Cyber-Physical Systems: Metrological Issues

S. Yatsyshyn
B. Stadnyk
Editors

Cyber-Physical Systems: Metrological Issues

International Frequency Sensor Association Publishing

S. Yatsyshyn and B. Stadnyk, *Editors*
Cyber-Physical Systems. Metrological Issues

ISBN: 978-84-608-9962-4
BN-20161105-01
BIC: TBM

The Monograph is printed upon the decision of the Scientific Council of National University 'Lviv Polytechnic' (Protocol No. 24, 16.06.2016).

Content

About the Authors **11**

Acknowledgement **13**

List of Abbreviations **15**

Introduction **17**

1. Cyber-Physical Systems. Structure, Operation and Problems **19**
   1.1. Metrology as a Basis for Designing CPS Models 23
     1.1.1. Typical Examples of CPSs and Experience of Metrological
      Maintenance *24*
     1.1.2. Research on the Nearest Future Generation Communication
      Technologies *28*
   1.2. Metrology to Support the Advanced Communication Technology 29
   1.3. Measurement Technology for both Development and Production 31
     1.3.1. Flexible Metrological Subsystems with Special Simulating Procedures .. *31*
     1.3.2. Metrological Subsystems with Special Standardization Facilities *32*
     1.3.3. Dispersed Measuring Subsystems *33*
   1.4. Spatial Problems, Volatility and Ageing Problems 35
   1.5. Conclusions 36
   References 37

2. Smart Measuring Instruments **41**
   2.1. Smart Sensors & Transducers and their Grids 42
     2.1.1. Distributed Sensor Network *45*
     2.1.2. Smart Grid Sensor *46*
     2.1.3. Middleware and Wireless Sensor Networks *46*
   2.2. Metrological Problems of Smart Measuring Instruments 48
     2.2.1. Specificy of Measuring Instruments Due to Their Smartness *48*
     2.2.2. Verification of Metrological Software and Middleware *50*
     2.2.3. Fractional Processes in Measuring Techniques *56*
   2.3. Metrology and Modelling for Additive Manufacturing 57
     2.3.1. Metrological Problems of Additive Manufacturing *58*
     2.3.2. Computation Method of Segmental Approximation for Random
      Topology Single-Surface Reconstruction of Interferometric Patterns *60*
     2.3.3. Model-based Diagnostic for Heavy Operating Conditions *68*
   2.4. Smart Sensors Grids Implementation 77
     2.4.1. Smart Energetics and Information Support *77*
     2.4.2. Smart Energy-Efficient Buildings *78*
     2.4.3. Metrological Study of Heat Energy Balance *92*
   2.5. Conclusions 98
   References 99

3. Embedded Measures as the Measuring Instruments..........105
3.1. Checked Instrument Based on the Inverse of Conductance Quantum 106
3.2. Study of Quantum Unit of Temperature and Temperature Measuring
Standard 114
3.2.1. Primary Thermometry and Quantum Units of Temperature 114
3.2.2. Promising Methods of Thermometry.. 117
3.2.3. Investigation in Creating the Quantum Unit of Temperature............... 119
3.3. Mass Measures with Coded Remote Access.. 123
3.4. Conclusions 128
References 130

4. Code-Controllable Measures for Correction of Measuring Channels 133
4.1. Quality Assurance in Measuring Instrument Design 133
4.2. Additive Error Correction for Measuring Instrument 142
4.3. Remote Error Correction of Measuring Channel.. 156
4.4. Conclusions 171
References 173

5. Techniques for Accuracy/Trueness Improvement.....................177
5.1. Major Metrological Characteristics of CPSs Units within Different
Approaches.. 177
5.2. Reliability and Accuracy/Trueness of Measurements 181
5.2.1. Methods to Improve the Accuracy, Errors and Examples of Reduction. 182
5.2.2. Duration of Noise Signal Gauging... 190
5.3. Metrological Problems of Raman Thermometry 191
5.4. Temperature Dependent Precision Threshold... 194
5.5. Dynamic Error and Instrumental Error 196
5.6. Methods of Correction and Statistical Minimization of Errors...................... 200
5.7. Measurement Inexactness in (Nano)thermometry 211
5.8. Coriolis Mass Flowmeter and its Uncertainty Management.......................... 213
5.8.1. Accuracy Problems on Example of CMF with Straight Tube............... 215
5.8.2. CMF Performance Improvement Under the S-type
Thermocouples Usage.. 220
5.9. Conclusions 221
References 222

6. Frequency, Noise and Spectrum Metrology 225
6.1. Frequency-Phase Techniques in Thermometry... 226
6.2. Noise Metrology 230
6.2.1. Fluctuations and Thermodynamics, Proper Noise and Thermometry... 231
6.2.2. Problems and Methodology of Noise Measurements............................. 242
6.3. Non-Invasive Diagnostics... 258
6.3.1. 1/f Noise Studying for Diagnostics... 258
6.3.2. Passive Method of Electronic Elements Quality Characterization......... 265
6.4. Raman Method in Metrology and Thermometry 269
6.4.1. Metrology of Raman Thermometer with Universal Calibration............. 270
6.4.2. Elaboration of Raman Method... 280
6.5. Conclusions 287
References 287

7. Qualimetric Estimation of CPSs and Their Products........................ **293**
  7.1. Qualimetry of Natural Gas as Energy Source................................ 295
  7.2. Foundations of Objective Qualimetry 311
  7.3. Objective Qualimetry on the Basis of Thermodynamics............................ 316
  7.4. Conclusions.. 319
  References.. 319

Index **321**

About the Authors

Svyatoslav Yatsyshyn, As. Member of International Thermoelectric Academy, Dr. Sc., Prof., Eng. Member of the Ukrainian Academy of Metrology. Areas of scientific and engineering research: theory and practice of temperature measurements; nanothermometry; thermodynamics and nanothermodynamics, standards of physical units.

Author of 3 books and 149 scientific works in metrology and thermometry, and other branches, including Cyber-Physical Systems, in English, Ukrainian, Russian, and Polish.

Bohdan Stadnyk, Member of International Thermoelectric Academy, Dr. Sc., D.H.C., Prof., Eng. Areas of scientific and engineering research: metrology and nanometrology, theory and practice of temperature measurements, including in atomic energy plants and engines; nanothermometry (methods and means).

Author of 18 textbooks, tutorials and handbooks, 533 scientific works in thermometry, thermoelectricity and other branches, in English, German, Ukrainian, Russian, Polish and other languages.

Yaroslav Lutsyk, Dr. Sc., Prof., Eng., Member of the Ukrainian Academy of Metrology. Areas of scientific and engineering research: the development of acoustic methods for measuring the temperature, temperature instrumentation and metrology for temperature measurement, nanothermometry.

Author of 7 books and 123 scientific works in metrology, thermometry and ultrasonic thermometry in English, Ukrainian, Russian, and Polish.

Yury Bobalo, Dr. Sc., Multi-D.H.C., Prof., Eng. Areas of scientific and engineering research: radio engineering and radiometry.

Author of 3 monographs, 11 textbooks and tutorials, 220 scientific works, in English, German, Ukrainian, Russian, Polish and other languages.

Pylyp Skoropad, Dr. Sc., Prof., Eng. Areas of scientific and engineering research: metrological instruments of contact thermometry, materials science.

Author of 1 monograph, 2 textbooks and 82 scientific works, in English, German, Ukrainian, Russian, Polish and other languages.

Mykola Mykyychuk, Member of the Ukrainian Academy of Metrology, Dr. Sc., Prof., Eng. Areas of scientific and engineering research: methods of verification of the measuring tools, solar energetics and its control, means of ensuring the measurements.

Author of 1 monograph, 2 textbooks and 52 scientific works, in English, Ukrainian.

Vasyl Yatsuk, Member of the Ukrainian Academy of Metrology, Dr. Sc., Prof., Eng. Areas of scientific and engineering research: automation and metrological assurance of measurements.

Author of 4 textbooks, 120 scientific works, in English, German, Ukrainian, and Russian.

Ihor Mykytyn, Dr. Sc., Prof., Eng. Areas of scientific and engineering research: noise thermometry, metrological engineering and verification of metrological software.

Author of 2 textbooks, 59 scientific works, in English, Ukrainian, Russian, Polish.

Zenoviy Kolodiy, Dr. Sc., Prof., Eng. Areas of scientific and engineering research: metrology and nanometrology, theory and practice of temperature measurements; nanothermometry (methods and means).

Author of 75 scientific works in English, Ukrainian, and Russian.

Oleh Seheda, Ph.D, As. Prof. Areas of scientific and engineering research: Raman thermometry.

Author of 18 scientific works in English, German, Ukrainian.

Vasyl Motalo, Dr. Sc., Prof., Eng. Areas of scientific and engineering research: methods and means of measurements.

Author of 4 textbooks, 72 scientific works in English, German, Ukrainian, Russian, Polish and other languages.

Acknowledgement

The scientific results, presented in this book, were obtained in the frame of research project number 0114U001243, 01.01.2014 - 31.12.2016, financially supported by the Ministry of Education and Science of Ukraine.

List of Abbreviations

AM	Additive Manufacturing
ASIC	Application-Specific Integrated Circuit
CC	Channel Commutator
CCM	Code-Controlled Measure
CL	Connection Line
CMF	Coriolis Mass Flowmeter
CNT	Carbon Nano Tube
CoE	Component of Error
CPS	Cyber-Physical System
DAS	Data Acquisition System
FFS	Flip-Flop Switch
FN	Flicker Noise.
IA	Isolation Amplifier
IAB	Input Amplification Block
ID	Isolation Device
LNG	Liquefied Natural Gas
LPG	Liquefied Petroleum Gas
MI	Measuring Instrument
MT	Measuring Transducer
NG	Natural Gas
NT	Noise Thermometer
NZE	Net-Zero Energy (building)
OA	Operation Amplifier
OA	Operation Amplifier
PS	Polar Switch
PSD	Power Spectral Density
QoP	Quality of Product
QoS	Quality of Service
SE	Sensitive Element
SEEH	Smart Energy-Efficient House
SST	Standard Signal Transducer
TF	Transformation Function
TT	Thermodynamic Temperature
UW	Ultrasonic Wave
WLI	White Light Interferometry
WSN	Wireless Sensor Network.

Introduction

The purpose of this book is presentation and consideration of main trends in the branch of metrology of Cyber-Physical Systems which are becoming a key element of everyday life. At the first place it destined for engineers, lecturers, students, persons who are not acquainted enough with specificity of Cyber-Physical Systems and their Metrology but are interested in it. The authors tried to highlight emergence and development of these systems, combined with the study of their metrology provision and support. Authors also presented their achievements respectively in the set of 15 works, 2015-2016, Sensors & Transducers, Issues 3 (2015) – 3 (2016), common title "Metrological Array of Cyber-Physical Systems", and in a similar set of articles, 2013, Sensors & Transducers, Issues 3-12 (titled "Development of Noise Measurements").

Day after day the globalization of industry, agriculture, transport, health-care and so on becomes more total. Certainly it contributes to continual development of Internet technologies, one manifestation of which is the occurrence of Cyber-Physical Systems. Objective quantitative information on the progress of technological processes is obtained by measuring their parameters. Measurement should be considered as a holistic process that starts from perception and transformation of object measurement data to its processing, storage, transmission and application for developing retroactivity in controlled technological objects. Therefore, in current conditions, one of the most important CPSs' parameters is their general and metrological reliability. The latter defines validity and significance of the obtained information that becomes especially important for CPSs operation.

Indeed, the manufacturing CPSs should not cause environmental damage, greater from the acceptable standards. The problem of preventing environmental and technogenic accidents and disasters should be noted also. Exploration of this book would give academic researchers and practitioners a novel insight into the complex problem of CPSs conjugated with unceasing development of metrology science, and powerful tools to analyse the arising daunts.

The presented monograph includes several arrays of metrological studies based on studying:

- Verifying and validating the metrological tasks for parameters determining the controlled equipment, process, materials through the development, implementation and realization of specific metrology and standardization methods, instruments, and equally the same as the facilities and etc. that is successfully described by the terms "metrological hardware, software, and middleware". Then, hardware is necessary to provide at the design stage and to install at manufacturing this component of CPS or at its operation. Software is usually installed while training the hardware. Middleware has to be transferred via Internet base stations, and should be installed automatically and provided the set of automatic calibration actions, measurement and characterization of objects in-situ. Calibration could be performed remotely under condition of code access to CPS LAN with implemented appropriate software and middleware.

- Aspects of metrological reliability, including its prediction, particularly of CPS integrated metrological subsystems remote nodes. This includes not only microwave research essential for the operation at optical wavelengths, but low and ultralow frequency methods in which it can be detected successfully the hidden and latent defects of complicated systems.

- The necessary metrological reliability of information and measurement system that, as a separate subsystem is the part of CPSs, in practice, constantly supervise the measurements. Reliable measurement information of required accuracy or trueness can be obtained only through technically informed choice of measuring instruments and includes the following data: availability of measured or monitored parameters of object; tolerance for deviations of these parameters and allowable measurement uncertaintics; allowable probability of false and unidentified rejections for each of monitored and controlled parameters and the values of confidence for them; distribution laws of measuring (controlled) parameters and their measurement errors that can arise while using the measuring instruments; measuring conditions: mechanical loads (vibration, shock, acceleration, etc.), climatic impacts (temperature, humidity, pressure, etc.), and so on.

We are very grateful to our reviewers for well-wishing criticism.

1.

Cyber-Physical Systems. Structure, Operation and Problems

B. Stadnyk, S. Yatsyshyn and Ya. Lutsyk

Cyber-physical systems is a system of collaborating computational elements controlling physical entities. Rather the similar seems to be definition of SEI of Carnegie Mellon University which concludes the next: CPSs are "engineered systems that are built from, and depend upon, the seamless integration of computational algorithms and physical components". Their objective is to enable efficient development of high-confidence distributed CPSs, whose nodes operate in a provably correct manner in terms of functionality and timing, leading to predictable and reliable behavior of the entire system.

Today, a precursor generation of CPSs can be found in areas as diverse as aerospace, automotive, chemical processes, civil infrastructure, energy, healthcare, manufacturing, transportation, entertainment, and consumer appliances. This generation is often referred to as embedded systems. In embedded systems the emphasis tends to be more on the computational elements, and less on an intense link between the computational and physical elements [1]. For embedded systems it is assumed that they are always ready to operate, and such systems built into the cars, aircrafts and other complex products are responsible for security. The reliability of such systems is ensured by maintaining the reliability of such systems is provided by maintaining the operability of its components due to redundancy, regulatory replacement of components etc. Unlike more traditional embedded systems, a full-fledged CPS is typically designed as a network of interacting elements with physical input and output instead of as standalone devices. Unfortunately the metrological problems of such systems are not considered here since it is the next stage of their existence and maintenance.

On the basis of available experience of metrology of the mentioned units operation, it seems appropriate to extend further the understanding of aforesaid systems in the direction of natural affiliation, to the next concept: a full-fledged CPS with the metrological assurance of group of operating parameters as well as the basic characteristics of the intermediate product has to be designed as a network of interacting elements with physical input and output of every element that is controlled at each stage of operation providing a qualitative final product. Furthermore, the CPS can change over time, and a priori is known that the components and connections between CPS units are not 100 per cent reliable.

Idea of Smart Manufacturing as the kind of CPS agrees with the general trends of present century that can be observed in the industry [2]. For example, to realize mass production on individual orders at the price of mass products is fully possible only if there exist the flexibility and adaptability provided by CPS means. Roughly the same can be assumed for the AM that develops the pillars of 3D-printing for release the finished goods. On the one hand, it better meets the requirements of mass customization than current technologies, but on the other - it can be fully automated and linked into a single process of conceptual design and manufacturing to get Direct Manufacturing as a division of Smart Manufacturing. Hence it emerges the need for theoretical works and engineer modeling related to the establishment of methods for generating models of the physical component of CPSs, interfaces of these models with models of computing systems, methods, certification and verification of models. And the basis of aforementioned has to become a set of Smart Sensors, on the one hand, and the Smart Actuators on the other hand. The operation of the last is served by their own aggregate of models which should deal with previous models and concepts.

The existing CPSs development programs were suggested by scientists in automatics, software, networks and their security. However, their effective development is impossible without taking into account the metrological aspects of designing, constructing, and operating CFS. Therefore current NIST [2] program focuses on involving metrological science to resolve some AM-problems. For instance, this program focuses on four areas which are closely interrelated: (1) Material characterization; (2) Real-time process control; (3) Process and product qualification, and (4) Systems integration. In the real-time process control area, the program firstly focuses on innovative process metrology to provide traceable and quantitative data for validating process models, calibrating in-process sensors, and determining optimal process conditions. The program has to establish metrics and test methods for assessing the performance of process metrology sensors and instruments. Validated physics-based models are used to develop reduced-order analytical models for use in the development of real-time control algorithms. Both iterative open-loop controls with post-process feedback and real-time closed-loop control approaches have to be tested. Performance metrics for both types of control methods are developed. Open architecture test bed is recommended to incorporate various process metrology instruments and control systems as a validation platform.

In the process and product qualification area, the program will establish foundations for equivalence-based qualification of materials, processes, and parts used in AM by developing novel test methods and protocols for round robin testing, as well as generating trusted data for sharing among the AM stakeholders. To enable model-based qualification, validated process models are needed. Validated temperature models will feed material models that predict AM material and part performance. Methods will be developed for integrating pre-process, in-process, and post-process measurements to demonstrate that a part will perform to specifications.

Similarly, the CPS technologies company utilizes the sophisticated metrological equipment for material and geometric characterization [3]. CPS uses Solidworks for 3-D design, AutoCAD for 2-D design and CAMWorks to generate NC Code to produce all process tooling. This enables CPS to control the key product tolerances of all dimensions. In addition to design control, the CPS machine shop, incorporating state-of-the-art machining and inspection equipment, fabricates all necessary molds, casting dies, etc. This capability provides control over key product tolerances and permits rapid product development and production scale-up. Verification and validation becomes key during new product development.

However, the continuous development of CPSs, designated schematically, comes up against a number of difficulties due to the following circumstances. Firstly, assign a priori that CPS creating (Industrial Revolution 4.0 [4]) on the basis of industrial Internet opportunities envisages the free conjunction and matching of wares from different countries. But, the current set of standards in different countries is insufficient to describe successfully the means and tools of modern scientific technology, and actually it is constantly evolving and improving [5]. Secondly, laboratories and leading research centers based on modern productive machinery are not able to be tested by virtue of their own complexity and problems of delivery to certified laboratories. Thirdly, unique and newly created machinery often requires self-verification and standardization of metrology facilities to ensure the qualitative work. Raman installation for nanoobjects' temperature measurements can serve as a prominent example of the mentioned. Basing on results of previous studies for its efficient operation we have proposed the CNT blackbody as a standard artefact. Fourthly, the existing standards by several orders lose their values (accuracy characteristics) while uploading them to the end user, and it is actually considered a normal metrological practice. However, such practice cannot be deemed adequate for CPSs. Fifthly, it is ignored the transfer of the metrology standards of all physical quantities to new contemporary framework wherein the basis for determining their dimensions are assigned the fundamental properties and constants of matter [6].

Series manufacturing of products and components is facing a new "industrial revolution." Production systems are working in a way that is increasingly flexible and networked. Production processes are becoming ever more efficient through the use of CPS, but also more complex and multi-faceted. The choice of suitable metrological systems plays a key role in implementing new production concepts. Measurement technology capable of keeping pace with the most modern production systems is the only way to ensure seamless quality control and efficient, cost-saving production.

By supporting specified approach in the field of metrology and standardization, we consider in this paper the basic aspects of horizontal-vertical cooperation and integration in CPS research and operation. CPS with their specific metrological support we plan to consider previously below, and sequentially examine within the limits of publications in the subsequent Journal issues. Considerations carry out for different, but the kind of science and technology, which has already achieved a certain level of metrological status sufficient to transfer achievements to other industries. CPSs are discussed below according to the NIST classification [2] (Fig. 1.1).

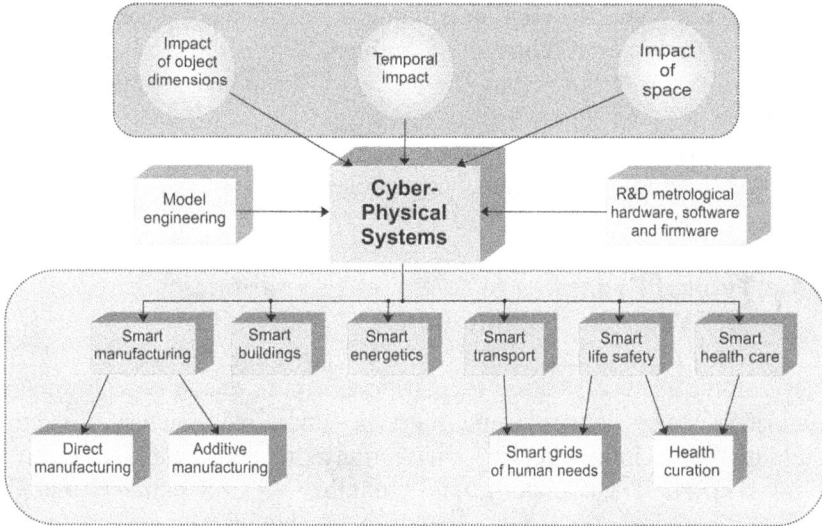

Fig. 1.1. Cyber-Physical Systems according to NIST classification.

1.1. Metrology as a Basis for Designing CPS Models

The basis of the construction of this discipline can be considered the metrology, which facilitates a precision adequate description of the external world and therefore adequate response by the CPS on deflecting impacts while operating. For this purpose metrology should continuously evolve to respond adequately to the challenges of today and to assimilate additional seemingly not involved scientific disciplines. As a result, emerge interdisciplinary relations [7], and the metrology is included in transdisciplinary field of contemporary technology.

In proposed approach there are considered the main features of model engineering of most common in the world of intelligent fire detectors, as the units of alarm systems which can be regarded as fire sensors smart grid. The embedded microprocessors allow implementing any algorithms without changing the element base. It simplifies the realization of algorithm and forms the basis of Life Safety model engineering.

Within specified approach we argue necessity of developing the heat kind of smart fire detector. There were investigated 2 algorithms of its operation with variable threshold level triggering. The first one ensures the operation of the detector without information about modes of process equipment. It forms the basis of a simplest model of operation that only considers the rate of change of ambient temperature. The second algorithm requires controlling the modes of the process equipment, in which the detector is installed. In this case, it limits the amount of heat value $(Q \cdot t)$, acquired by controlled units that is often considered an important technological parameter, especially in nanotechnology.

1.1.1. Typical Examples of CPSs and Experience of Metrological Maintenance

Application area of CPSs is partitioned into the following main segments: smart networks and services, smart manufacturing, smart buildings and infrastructure, smart transport and health care etc. (Fig. 1.1). NIST has expertise in a number of CPS domains and is uniquely positioned to achieve breakthroughs by adapting measurement science solutions across domains and, where possible, developing independent solutions.

Smart Manufacturing

Its main constituent seems to be the Manufacturing Intelligence. In industry domain the Cyber-Physical Systems empowered by Cloud technologies lead to new approaches that pave the path to Industry 4.0 as the European Commission IMC-AESOP project with partners demonstrated [4]. Smart Manufacturing includes Smart Machines, which differ from contemporary ones by multi-function, small sizes, the adaption ability to the needs of users that is realized by gathering the required functionality in one machine. By obtaining information on the changed requirements, it may itself make adjustments into the technological process. Subsequently we will try to consider one example of Direct Manufacturing, namely mass metal production with the use of precision measuring devices – balances. For their normal functioning should be provided calibration, executable in-situ. Thanks to man-in-loop technology is carried out remote accreditation of precision balances by bringing their performance in some operation ranges to values of working standards.

On this way the creation of remote horizontal-vertical structure of the metrological surveillance of intelligent instrumentation as a structural unit of CPS, envisages the remote control and verification actions with using code access to previously installed software and hardware. So it becomes necessary a new measurement science [8] that is critical to overcoming the serious limitations inherent to contemporary test methods and equipment. NIST fundamental waveform metrology must be greatly extended to provide the underpinning measurement science that can reproducibly characterize, compare complex waveforms with reduced uncertainties and thus providing quantitative descriptions of communications signal quality. Consequently, apply perspective measurement techniques to develop new or enhanced metrology required to support advances in high-speed both wireless and optical networks.

Smart Life Safety Systems and Smart Transport with Smart Sensors

Modern society depends on networks for the transmission of energy data, life safety data, antiterrorist defence etc. Therefore, the next goal is to create a CPS or Smart Grids more effective. For tasks that require more resources than are locally available, one common mechanism for rapid implementation of Smartphone-based mobile CPS nodes utilizes the network connectivity to link the mobile system with either a server or a cloud environment, enabling complex processing tasks that are impossible under local resource constraints.

Building the CPS requires a new science of characterizing and controlling dynamic processes across heterogeneous networks of smart sensors [9] and computational devices In spite of rapid evolution, authors [10] continue to face new difficulties and severe challenges, such as interdisciplinary integration and novel design methodology. Model engineering and maximal possible supply of sensors for data acquisition and processing enable to offer the best solutions to ensure efficiency, safety, and defence. For instance, protection against terrorist attacks may be performed as a high-speed system of explosion intensity redemption which occurred inside the plane at an altitude of 10 km and higher through application of optical sensors and the actuators of automatic depressurization/pressurization in a separate section of the plane [11].

Next stage of fire detectors development becomes emergency detectors. The latter include explosion detectors, acoustic detectors of destroying bridges, neutron sensors of petroleum wells [12], which become particularly important due to catastrophe of well in the Gulf of Mexico. These smart sensors/detectors should form particular global network systems. Including in control circuits of CPS will not only safely operate existing structures and prolong their life cycle, but also establish qualitatively new designs (Fig. 1.2).

Through transfer the measured data processing directly to the sensors and the use of fuzzy logics techniques (Sugeno [13] and Mamdani [14] methods), Smart grids built with the use of such sensors already significantly reduce response time and increase the probability of an emergency identifying, and also take into account the seasonal, daily, and other changes in working conditions, which in turn improves the defining characteristics of Smart systems. However, it exist a particular uncertainty for system's operation criteria in the event of a critical situation, dependent on the characteristics of the environment.

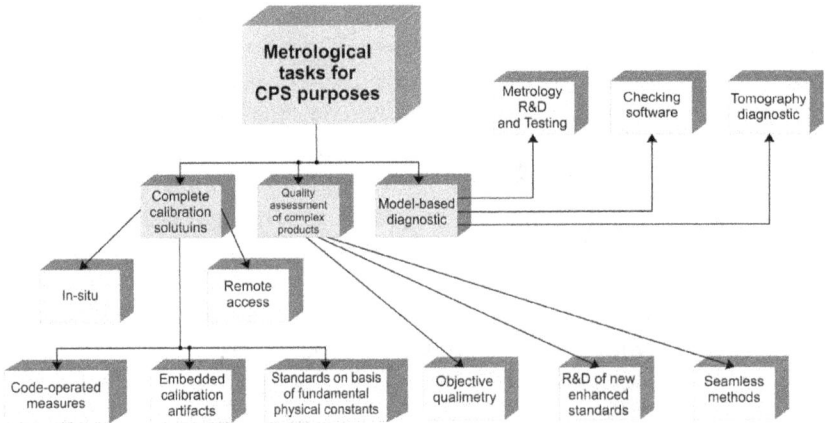

Fig. 1.2. Smart life safety systems with smart sensors.

Smart Buildings

Modern technologies allow the creation of Smart buildings, or constructions with minimal or zero consumption – Net-Zero Building [15]. But they need constant monitoring; therefore must be connected to Smart sensors grids and in appropriate way be controlled by CPS means in order to most expediently use those provided outside world resources and services. Combining data from different sources can achieve Smart buildings operating conditions close to optimal. The main demand to aforementioned objects is to achieve zero energy consumption. This is reached by a detailed study of thermal conditions with help of smart temperature sensors local net and provides adequate insulation in continuous multipoint temperature monitoring.

Some attention is paid to the preliminary evaluation of thermal conditions in the construction phase, eliminating bridges of cold etc., for which authors suggested a number of methods and tools of heat control. These are methods of infra-red thermal imager's diagnostics, method of quasi discrete point-wise studying the restrictive planes' temperature regime, as well as monitoring temperatures in time by using glued to the inner and outer surfaces of the wall cross section the chips with embedded temperature sensors and so on. Thus it becomes possible to detect and eliminate bridges of cold, to study the role of energy reflecting coatings with unknown blackness coefficient and to manage ultimately the operation of energy subsystems for supply, heating and ventilation.

Smart Health-Care Systems

NIST program [2] represents needs of medical implants development. As classical metrological task in the mentioned area it seems to be the research of adapted to person purposes smart health-care system of cancer treatment. Here the human cells are irradiated at the micro level by microwaves, and due to selective origins of absorption and power of irradiation which has been regulated through feedback provided by nanosensors [16].

Spectrum Metrology

In the field of Optical Metrology [17] a challenge is to develop or create new measurement techniques and standards, to meet the needs of next-generation advanced manufacturing, which will rely on nanometer scale materials and technologies. It is difficult to provide samples on which precision instruments can be calibrated on nanoscale. Calibration standards are important for repeatability to be ensured. It is difficult to select a universal calibration artefact with which we can achieve repeatability on nanoscale. At nanoscale, while calibrating care needs to be taken for the influence of external factors (noise, vibration, motion) and internal factors such as interaction between the artefact and equipment which can cause significant deviations. The most universal nanothermometry method, Raman method, apt for the direct temperature measurement of micro- and nano-objects within the range 100 nm –

100 μm, could be distinguished within this from cryogen till mid-high temperatures. In addition, it does not demand calibration; therefore the mentioned method is particularly considered in Chapter 7.

Note that this program can automatically mode be installed using the industry Internet being in the spirit of NIST program [2]: "With standardization and best practices to ensure data quality, usable formats and security, measured data could be made available alongside spectrum license data to improve the quality and quantity of information available for various decision makers. Further, in the spirit of the PCAST Spectrum Access System recommendation, this information fed into a compatibility algorithm could enable an automated dynamic spectrum access paradigm for select scenarios".

1.1.2. Research on the Nearest Future Generation Communication Technologies

According goal of NIST [2] "To measure and predict global behavior of CPS in their large information systems, which characterize future communication systems", we develop and evaluate methods and algorithms to predict behaviors that can lead to system failures as well as "deployment of successful methods and technologies form of measuring and predicting shifts in system behaviors".

We have researched [18] the noise method of studying the electronic element structure defectiveness. Low-reliable elements could be revealed with the help of proper noise power parameters. In comparison with other seamless methods of electronic element inner structure defectiveness diagnostics, the method of research by dint of electric noise power is multipurpose, and enables detecting the potentially low-reliable elements. With involving the dependence of flicker-noise parameters on the controlled object structure, we could diagnose the electronic element state and its evolution, especially at the primary stages of defects formation. So, electric noises are the manifestation of fluctuation-dissipation processes taking place inside the researched substances. Moreover, the method of studying proper noise as method of non-destructive testing could lie into the basis of passive noise spectroscopy. It could be realized the most precisely owing to the usage of noise thermometers as calibrated measuring devices.

"To predict behaviors that can lead to large-scale system failures in most spheres of science and technics", the measurements of determined signals are conducted, whilst in the case of noise values the random signals are of interest. The interference could be both determined and random. Besides, taking into consideration the fact that measurements are performed within the wide bandwidth, the determined interference could reveal itself at the various frequencies. Therefore we should recommend the filters of complicated configuration that combine a band filter for forming the work-bandwidth of a measuring device, and rejecter filters for some frequencies which could be numerous. Taking into account the specific conditions of measurement as well as difficult construction and working principles of such filters, the synthesis of digital intelligent filters with the usage of rapid Fourier transformation which meets the enumerated requirements is optimal. Nowadays we are ready to study large information subsystems of remote CPS by developing the appropriate metrological software, installed distantly.

1.2. Metrology to Support the Advanced Communication Technology

It includes the specific metrological tasks [2]. Some of them are outlined below:

- Calibration artefacts with full point-by-point uncertainty characterization, which will enable a wide range of calibrated time- and frequency-domain measurements to be performed in industrial laboratories, thus accelerating product development by removing the need for intermediate calibration steps.

- Measurement methods for industry to directly verify the over-the-air performance of mobile devices.

- Methods to quantify and validate large signal network analyzer measurements to enable communications networks with greater dynamic range and greater channel capacity to be achieved using base stations and mobile devices.

Thereby, it seems to be demanded the encoded manageable measures that especially important for security and integrity of information systems becoming a critical issue within most types of organizations [19]. Simultaneously developing measures and standards for metrology and whole industry needs face with analogous requirements as high-mentioned. We accepted such measures R&D to ensure their efficient remote operation in industry. For this purpose the measures, preset in complex information-measuring systems, are activated by a qualified metrologists from a distance by start-coded signal [20]. An example of quite simple and efficient solution of CPS metrological problems is the appearance of new approach for CPSs. In the opinion [21] the complete calibration solution for process calibrators and multifunction process calibrators is quite new area of science for CPSs, it's a combination of our already proven, high end pressure calibration service with our new electrical calibration capabilities.

Another way to improve quality of CPS products is suggested to be the embedded high-precision reference standards that are based on the fundamental properties of matter. We have considered [22] the implementation the state standard of electrical resistance. This standard is proposed to develop by applying the latest nanotechnology achievements, such as research of electrical conductivity of nanopatterns (f.i., carbon nanotubes). Being in superconductive state, the latter are inherent in the resistance value which corresponds to inverse of conductance quantum that is equal (12906.4037 ± 0.0020) Ω, due to transient resistance of contacts. Four-wires circuit is sufficient to carry out the measurement of this resistance, and the Wheatstone bridge inston bridge together with the Hamon network would be sufficient to transfer this value to values of current standards $(1.0; 100.0$ $\Omega)$ and then to adjust with high precision the values of the working resistors to the required denominations.

In many cases it is impossible to avoid checking the metrological applied and embedded software. While verification the latest, it arises the problem associated with access to executive program. Mostly is unable to submit the program of test sequences of numbers and verify it. Ways to overcome problems may lie in the modification of measurement means as CPS subsystems by installing an additional digital input.

1.3. Measurement Technology for both Development and Production

1.3.1. Flexible Metrological Subsystems with Special Simulating Procedures

Product developers and production planners are searching for essentially the same thing: flexible systems featuring easy integration of new technologies without users having to retrain each time. The challenge is to respond to market demands using available methods and tools. The two great pillars of support in this area are simulation and product design. However, these theoretically determined data points and components must still be tested with real prototypes under conditions closely simulating series production. That includes stress, function and fatigue tests. Components are subjected to mechanical stress and physical loading in all cases and a life cycle is simulated as close to real condition as possible. Mechanical loading includes application of forces, pressures, strains and torques. A test covering years can be simulated in a few hours with the effect of temperature [23].

One of NIST initiatives, associated with model-based diagnostics and prognostics and proposition of the leading German researchers on the implementation of high-speed method of diagnosing CFS structures by alternating loads [3], we support by our 30-year development of temperature means metrology at mechanical loading (stress, torsion and etc.). We have researched the impacts of mechanical loading at different temperatures on performance of temperature sensors in series of scientific works and support the mentioned work. Thus, at the rapid heat of the sensors of noise thermometers the transient noise process caused by thermodynamic disequilibrium has been detected. Such a behavior of a noise signal could be revealed at the expense of internal changes intensified consequently of applying an uneven temperature gradient on the substance with inner defects. The research on the behavior of a noise signal in the dynamic temperature mode has been made following the methods of rapid transferring of a sensor from the medium with one temperature into that with higher temperature.

1.3.2. Metrological Subsystems with Special Standardization Facilities

Therefore, to solve metrological problems of CPS we proposed creating the local certification centers, wherein to be kept operating standards of physical quantities (Fig. 1.3). As mobile standards are characterized by significantly less accuracy compared to stationary ones, it should be given proper attention to them. Depending on the way of such standards application, the distance at which they are transmitted in electrical or optical manner varies: see in Fig. 1.3 - the measure 1 or measure 2 in the form of circles.

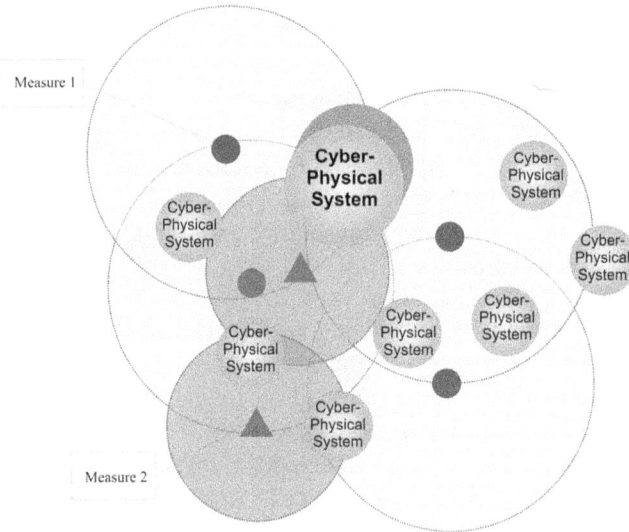

Fig. 1.3. Implementation of calibration procedures depending on CPSs and certification centers locations.

As a result, forms a complex spatial structure of current CPS with appropriate standard provision, based on the optimization of the quality of fabricated products considering metrological support.

In modern practice the measuring processes have to be constantly supervised, trying to assure required metrological reliability. Today normative documents recommend systematic and complete control of measurement process in order to guarantee the required quality level of goods, products and services produced by measuring means application.

This control is treated as separate and lengthy procedures. Frequency of metrology inspections is primary determined by metrological reliability of measurement. To improve the latter frequency would be increased. This characteristic mainly depends on the metrological performance drift, and control results have to be fixed specially that would help in future diagnostic of manufacturing cycle according to production quality [24].

1.3.3. Dispersed Measuring Subsystems

Frequently data acquisition system of measurement results are performed as a separate switching, measuring, computing and other operations in order to create, accumulate and elaborate measurement results of a wide range of electrical and non-electrical measurands. PC distribution resulted in the formation of so-called measuring circuits intended for multichannel conversion of measuring signals into digital form and vice versa - for the formation the analog output signals. The length of the lines there may be almost unlimited. However, in industrial conditions connecting lines, especially at considerable length can add quite significant and, more importantly, uncontrolled error in the measurement result. Dispersed measuring systems as the subsystems of dispersed CPSs can be composed of some spatially shared technological elements (Fig. 1.4). Measured values are controlled via separate measuring channels (channel + signal conditioning unit + ADC + MC), which can be regarded as smart measurement means. The measurement results are transmitted via interfaces to local measurement stations, wherein the processing and analysis are performed, then control and correction signals are produced and etc. From thence through the communication nodes and channels (radio and optical channels) or networks they are transmitted to a central management point based on the PC with the appropriate software. Thus length of the lines is practically unlimited.

Thus, the assessment of products quality is a complex multifactor problem, within which is difficult to evaluate the role and relative weight of each factor, as well as to expressed in physical units the objective numerical value of it or to validate this factor, and finally to determine certain characteristics. At best, the result is expressed as the number reasonably combining all these impact-factors. That is a subjective Qualimetry. It is exploited while comparison the similar products of the same destination from different manufacturers. However, no one can prove conclusively the correctness of the choice of those or other factors that affect the same assessment. From the metrological point of view, no one can guarantee absence of correlation for separate factors among themselves that negative impacts the obtained results. On the basis of experience of studying factors influencing the performance of thermoelectric sensors, to evaluate the quality of new products we suggested to carry out methodologically correct selection of uncorrelated factors. Underpinning for this should serve an objective Qualimetry built on the basis of thermodynamics. The latter ensures correctness of choice the determining factors - characteristics – of being evaluated product as a set of unrelated variables (measurands).

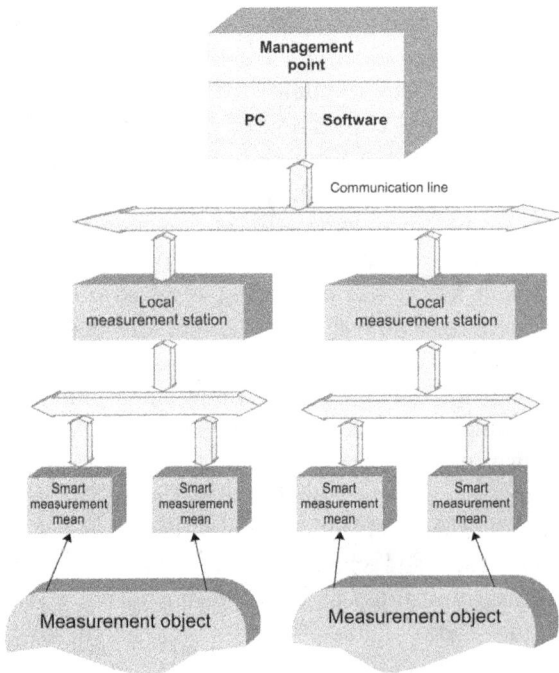

Fig. 1.4. Dispersed measuring systems.

While using the embedded systems a priori assume that they are always available to operate; such systems are built into vehicles and other complex products wherein may be responsible for security. The reliability of such systems is ensured by maintaining the operability of its components due to redundancy, regulatory replacement and so on.

Everything complicates in CPSs. Their system is changeable over time that is based on initial setup, and certainly aware that components and links between them are not entirely reliable. The mentioned concerns the overall reliability. Similar refers the metrological reliability especially of information and measurement subsystems of CPSs. Here the specified reliability manifests a direct impact on final products, or on finishing results of CPS operation. For instance, the pressure sensor works not quite satisfactory in health-care CPS, or emerges the indirect impact on interim results. As example, due to aging effect, it emerges the drift of temperature sensor readouts, undetected on time. That results in deviations from optimum mode at manufacturing of quantum dots.

1.4. Spatial Problems, Volatility and Ageing Problems

One of the most important tasks in the field of design, development and management of CPSs is a question of teamwork and users. Issues that are important in the context of this collaboration is to identify and model the "situational awareness", the human experience of these systems and the environment as well as initiated by these systems changes in general conditions or individual parameters. This can be crucial in deciding. For instance, smart house grids may be not only power generating element of energy-savings society, but also due to energy dissipation they alter the climate in neighborhood [25].

In this regard, traditional borders between human and machine become quite blurred. Communication in these networks are carried out through the exchange and sharing of information. It is becoming more and more applicable in human being. Own example of global type CPSs wherein information would be considered as the end product of world cognition is also self-customizable and self-regulatory network of autonomous moored and drifting weather buoys. Each of them equipped with a range of different sensors with their own processing powers able to form together a global information network concerning state of weather, storms, tidal waves, etc. They can be made without actuators nevertheless form the final information product.

At the same time, advances in science and engineering improve the connections between computational and physical elements of smart units design, dramatically increasing the adaptability, autonomy, efficiency, functionality, reliability, safety, and usability of CPSs. This expands the field of CPSs application by including new areas, such as transport;

machinery especially of precise types (robotic surgery and nano manufacturing); operation in dangerous or inaccessible environments; coordination (air traffic control); efficiency (smart energy-effective houses); and so on.

Mobile CPSs are a prominent example of considered systems. Usually Smartphones are referred to mobile CPSs, and Smartphone platforms make ideal mobile CPSs for a number of reasons, including: significant computational resources, such as processing capability, local storage; multiple sensory input/output devices, such as touch screens, cameras, GPS chips, etc.; a row of communication mechanisms, such as Wi-Fi, 3G, EDGE, Bluetooth for interconnecting devices to either the Internet, or to other devices; high-level programming languages that enable rapid development of mobile CPS node software, such as Java and others; end-user maintenance and upkeep. Really the high-mentioned type of CPSs can be attributed to the mobile one only conventionally as they are deprived of independent means of transportation or movement in space.

However, we are of great interest in technical types of actually mobile CPSs for tracking and analyzing CO_2 emissions, measuring traffic, monitoring oil and gas pipelines, etc. Then for tasks that require more resources than are locally available the considered CPSs utilize the network connectivity to link them with cloud environment, enabling complex processing tasks. Especially, it concerns the automatic pilot avionics, drones and other sensor-based communication-enabled autonomous systems. A real-world example of such a system is the Distributed Robot Garden in which a team of robots tend a garden of tomato plants [26]. This CPS combines distributed sensing (each plant is equipped with a sensor node monitoring its status), navigation, manipulation and wireless networking [27].

1.5. Conclusions

Conjugations of networking and information measuring technologies with manufactured products as well as with associated services are underlying a new generation of smart and flexible cyber-physical systems. The latter can only be implemented with smart sensors and actuators, powerful high-precision measuring equipment and technologies.

CPS study envisages attention to metrological and standardization problems. Their solution impacts not only on estimation of failures or unplanned disorders of production cycle, but the quality of assigned tasks execution (product manufacturing and etc.).

Proper attention in further CPS development is impossible without solving the specific metrological problems, the core of which lies in the triangle «embedded metrological hardware – Internet-borrowed metrological software – automatic implemented metrological middleware» providing sufficient level of metrological assurance. In particular, it facilitates the implementation of measurement methods for industry to verify directly the over-the-air performance of standard artefacts and/or mobile metrological instruments.

The development of simulation models for the purposes of the above triangle and methods for assessing the reliability of CPS, methods of nondestructive control and other test methods that include the Internet-borrowed noise and other frequency methods, able to provide improving the reliability and perfection of CPS operation.

Forecasting of CPS's metrology subsystems performance and CPS operation ensuring is partially achieved by development of Standards of physical quantities, basing on the fundamental constants of matter and exploring their transformation in the frequency characteristics, that could gain 20-fold enhancing of measurement accuracy/trueness and consequently the quality of CPS's operation. The first such Standard, enables to improve by 3-5 orders the electrical measurements exactness seems to be the intrinsic Standard of electrical resistance, based on superconductive CNT.

References

[1]. Lee, Edward, Cyber Physical Systems: Design Challenges, Berkeley Technical Report No. UCB/EECS-2008-8, *University of California*, January 23, 2008, Retrieved 2008-06-07.
[2]. NIST National Technical Information Service, FY 2014, Budget Submission to Congress, 2014.
[3]. CPS Technologies, Reliability with Smart Composite Products, http://www.alsic.com/alsic-design.html
[4]. Industry 4.0: Revolution in Production, By SAP Guest, (http://blogs.sap.com/innovation /innovation/industry-4-0 revolution-in-production-019559).

[5]. M. Stock, The Watt balance: determination of Planck constant and redefinition of the kilogram, *Royal Society Discussing Meeting*, Jan. 2011 (www.bimp.org.).

[6]. Ia. Mills et al., The New SI: units and fundamental constants, *Royal Society Discussing Meeting*, Jan. 2011. (www.bipm.org/utils/common/pdf/ RoySoc/Ian_Mills.pdf).

[7]. Y. Wang, Cognitive Informatics: a new transdisciplinary research field, *Brain and Mind,* Vol. 4, 2003, pp.115-124.

[8]. ASTM International, Technical Committee F42, Additive Manufacturing Technologies, http://www.astm.org/COMMITTEE/F42.htm

[9]. S. Yurish, Sensors: Smart vs. Intelligent, *Sensors and Transducers,* Vol. 114, Issue 3, March 2010, pp. I-VI.

[10]. J. Wan, H. Yau, H. Sau, F. Li, Advances in Cyber-Physical Systems Research, *KSII Transactions on Internet and Information Systems,* Vol. 5, No.11, 2011, pp. 1891-1908.

[11]. S. Yatsyshyn, A. Kushnir, Pyrometric converters in automated installations of explosion suppression, *Questions of Defense Equipment. Ser. 16: Technical Means of Terrorism Countering,* No. 1-2, 2008, pp. 80-81.

[12]. GE's Drilling Measurements Catalog (https://www.geoilandgas.com/ drilling/drilling-measurement-solutions).

[13]. M. Sugeno, T. Yasukawa, A fuzzy-logics-based approach to Qualitative Modeling, *IEEE Transactions on Fuzzy Systems*, Vol. 1, No. 1, 1993, pp. 7-31.

[14]. C. Pappis, E. Mamdani, A Fuzzy Logic Controller for a Traffic Junction, *System, Man and Cybernetics,* Vol. 7, Issue 10, 2007, pp. 707-717.

[15]. P. Torcellini, S. Pless, M. Deru, D. Crawley, Zero Energy Buildings: A Critical Look at the Definition, in *Proceedings of the ACEEE Summer Study on Energy Efficiency in Buildings Conference*, Pacific Grove, California, Aug. 14-18, 2006, pp. 1-15.

[16]. V. Khanna, Frontiers of Nanosensor Technology, *Sensors and Transducers,* Vol. 103, Issue 4, April 2009, pp. 1-16.

[17]. Understanding the Spectrum Environment: Using data and monitoring to improve spectrum utilization, *NITRD Wireless Spectrum R&D Senior Steering Group Workshop V Report,* 2014, Arlington, VA (https://www.nitrd.gov/nitrdgroups/images/8/80/WSRD_Workshop_V_R eport.pdf).

[18]. Z. Kolodiy, S. Yatsyshyn, B. Stadnyk, Metrological Array of Cyber-Physical Systems. Part 4. Non-Invasive Diagnostics, *Sensors and Transducers,* Vol. 187, Issue 4, April 2015, pp. 108-112.

[19]. R. R. Martin, Making Security Measurable and Manageable (http://measurablesecurity. mitre.org/index.html).

[20]. V. Yatsuk, P. Malachivski, Methods of improving the measurement accuracy, *Beskyd-Bit,* Lviv, 2008 (in Ukrainian).

[21]. The Total Calibration Solution, *CPS. Instrumentation & Calibration Experts,* 24 Sep 2014 (http://www.cps.co.nz/blog-display.aspx? ArticleId=98&categoryId=63).

[22]. B. Stadnyk, S. Yatsyshyn, State standard of electrical resistance on the basis of von Klitzing constant, in *Proceedings of the 58th Internationales Wissenschaftliches Kolloquium,* Technische Universität Ilmenau, 08 – 12 September 2014, pp.138-141.

[23]. M. Guckes, Product Manager Industrial Amplifiers and Software (http://www.hbm.com/en/4112/industry-4/).

[24]. V. Yatsuk et al., Possibilities of Precision Ohmmeter Calibration at the Exploitation Condition, in *Proceedings of the 7th IEEE Internat. Conference on Intelligent Data Acquisition and Advanced Computing Systems: Technology and Applications,* Berlin, Germany, 12-14 September 2013, pp. 24-28.

[25]. S. Karnouskos, Cyber-Physical Systems in the Smart Grid, in *Proceedings of the 9th IEEE International Conference on Industrial Informatics (INDIN),* 2011, July 2011. Retrieved 20 Apr 2014.

[26]. L. Sanneman, D. Ajito, J. DelPreto, A. Mehta *et al.* A Distributed Robot Garden System, in *Proceedings of the IEEE International Conference on Robotics and Automation,* Seattle, WA, 26-30 May 2015, pp. 6120-6127.

2.

Smart Measuring Instruments

S. Yatsyshyn, Yu. Bobalo, B. Stadnyk and P. Skoropad

Smart MIs are the prerequisite for CPS design as they constitute the essential units of information and measurement subsystems. There is a set of smart MIs which is divided into the following subsets: smart sensors, smart transducers, their grids and etc. that can be joined together in modern WSNs. Emerging field of cheap and easily deployed sensors offers an unprecedented opportunity for a wide spectrum of various applications. When combined, they offer numerous advantages over traditional networks. These include a large-scale flexible architecture, high-resolution data, and application-adaptive mechanisms as well as a row of metrologically specific features and performance (self-check, self-validation, self-verification, self-calibration, self-adjustment, etc.).

For, instance, special attention according to opinion of Carnegi Mellon University scientists can be given to verificating the WSNs in full compliance with well-known methodology. Timing verification has to guarantee the execution of assigned tasks in real-time WSNs within planned duration. In this area our efforts are directed in developing the schedulability techniques for multicore platforms, where new challenges of shared resources, such as memory, different and changeable algorithms of signal processing and storing, transmission and aggregation have to be considered. Often it would be needed for decision-making in real-time systems, especially in health-care and life safety.

Main milestones in everyday work aiming the insurance of reliable WSN reliable operation lie in the direction of functional and probabilistic verifications. We provide the software and middleware development aiming to reach predetermined behavior. The easiest way to achieve this may be demonstrated on the example of widespread wireless fire detectors networks. They are characterized by a number of special algorithms directed on as fast as possible and accurate triggering and actuating the automation of higher level. So, it becomes the necessity of researching and implementing the original operation algorithms for fire sensors and also the checking algorithms for periodic real-time software examination. Considering their structural complexity (presence of smoke and heat sensitive elements, various principles of elaboration of the received signals, their drift of characteristics, and pollution of translucent elements, etc.) development of such algorithms is a daunting task. Herein, human life may be the price of bug. Equally important seems to be probabilistic verification that is assigned to boost the probability of reaching WSN declared goals (estimation of their chances being achieved). This can be fulfilled in two known ways: a) Probabilistic model based on Markov approach, and b) The similar one based on Monte Carlo simulation techniques.

Significant attention should be given to collaborative autonomy of WSNs designed from the autonomous changeable groups of dissimilar sensors. We try to optimize performance and scalability of them by designing the decentralized operating environment. It may be open- and closed- sourced; the latter is proposed for WSNs that perform mission-oriented tasks. New challenge of CPSs operation is their ability of rapid adaptation to continuous environment alters even unto malfunctions. "A self-adaptive system is a system capable of changing its behavior and structure to adapt to changes in itself and its operating environment without human intervention" [1] while considering the WSN's approaches.

We obligate to develop not only overall adaptation strategies including predictions of environment evolution but also the specific metrologic proactive adaptive mechanisms caused by mentioned changes. If it is the MIs, the arrangements in particular are bound with the change in metrological characteristics and as the consequence with CPS dysfunction. They are examined further in Chapters 4-5. Additionally due to modelling (p. 2.3) the adaptations can be performed proactively outpacing the happening drifts and other kinds of performance changes. As result, it spares the time necessary for adaptation strategy that has to be fulfilled.

2.1. Smart Sensors & Transducers and their Grids

Smart sensor, according to generally accepted industry definitions, combines a sensing element, analog interface circuit, ADC, and bus interface; all in one housing [2]. Except it, smart sensor includes the microprocessor that conditions the signals before transmission to the control network. Also it filters out undesired noise and compensates for errors before sending the data. Making the grade against the newest generation of smart sensors, however, means that additional functionality must be included, such as self-testing, self-identification, self-validation, or self-adaptation. Of particular interest and importance to designers are such smart sensor capabilities as self-calibration and self-diagnosis, the ability to use signal processing, and multi-sensing capabilities.

Smart Temperature Sensor is the analog or digital primary thermosensitive transducer combined with a processing unit and a communicating interface [3] and able to perform a row of smart metrological functions due to installed metrological software. This is intelligent temperature sensor with a number of specialized algorithms provided in the design or installation stage, i.e. a sensor with such embedded algorithms that are necessary to provide implementation of following specialized metrology functions. Namely, such functions include, f.i. the ability to realize automatic switching of sub-range of measurement, depending on input signal value; automatic self-validation, self-check, self-diagnostics and etc.; the introduction of adjustments when the action of impact factor takes place; linearization of metrologic characteristics; compensation of cold-junction temperature for thermocouples and so on.

The high-mentioned autonomous smart sensors being spatially distributed are frequently combined in network assigned to monitor preset or environmental conditions (T, p, c, etc.) and to cooperatively send received and processed data by network to the end user. A set of such sensors is named WSN if the specified sensors are interconnected and distributed in space. Furthermore, the most modern WSNs can operate bi-directionally which means that the control of smart sensors becomes enable as well as their software and middleware updating.

Each WSN consists structurally from a large number (up to even thousands) of nodes which are the individual sensors that have ability to radio communicate with one or several neighboring units. Most common WSN is the fire alarm sensors network each branch of which has up to 2^6 sensors that was caused by limiting the length of microcontroller register (8 bits) [4]. Topology of every WSN may differ: star, cluster tree, mesh, up to advanced multi-hop mesh WSN. Propagation technique between the hops of network can be routing or flooding. Nowadays, the problem arises to adapt the traditional network topologies to the contemporary communicating conditions. Its significant segment is described in [5] and considered below.

Early sensor networks used simple twisted shielded–pair implementations for each sensor. Later, the industry adopted multidrop buses (e.g., Ethernet). Now we can apply web-based networks (e.g., the World Wide Web) implemented on the factory floor. Three classic network topologies (point-to-point, multidrop, and web) are well-known. In the more reliable first one, each sensor node requires a separate twisted shielded–pair wire connection. In a multidrop network, each sensor node puts its information onto a common medium; this requires careful attention to protocols in hardware and software. Afterwards, few networks have provided frequency-modulated signals on wires to carry multiple sensor readouts on FM channels. Furthermore, early WSNs were simple radio-frequency implementations of this topology. They applied RF modems to convert the RS-232 signal to a radio signal and back again. Due to simple FM coding and interference impact, the reliability of such networking does not satisfied principally. Only when it was selected the particular frequency ranges of network operation the complete wireless local area networks were emerged. Remote data acquisition systems of similar topology are frequently implemented with in-field data concentrators and radio transmitter aiming the hosts, where the signals are demultiplexed into original sensor signals.

Once the industry began the migration to multidrop buses, problems associated with digitization began to emerge [6]. Especially, it occurs with point-to-point systems, where a single clock of host could be used to time stamp when the analog signals from multiple sensors were acquired. With the distributed smart sensors required to implement a multidrop network, synchronization of their clocks becomes the critical issue and important design parameter. Somewhat different problems are inherent in a web topology which all nodes are potentially connected to all other nodes. Some of them can be eliminated by applying the repeaters and routers to make virtual connections.

A significant interest is emerged to the ad hoc networks. They are multi-hop networks consisting of wireless autonomous hosts, where each host may serve as a router to assists traffic from other nodes [7]. Wireless ad hoc networks cover a wide range of network scenarios, including sensor, mobile ad hoc, personal area, and other networks. The research activities in the specific field of WSNs include sensor training, security through smart node cooperation, sensor area coverage with random and deterministic placement, object location, sensor position determination, energy efficient broadcasting and activity scheduling, routing, connectivity, data dissemination and gathering, path exposure, tree reconfiguration, and topology construction.

Network topology involves not only the sensor nodes, but also the base stations and cross-layers for normal operation. Cross-layering becomes quite important for WSNs. So cross-layer can be used to make the optimal modulation to improve the transmission performance, such as data rate, energy efficiency, QoS, etc.

2.1.1. Distributed Sensor Network

In accordance with [7] the ideal wireless sensor is networked and scalable, smart and programmable, capable of fast data acquisition, reliable and accurate over the long term, and requires the minimal maintenance. Selecting the optimum sensors and wireless communications link requires knowledge of the application and problem definition: battery life, sensors update rate, and size of all major design considerations. Examples of low data rate sensors include temperature, humidity kinds. Examples of high data rate sensors are the sensors of strain, acceleration, and vibration. Contemporary examples of extra high data rate sensors are thermal and optical imagers, remote control sensors, etc.

Advance in microelectronics have resulted in the ability of sensors, based on digital electronics, and radio communications to develop the integrated units. Such sensors acquire the ability to communicate with each other with help of wireless data routing protocols. The WSN preferably has to consist of a base-station ("gateway") that can communicate with a few of wireless sensors via radio-waves. Data is accumulated at the sensors node, compressed and transmitted to the main or to the intermediate base station. We are not considered the architecture of WSNs below as it is traditional (star, mesh, hybrid star - mesh). Similarly it is not the special purpose to study ordinary network standards (IEEE802.11x that is used for LANs at high bandwidth data transfer; IEEE802.15.1 and 2 that are known as Bluetooth that is of lower power and applies for personal area network; IEEE802.15.4 that is specifically designed for the requirements of WSNs, here the worldwide licence-free band – 2…4 GHz – is applied; ZigBee which alliance encompasses the IEEE802.15.4 specification and expands on the network specification and the application interface; and one of most modern standard is IEEE1451.5 that aims the operating of smart sensor working groups in their wireless option.

If a centralised architecture is used in a sensor network and the central node fails, then the entire network will collapse, however the reliability of the sensor network can be increased by using distributed control architecture. Distributed control is used in WSNs for the following reasons: sensor nodes are prone to failure, for better collection of data, to provide nodes with backup in case of failure of the central one, resources have to be self-organized.

2.1.2. Smart Grid Sensor

A smart grid sensor is a small, lightweight node that serves as a detection station in a sensor network. Smart grid sensors enable the remote monitoring of equipment such as transformers and power lines and the demand-side management of resources on energy smart grid [8]. Smart grid sensors can be used to monitor weather conditions and power line temperature, which can then be used to calculate the line's carrying capacity. This process is called dynamic line rating and it enables power companies to increase the power flow of existing transmission lines. Smart grid sensors can also be used within homes and businesses to increase energy efficiency [9]. According to Nano Markets, companies like GE, LG and Whirlpool have already announced their commitment to building home appliances that are smart-enabled. Smart grid sensors will link these appliances with smart meters, providing visibility into real-time power consumption. Power companies can use this information to develop real-time pricing and consumers can use the information to lower their power consumption at peak times, during the high electricity prices. Specific electronic devices can automatically change behavior in response to real-time information from electric grid smart sensors [10].

2.1.3. Middleware and Wireless Sensor Networks

Middleware layer is a novel approach to fully meeting the design and implementation challenges of WSN technologies. We are considering the WSN middleware as a software infrastructure that glues together the network hardware, operating systems, network stacks, and applications.

A complete middleware solution should contain a runtime environment that supports and coordinates multiple applications, and standardized system services such as data aggregation, control and management policies adapting to target applications, and mechanisms to achieve adaptive and efficient system resources use to prolong the sensor network's life. Middleware should provide low-level programming models to meet the major challenge of bridging the gap between hardware technology's raw potential and the necessary broad activities such as reconfiguration, execution, and communication.

The relevant middleware projects for WSNs were studied in [11]. In cases where physical contact for replacement or maintenance is impossible, wireless media is the only way for remote accessibility. Hence, middleware should provide mechanisms for efficient processor and memory use while enabling lower-power communication. Major WSN properties studied below are the next. Scalability is the first one and is defined as follows: if an application grows, the WSN has to be flexible enough to permit this growth anywhere and anytime without affecting network performance. Efficient middleware services must be capable of maintaining acceptable performance levels as the network changes its dimensions and topology. The latter is subject to frequent changes owing to factors such as malfunctioning, device failure, mobility, and interference impacts and so on.

Most WSNs' applications are real-time phenomena; so, middleware should provide real-time services to adapt to the changes and provide consistent data. Application knowledge's design principles dictate another important and unique property of WSN middleware. Middleware must include mechanisms for injecting application knowledge of WSN's infrastructure. This lets developers map application communication requirements to network parameters, which enable them to fine-tune network monitoring. WSNs are being widely deployed in domains that involve sensors information for example, in healthcare and rescue areas. The untethered and large deployment of WSNs in harsh environments increases their exposure to malicious intrusions and attacks such as denial of service. In addition, the wireless medium facilitates eavesdropping and adversarial packet injection to compromise the network's functioning. All these factors make security extremely important. Furthermore, sensor nodes have limited power and processing resources, so standard security mechanisms, which are heavy in weight and resource consumption, are unsuitable. These challenges increase the need to develop comprehensive and secure solutions that achieve wider protection, while maintaining desirable network performance. Middleware efforts should concentrate on developing and integrating security in the initial phases of software design, hence achieving different security requirements such as confidentiality, authentication, integrity, freshness, and availability [12-13].

2.2. Metrological Problems of Smart Measuring Instruments

CPSs' development foresees, in particular, the progress in modeling tools and data sets that enable researchers accurately model the scale and dynamics of current and future Internet control systems. The same holds for models and tools for measuring and predicting the performance of secure network-based CPS. In this direction the method of non-destructive noise spectroscopy is considered as perspective one [14].

2.2.1. Specificy of Measuring Instruments Due to Their Smartness

Analog interface is a set of MIs that are an integral part of measuring channel between the primary measuring transducer and ADC. It performs the following functions: scale transformation of measurement signals, their filtering, temperature compensation of thermocouples cold junctions, galvanic separation of transferring and receiving parts, linearization of characteristics and initialization of passive MTs, measuring signals multiplexing, measurement and service information transmission.

Smart sensors are supplied with digital information transmissive means by equipping them with built-in digital controllers to match the universal network interface or by combining technology of analog and digital transmission in a single CL. By the structure all smart sensors are divided into 4 groups: sensors of centralized and decentralized types, as well as sensors with digital and analog buses. By correction methods the analog interfaces with smart sensors are divided into the groups: with manual error correction, with auto correction of errors in analog-digital form, and with digital correction of errors, that are considered in detail in the Chapters 4-5.

It is imperative the further development of R&D works regarding the assessment of accuracy in the field of metrology in particular while single measuring the multivariable values by using multiparameter smart MIs, for instance CMFs [15]. They can be considered to flexible tools and equipment to be able to quickly re-engineer existing procedures. Resultantly, by choosing the right measurement technology solution we can efficiently supply both development and production with relevant data.

CMF transformation function is the dependence of the liquid mass or volume on its flowing velocity through the specified cut. It is determined by the comparison of the time characteristics of two identical sensors in the CMF input and output. More phase difference of mentioned characteristics corresponds to faster controlled environment flowing. So there is a dependence on the hydrodynamic regime of the current environment flowing through the CMF, its viscosity, the temperature etc. Within the current metrological conception the CMF consists of transducers with appropriate sensors and peripheral devices and the microprocessor unit of received signals processing. The CMF sensors determines the flow velocity, temperature and provide information in form of output signals to the microprocessor that carries out the function of the brain of the measurement device and system in total providing access to the display, main menu and output device of processed information for the interaction with other systems, for instance, the filling system. Peripheral devices provide monitoring, warning signalization and other functions, for instance, periodic processes management and the function of liquid density more accurate determination etc. The CMF transfer function error temperature component is the error specified by the temperature regime of liquid/gas flowing. It depends on the temperature of the control environment (on temperature dependence of the liquid flowing regime through the CMF); on the CMF body outlet temperature that is provided with the help of temperature detecting means for detecting a temperature of the inner tube, and temperature detecting means for detecting a temperature of the outer tube; temperature correcting means for compensating an instrumental error according to a change of temperature of the inner tube; temperature difference correcting means for compensating an instrumental error according to a difference between a temperature of the inner tube and a temperature of the outer tube.

Next example of the necessity to develop smart MIs could be industrial tomography systems. Their development consist in development and implementation of methods and algorithms to process the results of measurements which would enable for a limited amount of gauges exactly and quickly obtain high-quality images of the distribution of the studied variables. Mainly it would be the spatial distribution of the electrical conductivity of the medium. Improvement of the exactness and rate of tomography measurement have been reached recently [15] by reducing additive gauge errors through the use of the difference measurement method; by elaborating the particular methodology for calculating the sensitivity matrix without of methodical errors inherent in finite-difference methods and based on only a unitary solution of the direct task; ensure the stability and convergence of the iterative procedure of the spatial distribution determination.

2.2.2. Verification of Metrological Software and Middleware

The presented in this book approach substantially differs from the generally accepted one. The latter is described in the following subsection. Commonly as the software metric is considered a standard of measure of a degree to which a software system or process possesses some features. Since quantitative measurements are essential in all areas, the goal is obtaining objective, reproducible and quantifiable measurements, which may have numerous applications in schedule and budget planning, cost estimation, quality assurance testing, software debugging, software performance optimization, and so on.

Our specific measurement consists in evaluation of MIs performance reliability, trueness, and other metrological properties, due to the quality of the certain kind of metrological software, or the software linked with metrological features of MIs.

Design and development of MIs involves the reduction of analog units and expansion of digital ones due to the advances of last. These include the digital microcontrollers, programmable logic arrays and more. Trend of modernity consists in growing the "weight" of software in measurement means. Inconsistency of software to the measuring tasks of certain MI, accidental or intentional alteration of its functions can lead to incorrect measurement results. It is therefore advisable to conduct verification of software to derive its impact on the metrological characteristics of MI and the possibility of further running the studied software as a consisting part of the device, tools, or metrological mean [16].

Under the MI's software we understand a set of programs and procedures designed to register, aggregate, process, display and save, and post the measurement results. Such software as a functional part of MI is delivered jointly with hardware [17-18]. According to [18] software that may affect the metrological characteristics of measurement instruments, concerns: a) Programs and program modules that participate in processing of measurement results; b) Involved in the calculations the software parameters that affect the measurements result; c) Programs and program modules that carry out presentation of measurement data, its storage and transmission, software update and identification, secure software development [19] and data protection; d) Components of protected interface for data exchange between software modules of CPS's units.

The last 2 items are appropriately attributed to overall not metrological verification of software, since no measuring or computing actions that affect the measurement results, have been fulfilled. These points characterize the correctness of the software functioning in general, especially while exploiting the set of different compilers, build utilities, debuggers, code analyzers, and software versioning control systems [20]. So, we have the expertise to incorporate security practices – authentication, authorization and auditing – into each phase of the verification, from software design and implementation to testing and deployment.

Secure software engineering has become an increasingly important part of software quality, particularly due to the development of the Internet. While IT security measures can offer basic protection for the main areas of our IT systems, secure software is also critical for establishing a completely secure CPS's environment. Every single software MI developer must care about security, because users need to be able to trust the proposed software including the MI software and middleware. Entire cover cycle of the aforementioned wares development includes requirements engineering, trust & threat modelling, secure coding, security testing, and security response to code protection [21].

MI software metrological verification raises the problem of choice the appropriate methods of software and middleware assessing, testing, and certifying. Result of the metrological validation must be the confirmation or negation of studied ware to the requirements noted in normative documents. Procedures and methods of checking software, and determining its disadvantages are considered below. Software study include first of all the fulfilment the procedures of unambiguity ensuring the operating functions for generated data. Selection of the procedures is defined by regulation requirements, as well as by software developer or user desires to confirm its compliance with target specification.

To validate the type of MI, test procedure should envisage an identification and evaluation of software impact on metrological characteristics as well as to prevent unauthorized software reconfiguration and interference that can degrade the trueness of received measurement results. Therefore the developed under tests, project description of MI type should contain additionally to metrological characteristics the description of software, identity, impact assessment, and the level of protection against unintentional and intentional changes. Certification of MI software and middleware is the research that aims the determination of characteristics, features, identification data, and confirmation of requirements compliance. In accordance with methodology of software certification to determine one or more characteristics (analysis of documentation and source code, functional inspection under controlled conditions, etc.) the tests are carried out [22].

Distinguish two kinds of certification: general and metrological. The object of general appraisal is complete software, study of which is conducted to justify the application of algorithm (program) within specific tasks. To assess the impact of software on the data inexactness or untrueness, the metrological certification examines merely that software which is an integral part of complete software of the particular MIs. Software certification is mostly voluntary except the software that performs especially responsible functions, where the lack of quality, mistakes or failures can severely disrupted or are dangerous for life and health (aviation, nuclear power, management of authorities and banking systems, etc.) [23].

It is important to choose the correct verification method for MI software and middleware. There are considered the possibility and cost of its implementation, the checking quality and more. Methods of software verification are considered below (Fig. 2.1).

Method of comparative testing with applying the reference software is used for the certain software which helps to identify its features being checked. As the reference software can be studied: attested and/or certified software, functionality of which is similar to tested software features; specially developed software with functions that are identical to functions of software that has to be checked; software for assignments of computing tasks (e.g. spreadsheets, software for mathematical and statistical calculations, etc.).

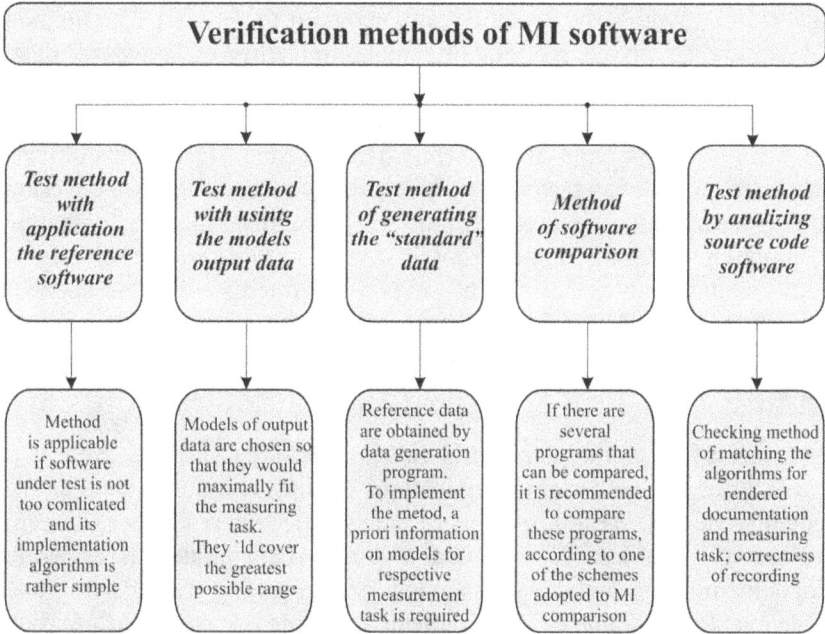

Verification methods of MI software

Test method with application the reference software	Test method with usintg the models output data	Test method of generating the "standard" data	Method of software comparison	Test method by analizing source code software
Method is applicable if software under test is not too comlicated and its implementation algorithm is rather simple	Models of output data are chosen so that they would maximally fit the measuring task. They 'ld cover the greatest possible range	Reference data are obtained by data generation program. To implement the metod, a priori information on models for respective measurement task is required	If there are several programs that can be compared, it is recommended to compare these programs, according to one of the schemes adopted to MI comparison	Checking method of matching the algorithms for rendered documentation and measuring task; correctness of recording

Fig. 2.1. Verification methods of software for metrological instruments.

To development of reference software, somebody may resort in cases when the software under test is not too complicated and its implementation algorithm is rather simple. Reference software should not reproduce all the functionality of software that is tested and could only contain functions and parameters that affect MI metrological characteristics. In certain cases, the peculiarities of graphic user interface and functions that do not participate in the handling of measurements (e.g., display functions, data storage, etc.) are not taken into account. Given method enables to consider the metrological characteristics and maximum peculiarities of software that is tested. Drawback of this method is that often the complexity of software implementation makes it inexpedient due to high costs of developing the reference software.

In absence of reference software, the priority is given to the method of comparative trials by using output data models and comparative tests with the generating the "standard" data. The latter is recommended for certification of data processing algorithms of measurement results. Method enables to evaluate the algorithm possibilities by comparing of processed results of output data model obtained with help of mentioned algorithm, towards the specified parameters of this model. The method of output data models is a kind of generation method of "standard" data, not just data generated by specially developed programs; only data are not generated by the specially developed program, and programmatically they are set at the input of software which is tested. Output data models are selected in a way that they fully regard the measuring tasks covering the greatest range of possible values. These may include: data that completely cover the range of possible values; data close to the largest and smallest values as well as several intermediate values; specific values of input variables – points of the sharp rise or rupture of derivatives, zero, single, and extremely small numerical values of variables, etc.

If values of a variable depend on values of another variable, the test is carried out for the certain combinations of these variables, such as the equality of two variables, or large and small their difference, or zero values of variables. Test method of output data models is easier to implement than the classic method of generating "standard" data. However, development of this method requires a priori information about software algorithms and their program realization, which it is not always known.

Method of generating "standard" data, as the method of output data models, is applied alternatively to the reference software method, or in the case of inability to check the particular functions of implemented software. One of the prerequisites of method of generating "standard" data is the availability of a priori information about the corresponding measuring task. "Reference" data is produced by specially developed generating program, i.e. the generator of "standard" data based on specified output data. The last realize in one of programming languages or by using the standard mathematical (statistical) software package. Initial data for testing, including for the generation of "reference" data, are formed considering the properties of software algorithms that are checked. Method of generating "standard" data is an alternative one to the method of using the reference software. On the other hand, development generator of "standard" data is advisable when it is cheaper than other methods of implementation [24].

If there are several program realizations of the same measurement algorithm at absence of the reference software, is expedient to carry out checks by comparing method implementation. According to it, the same set of "standard" data is submitted to inputs software products, and comparison of relevant test results is performed. Comparison method is simple and requires no the extra programs. However, programs with the same functions are quite rare.

When testing on an analysis of source code of software is checked the following items: conformity of algorithms structure to the rendered documentation; correctness of recorded algorithms to the chosen programming language; matching the selected algorithms up measuring tasks (detection of unstable algorithms).

When checking the compliance of algorithms structure to the rendered documentation, according to text of program the algorithms block diagram can be composed and compared with the algorithms described in documentation. In the event of differences between the structures of algorithms, the additional analysis elements block designs is conducted. When checking the correctness of algorithms to the chosen programming language, the compliance of code to programming regulations, the presence of uncertain variables and operators, the correct organization of cycles and so on are established. Conformity of selected algorithms with measuring task can be estimated by mathematical analysis of implemented software algorithms. It can be explored algorithms of realized different characteristics, in particular, may be executed an optimal analysis of numerical methods for solving measurement tasks. This method provides an opportunity for detailed assessment of metrological software. However to implement test method on analysis of source code, the experts in software industry and in metrology must be engaged simultaneously. So, costs for its implementation are much higher compare to other methods [25].

Middleware [11-13], being downloaded from the Internet and installed in MIs, is the software that provides services to software applications beyond those available from the operating system of programmable block of MI. It is often can be presented as "software glue". Middleware makes it easier for software developers to fulfil communication and input/output, so they can concentrate the attention on their own problems. Middleware is the software layer that lies between the operating system and such an application as WSNs. As a result, it supports complex, distributed business software applications.

Middleware includes Web servers, application servers, content management systems, and similar tools that support application development and delivery. It is especially integral to information technology based on Extensible Markup Language (XML), Simple Object Access Protocol (SOAP), Web services, SOA, Web 2.0 infrastructure, and Lightweight Directory Access Protocol (LDAP), which is commonly used for communicating and managing in distributed applications.

In simulation technology, middleware is generally applied in the context of high level architecture to a number of distributed simulations. This is a layer of software that lies between the application code and the run-time infrastructure. Middleware mainly consists of a library of functions, and enables a row of applications—simulations or federates in HLA terminology—to page these functions from the common library rather than re-create them for each application. Wireless networking developers can use middleware to meet the challenges associated with WSNs. Implementation of middleware allows developers to integrate operating systems and hardware with the wide variety of various applications that are currently available. For instance, radio-frequency software toolkits provide middleware to filter noisy and redundant raw data.

2.2.3. Fractional Processes in Measuring Techniques

Fractional processes are widely found in science, technology and engineering systems. In fractional processes and fractional-order signal processing, some complex random signals, characterized by the presence of a heavy-tailed distribution or non-negligible dependence between distant observations (local and long memory), are introduced and examined from the 'fractional' perspective using simulation, fractional-order modeling and filtering and realization of fractional-order systems. These fractional-order signal processing techniques are based on fractional calculus, the fractional Fourier transform and fractional lower-order moments. Fractional processes and fractional-order signal processing: presents fractional processes of fixed, variable and distributed order studied as the output of fractional-order differential systems; introduces these techniques and the fractional signals and fractional systems point of view [26].

Currently, wireless communications networks are developing quite quickly and utilizing more spectral resources. Despite all the efforts to improve the wireless network and energy-efficiency, spectrum is still overloaded. Problem results in decreasing the signal intensity and powering the interference impact. So, it is highly important to develop new methods of data transmission in wireless networks, including the WSNs. Better energy- and spectral- efficient methods have to be provided recently. To reach such the aim, we have proposed the techniques, signal models and hardware for new types of fractional comb-structured signals generation and transmission (on PIC16F873 pulse generator basis with the shortest pulse duration 20 μs at 4 MHz clock rate and 4 μs at 20 MHz rate). Obtained results envisaged that proposed signal model can be implemented in wideband wireless communications networks, providing the satisfactory characteristics. Such type of signals, namely the fractional signals of fractal orders 0, 1, 2, 3, may be recommended to encoding the digital information to be transmitted via wideband communication channels [27].

2.3. Metrology and Modelling for Additive Manufacturing

Increasing demands for precision measurement raises new problems of optimization of mathematical models of measuring transformations, and adequate processing of experimental data. This problem in modern MIs is particularly relevant due to the capabilities of inexpensive hardware implementation in basis of modern microelectronic components that opens the possibility of computing realization directly in the measuring path.

Obtaining the necessary precision in many cases is only possible in the case of optimal mathematical models that provide in a certain sense the best approach of MI general transformation function. Rational choice of mathematical model of transformation function in many cases improves the measuring accuracy or expands the measurement range of preset accuracy.

In this regard, becomes important choice of suitable criterion during processing the experimental data. Normal criteria mostly applicable for analyzing experimental data, is the most common method of mean squared errors, which consists in minimizing the sum of squares of the errors and computing the average of these squares. Unfortunately, the root-mean-square approximation does not provide achievement of the lowest difference between the estimator and function that is estimated at all points of observation, which is desirable during the precision processing of experimental data. Therefore, for solving the calibration tasks should be used the minimax criterion which ensures the minimum possible errors of reproducing the experimental calibration characteristics.

Physical modeling is an experimental method of scientific research, which implies the substitution of the studied physical process by other similar to it of the same physical nature - by model.

Physical model is a smaller or larger physical copy of an object. The geometries of model and object are often similar in the sense that one is a rescaling of the other; in such cases the scale is an important characteristic. Geometrically similar to the original the model can be both reduced and increased in the comparison with original sizes, and the model of process or phenomenon may differ from the real process by the quantitative physical characteristics such as power, energy, process pressure etc. In a broad sense, any physical experiment conducted in laboratory, including an experiment with natural object or part of it, is a physical modeling. The latter is based on the similarity theory and dimensional analysis, establishing the similarities criteria. The identity of the latter for a nature and the model provides the ability to transfer the experimental results obtained by physical modeling, in natural conditions. With the implementation of relevant conditions of physical modeling, i.e. the identity of similarity criteria, the values of variables that characterize a real phenomenon of proportionality of the similar points in space and at similar moments of time, become to be proportional to values of the same variables for the model. Presence of such proportionality allows perform recalculation of experimental results that were obtained on a model by multiplying the value of each of the identified variables on a constant for all values a given dimension set factor - the similarity factor [28].

2.3.1. Metrological Problems of Additive Manufacturing

Advances in AM processes and systems caused by the necessity of design-to-product transformation have resulted in evolution of measurement science, namely in the conjugated material characterization and their performance qualification as well as in process sensing, monitoring and model-based controlling of particular stages of CPS's operation. Since the AM provides "the agility needed to rapidly make innovative customized complex products and replacement parts that are not realizable by more traditional manufacturing technologies or are required to be produced in low volumes" [29], it is especially demanded by CPS-based technology. Therefore NIST Program focuses on overcoming the hurdles, for instance, of 3D-printers wide-spreading resulted from metrology insufficient development in aforementioned area. Such an optimized system requires the tightly integrated software and hardware components with well-understandable interactions. That is why it should be studied below the contribution of these components into the total efficiency. Unfortunately, 3D-printers, especially printers that perform the units from unique compositions, are currently supplied to market as closed "black boxes".

We, as the process developers, try to access the internal hardware and software components with integrated new sensors, models, or control algorithms oblivious to AM vendors [30].

Researches of Selective Laser Sintering technology have been conducted (Fig. 2.2) in this way [31] basing on their own measurements and on EOS GmbH data. Here the Quality control is one of the basic requirements. Monitoring the key parameters of laser melting of metallic powders - temperature, oxygen content, power of the laser, and quality of powder - have become a part of modern standards. However, it is impossible to describe exhaustively the product quality on the basis only of equipment parameters. Systems of controlling the local processes in-situ, created by coaxial principle, are desperately needed. In such away it could be received the information about defects that arise during manufacturing.

For monitoring the processes system QMmeltpool 3D [28] in real time provides data that determines the products quality, and records parameters of the melt zone with spatial reference, ensuring the 3D-visualization for further analysis which is similar by HD-resolution to computer tomography. The received result is exceptionally accurate 3D-description of the product, or more truly the definition of characteristic features of deposited layer (measured by photodiode and camera data on area and temperature of the melt zone that are conjugated with relevant data on the laser beam location). General data are visualized in 3 coordinates immediately, gaining the opportunity to track the process of creating each detail in space. This approach simplifies the identification of individual defects in obtained product. Due to the high resolution and frequency of digitization (every 0.1 mm, depending on the speed of scanning), the mentioned system simplify the identification of defects at early manufacturing stages as well as avoid them in the future. Consequently, there arises the possibility to optimize production and technical developments through iterative parameter variation. It can be reached through a two-way process, firstly, by software development on correct description of the details' surface while manufacturing or reconstructing, or, secondly, by further technology development of layerwise manufacturing of products of given shape from specific, for example, a composite material.

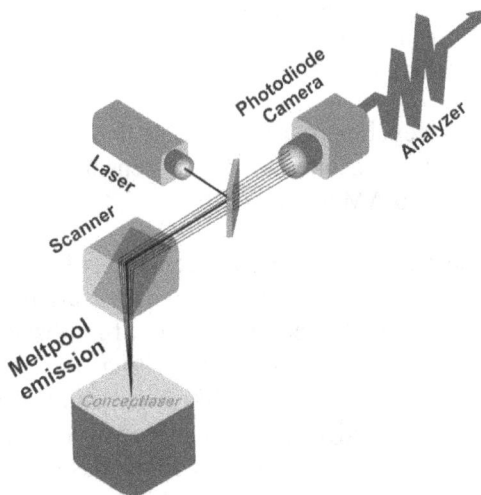

Fig. 2.2. Local monitoring the melt area with help of system QMmeltpool 3D: photodiode and camera are responsible for coaxial monitoring [31].

There exists a number of problems in aforementioned items, for instance in software reconstruction of complex surfaces. Let us examine below the way they are reached in more advanced science and technology.

2.3.2. Computation Method of Segmental Approximation for Random Topology Single-Surface Reconstruction of Interferometric Patterns

Measurement of topology or surface profile of objects is important task in many areas, for instance, in medicine [32-33], industry [34-35], scientific researches [36]. One of the methods of surface topology measurement is the WLI [37]. Owing to absence of physical contact with researched surface, high resolution and scanning speed, ability to analyse objects of big geometrical size, this technique has been used extensively in tasks of surface research. Moreover WLI unlike monochromatic interferometry enables to measure stepped surfaces and surfaces with significant curvature [38-39].

Mentioned advantages are wider presented in applications of WLI different techniques for various nanotechnological plants and cyber-physical systems. The most common examples of such usage is measurement of roughness and microstructures profile with nanodimensional Z-resolution, material standards calibration, and also integration with different non-optical measurement techniques for more thorough and versatile analysis of researched object nature [40-41]. For instance, modern 3D profilometer Talysurf CCI 6000 which operates on the basis of WLI and provides surface profile measurement of objects of linear dimension up to several mm with Z-resolution 0.01 nm and XY-resolution 360 nm [42-43].

However the use of white light complicates the analysis of interference signal due to envelope function of Gaussian spectrum [44-45]. Today there are many methods of WLI data processing: fringes tracing, phase shift methods, determination of maximum intensity of interferogram, direct phase demodulation [46-47]. But effectiveness of these methods reduces during complex surfaces reconstruction, especially at analysis of dynamic (variable in time), nonlinear (spherical) surfaces [48]. For this purpose it is urgent to develop new approaches which ensure improvement of metrological characteristics of surface reconstruction methods, and are efficient in realization.

The interferometer of classical variant includes light source, beam splitter, reference mirror, researched surface and screen (CCD-camera). The light wave of light source is divided into two paths directed to reference mirror and researched surface respectively. Image is formed in the screen as result of superposition of waves that reflected from respective surfaces. Strengthening and weakening of resultant waves is formed by means of optical difference of light rays paths.

Mathematical model of white-light interferogram is considered as follows [49]:

$$I(T) = I_0 + E(T)C(T)$$

$$E(T) = I_M exp - \left(\frac{4\Delta\lambda^2 T^2}{\lambda_0^4}\right) \tag{2.1}$$

$$C(T) = cos\left(\frac{4\pi}{\lambda_0}\right)T,$$

where I_0 is the constant component; E(T) represents the interferogram envelope; C(T) denotes the carrier; IM is the modulation amplitude; T is the optical path difference; λ_0 represents the central wavelength; $\Delta\lambda$ is the spectral bandwidth of the light source. The sense of reconstruction is to determine the optical path difference T from interferogram that is proportional to height displacement h (or profile) at every point of researched surface:

$$h = \frac{T}{v}, \tag{2.2}$$

where v is the refractive index. We study the tilted and spherical surfaces. The first one is used in measurement of small inclination angles of flat objects (wedge) and spherical surface corresponds to membrane model in pressure sensors [35].

By applying mathematical model it is possible to synthesize interferometric patterns. Interferogram view depends on light source parameters and surface topology. In the Fig. 2.3 there are interferograms of tilted (a) and spherical (b) surfaces for light of central wavelength $\lambda_0 = 620\ nm$ and bandwidth $\Delta\lambda = 52\ nm$. Interferograms contain 1000 pixels on vertical and horizontal axes.

Synthesis of surfaces, interferograms and data processing with the aim of reconstruction of surface profile is done in MatLab package, because it contains developed library of functions which are necessary to conduct researches.

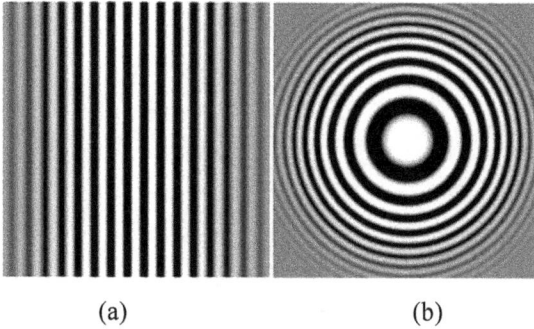

(a)   (b)

Fig. 2.3. White light interferograms of tilted (a), and spherical (b) surfaces.

Interferometric pattern is created as set of pixels of different intensity and every line of interferogram can be considered as discrete signal I(n). Dependence of intensity signal on sample number n is different significantly for flat and spherical surfaces.

In contrast to this, the dependence between intensity signal I and optical path difference T in accordance with (2.1) is identical regardless of researched surfaces form. This aspect of invariance is put in the basis of developed method of segmental approximation.

Segmental Approximation Method. Since mentioned above the intensity signal as function of optical path difference is invariant to the profile of researched surface. So the shape of the intensity signal obtained on the linear surface can be transferred on spherical surface and vice versa on conditions that values of optical path difference are close. If interferogram segment is approximated by polynomial:

$$\tilde{I} = \sum_{m=1}^{M} \alpha^m T^m \qquad (2.3)$$

So reconstruction of surface profile demands search of polynomial roots that is computing-complicated task. However the reconstruction is sufficiently simplified under application of the function which is reciprocal to I(T). Determination of informative parameter T demands approximation function value computing (Fig. 2.4):

$$\tilde{T} = \sum_{n=0}^{N} \alpha^n I^n \qquad (2.4)$$

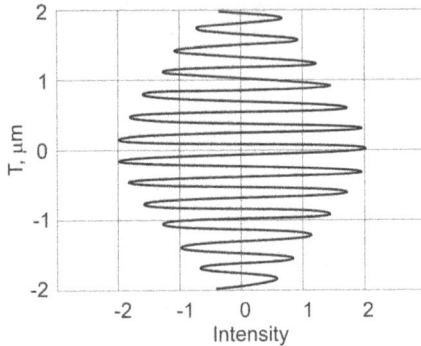

Fig. 2.4. Graph of informative parameter T on interferogram intensity I.

This approach enables to approximate profile fragment on calibration surface (with known parameters) and further to apply approximating results for any other surface. In practice it looks like as calibration of interferometer measuring channel, therefore light source parameters have to be identical at calibrating and measuring stages.

Given in the Fig. 2.4 function is multivalued with variable area of T dimensions at different segments. These conditions complicate the problem of polynomial approximation of T(I) function. Therefore approximation has to be performed for every segment of T(I) function. Segment is a stretch of function within change of I argument values. T(I) function segmentation is conveniently to bring to search of local extremes of its reciprocal version, i.e. to intensity of registered signal.

At every segment the optical path difference is approximated separately by algebraic polynomial with help of MatLab package:

$$\tilde{T} = a_0 + a_1 I_C + a_2 I_C^2 + \cdots + a_N I_C^N \tag{2.5}$$

Here I_C is the signal intensity on calibration stage; $a_0 \ldots a_N$ are the polynomial coefficients. Polynomial coefficients are determined from minimum condition of root-mean-square error of approximation:

$$\varepsilon_A^2 = min \sum_{k=1}^{K} (T_k \widetilde{T_k})^2, \tag{2.6}$$

where k is the number of signal samples. Basing on invariance of intensity signal concerning optical path difference, profile of researched surface can be approximated by polynomial with coefficients determined during calibration:

$$T = a_0' + a_1'I + a_2'I^2 + \cdots + a_N'I^{2N}, \qquad (2.7)$$

where I is the signal intensity of investigated surface; $a_0' \ldots a_N'$ are the polynomial coefficients obtained in calibration step. Coefficients values of approximating polynomial depend on intensity signal swing. As confirmation the coefficients of polynomial of 3rd degree for segments 1, 2, 15 and 16 are given in Table 2.1 received by calibration on the tilted surface.

Table 2.1. Coefficients values of approximating polynomials in different segments.

Segment	a_0, μm	a_1, μm/kd	a_2, μm/kd^2	a_3, μm/kd^3
1	-2.4	-0.17	-0.2	-1.75
2	-2.25	0.13	-0.11	0.81
15	-0.23	-0.02	$-3.96 \cdot 10^{-4}$	-0.004
16	-0.08	0.02	$-1.28 \cdot 10^{-4}$	0.0038

This condition makes to select the nearest segment by swing from set of calibrating segments. The criterion for selection is the coverage of analyzing surface segment by calibration one. The simplest selection algorithm is the comparison of extreme values of the analyzed segment with a set of calibration values towards the growth of their swing.

Thus surface reconstruction method by segmental approximation consists of two stages: calibration and, properly, reconstruction.

The interferogram, obtained for surface with known parameters, is registered at calibration of interferometer measuring channel. Then segmentation of received interferogram is performed and polynomial coefficients for every segment are determined by (2.5).

Reconstruction stage includes the next steps:

- Downloading of researched surface interferogram data;

- Segmentation of researched interferogram and searching identical "calibration" segment;

- Informative T parameter computing by (2.7) for every segment;

- Visualization of the reconstructed surface.

Methodology of accuracy research envisages the synthesis of tilted and spherical surfaces, simulation of interferogram (Fig. 2.3) on the basis of mathematical model, and surface reconstruction of the basis of developed method. For purpose of deep analysis of method performance the surface reconstruction accuracy was studied on the basis of some lines data. So far as in contrast to tilted surface, the spherical one depends on coordinate y (line number), and research concerned the central row No. 500 and remote from center row No. 150.

To estimate accuracy of reconstruction, the conventional error and RMS conventional error are determined:

$$\gamma(T) = \frac{|T_{rec} - T_{origin}|}{max(T_{origin}) - min(T_{origin})} \cdot 100\,\% \qquad (2.8)$$

$$\sigma = \frac{\sqrt{\sum_{i=1}^{N}[T_{rec}(i) - T_{origin}(i)]^2}}{N[max(T_{origin}) - min(T_{origin})]} \cdot 100\,\%, \qquad (2.9)$$

where $T_{rec}(i)$ is the height of reconstructed surface at the point i; $T_{origin}(i)$ is the height of original surface at the point i; N is a number of samples. The conventional error of tilted surface profile reconstruction is given in Fig. 2.5.

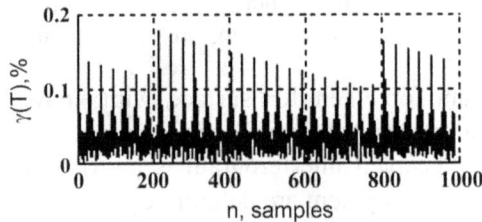

Fig. 2.5. Conventional error of tilted surface profile reconstruction.

As shown, the reconstruction maximum error doesn't exceed 0.2 %. Error graph is inherent in periodical character with brightly identified

66

spurs. Maximum errors are observed on the borders of segments which are specified by deviation of approximating polynomial from real interferogram. RMS error is equal to 0.05 %.

There are the profile reconstructions (Fig. 2.6) of conventional errors of spherical surface for central row No. 500 and remoted from the center row No. 150. The error character is not periodic for spherical surface, as continuance of interferogram segments of spherical surface is increasing with approaching to the center. Maximum values of conventional error do not exceed 1 %, and RMS error is 0.2 % for row No. 500 and 0.18 % for row No. 150.

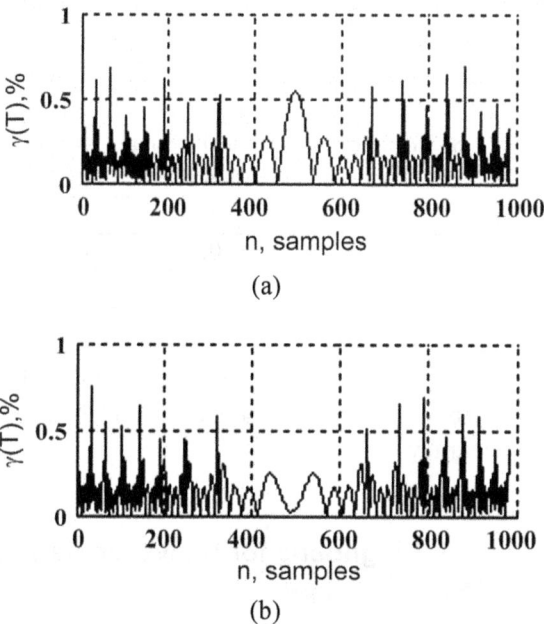

(a)

(b)

Fig. 2.6. Conventional error of spherical surface profile reconstruction for rows No. 500 (a), and No. 150 (b).

So far as the source of errors is discrepancy between real interferogram and approximation polynomial, especially on the borders of segments, it was studied the possibility of improvement of reconstruction accuracy by increase of approximation degree. With increase of polynomial degree the character of errors is not significantly changed; therefore only numerical values of errors are given for linear and spherical surfaces on condition of the use of approximation polynomials of 3, 5 and

15 degrees. Table 2.2 presents the maximum conventional errors of reconstruction, and Table 2.3 describes their RMS values.

Table 2.2. Maximum conventional errors of reconstruction depending on the polynomial degree.

Polynomial degree	$\gamma(T)$, % tilted surface	$\gamma(T)$, % spherical 500 row	$\gamma(T)$, % spherical 150 row
3	0.17	0.69	0.78
5	0.12	0.5	0.57
15	0.06	0.28	0.32

Table 2.3. RMS errors of reconstruction depending on the polynomial degree.

Polynomial degree	σ, % tilted surface	σ, % spherical 500 row	σ, % spherical 150 row
3	0.05	0.2	0.18
5	0.03	0.12	0.11
15	0.01	0.03	0.04

Research results envisage that the increase of approximation polynomial degree almost twice and in five times does not adequately improve the reconstruction accuracy.

2.3.3. Model-based Diagnostic for Heavy Operating Conditions

For modern manufacturing, with the motivation CPS development, a "coupled-model" approach is vigorously studied. Coupled model is a digital twin of the real machine that operates in computing (cloud) platform and simulates health condition with integrated knowledge from both data driven analytical algorithms. NIST has expertise in a number of CPS domains and is uniquely positioned to achieve breakthroughs by adapting measurement science solutions across domains and, where possible, developing domain independent solutions. This initiative focuses on three key problems. The first of them is associated with model-based diagnostics and prognostics [50]. Ultra-modern, flexible manufacturing processes can only be implemented with powerful measuring equipment. Product developers and production planners are searching for essentially the same thing: flexible systems featuring easy integration of new technologies without users having to retrain each time. The challenge is to respond to market demands using available methods and tools. The two great pillars of support in this area are simulation and product design [51].

Temperature precise measurement in the hostile environment can be considered as one of such tasks. Nowadays this measurement is very urgent due to intensive development of new directions especially in chemical, pharmaceutical and food industries. It's known that some stages of the technological process, which are in above mentioned branches, to provide corresponding characteristics of final product have to go under the certain temperature taking into account supporting of optimal conditions of chemical reactions running as well as technogenic safety. Since demands for temperature measuring instruments in such processes are very strict because they finally provide the product's quality and safety of long term producing.

Under such processes realization thermometric major instruments are mainly the resistance thermotransducers, and, as it is known, the hostile environment impacts negatively on the SE substance of thermotransducer [52]. Moreover, it is necessary to take into account the fact that under realization of continuous technological processes there is no possibility to change the thermotransducer whenever. Such temperature measuring instruments working in aggressive environment can be validated only under planned metrological verification, and in the period between scheduled metrological checks it emerges a possibility of worsening its metrological characteristics with corresponding outcomes.

Theoretically determined data points and components must still be tested with real prototypes of resistance thermotransducers under conditions closely simulating series production. The studied objects are subjected to mechanical stress and physical loading in all cases and a life cycle is simulated as close to real condition as possible. These are wire sensitive elements of resistance thermotransducers. They can be efficiently modeled in MatLab environment. SE cross-section consists of NxM clusters. In our model (CorrSim 2014 program) the wire's state is described by 4 matrixes and the order of every matrix corresponds to the amount of clusters NxM:

1) Mat is the matrix that includes relative current area of clusters. It means that every element of Mat matrix contains figure from 0 to 1 that characterizes the relative area of the cluster at this moment of time with regard to its initial area a·b. At the first moment, all elements which correspond to SE cross-section are given the value 1. And elements surrounding SE are assigned the value 0 that corresponds to the hostile environment. During modeling of the corrosion process the values of Mat matrix elements decrease from 1 to 0. Elements which have obtained the value 0 are considered to be absolutely corroded and their area is filled with the hostile environment. Consequently corroded cluster begins to impact on neighbor nonzero elements.

2) K is the matrix corrosion velocity, or rate. Every matrix element contains information about linear velocity of the current cluster corrosion given in mm per hour.

3) Matrixes SizeA and SizeB contain current cross-section areas of every cluster.

So, with the help of above mentioned four matrixes the state of every cluster of SE cross-section can be described at any moment of time. This model enables to set parameters separately to every cluster that provides the row of following advantages:

-   Allows modeling the heterogeneous resistance of certain clusters against corrosion that enables to model corrosive processes of SEs with different admixtures and heterogeneities.

-   Enables to model anticorrosive coatings of SEs of given reduced values of corrosion velocity for clusters around the perimeter wire.

-   Enables to specify various shapes of the thermotransducer SE cross-section, for instance, circular or rectangular shape.

Essence of the algorithm consists in the following. For every nonzero element of Mat matrix is performed the checking of 4 neighbor (above, below, left and right located) clusters concerning presence of zero value for these elements. As mentioned above elements with zero value are considered the hostile environment and effect on neighbor nonzero elements (clusters). For instance, if near nonzero element C_1 of the Mat matrix there is zero element C_2 located near the side b (cluster width) then nonzero element C_1 area decreases on the value of the corrosion velocity multiplied on the cluster width by which this cluster borders on

70

the element C_2 (hostile environment). And current cluster length C_1 decreases under this on the corrosion velocity and it will be preserved in corresponding element of SizeA matrix. So current area of every cluster at certain moment of time is calculated as product of corresponding elements of the matrixes SizeA and SizeB which contain proper clusters a and b sizes.

Cluster $R_{x,y}$ resistance at every moment of time is determined by the next equation:

$$R_{x,y} = \frac{\rho L}{ab \cdot Mat(x,y)}, \tag{2.10}$$

where ρ is the resistivity of the thermotransducer SE materials; ab is the area of cluster cross-section at the initial moment of time; Mat(x, y) is the cluster relative area at the current moment of time. Thermotransducer SE resistance at every moment of time is determined:

$$R = \sum_N \sum_M R_{x,y} \tag{2.11}$$

Taking into account (2.10) we received:

$$R = \frac{\rho L}{ab \sum_{x=1}^{N} \sum_{y=1}^{M} Mat(x,y)} \tag{2.12}$$

With the help of the CorrSim2014 program it has been conducted research of two major types of the thermotransducer SE corrosion: they are the pitting and uniform types. CorrSim2014 program enables to visualize in the shape of thermal map the process of corrosion as dynamic variable of the SE cross-section. Thermal maps are graphical data representation in which matrix individual values are presented in the form of colors. Moreover, CorrSim2014 allows to visualize SE resistance alters in time under the corrosion impact. The process of the corrosion of rectangular and circular thermotransducer SEs for uniform and pitting corrosion respectively is shown in Figs. 2.7-2.8. Sensitive elements are immersed in the hostile environment. Parameters of modeling: L=0.08 m; $\rho=150 \cdot 10^{-8}$ $\Omega \cdot$m. The column nearby the thermal map contains the range of numbers from 0 to 1, to which particular colors correspond.

For instance, in the Fig. 2.8, a, where uniform corrosion for SE with rectangular cross-section is modeled, 1 meets the saturated yellow color. It implies that SE cluster hasn't corroded yet. Number 0, which is given dark green color, corresponds to the hostile environment or the cluster that has corroded entirely. Next the resistance alter in time is shown for this SE as result of the hostile environment affect.

Correctness of results obtained by the means of CorrSim2014 program is confirmed by the model earlier developed in [52]. Simplifications adopted in the model: velocity and type of the corrosion are taken uniform within cluster; during cluster diminishing by the hostile environment the last one fills cluster's volume completely and continues to affect upon the next clusters; Only one SE cross-section area is analyzed, and the state of any other cross-sections of the SE is considered to be changed in the same way. For instance, if one of the clusters has entirely corroded then all clusters along the wire afore of this cluster have been corroded.

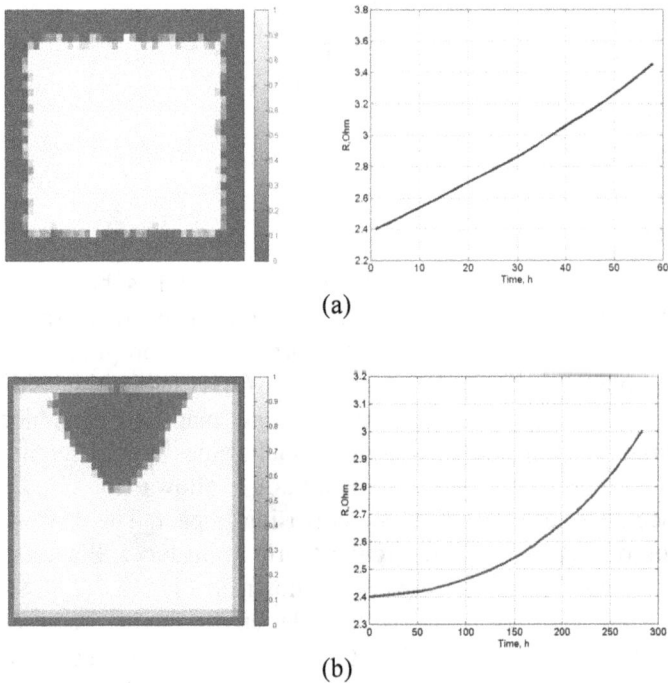

(a)

(b)

Fig. 2.7. Modeling of the sensitive element corrosion with rectangular cross-section area: a) uniform corrosion; b) pitting corrosion. Modeling parameters: A=0.25 mm; B=0.2 mm; N=40; M=30; corrosive velocity v=33·10^{-5} mm/h.

(a)

(b)

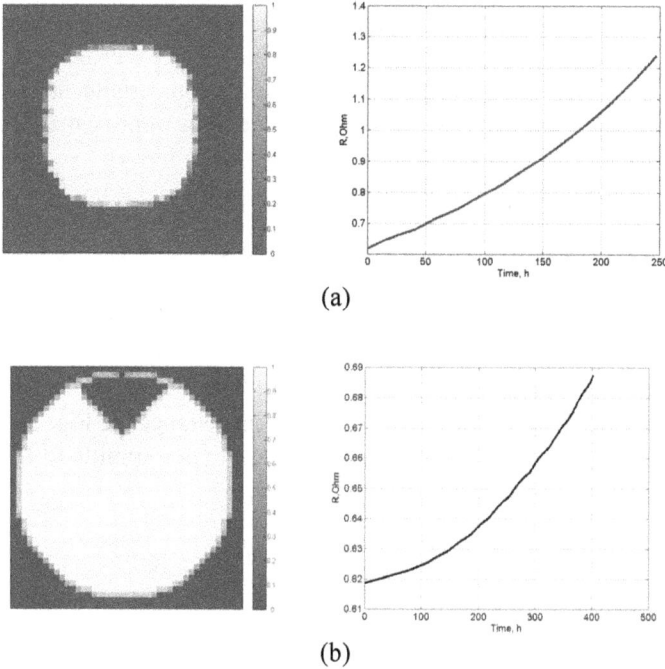

Fig. 2.8. Modeling of the sensitive element corrosion with circular cross-section area: a) uniform corrosion; b) pitting corrosion. Modeling parameters: A=0.5mm; B=0.2 mm; N=41; M=N; corrosive velocity v=32·10⁻⁵ mm/h.

Effect of mechanical strain on the SE resistance change velocity is examined due to influence of these strains on corrosive velocity (Fig. 2.9). It is considered one of the frequent cases when SE wire of R radius is wound uniformly on the core as shown in the Fig. 2.10. Here the mechanical strains are distributed in the cross-section area under such wire wounding. Herewith the stresses are the same over the entire length.

Stress gradient on x coordinate of wire longitudinal section, in the case of elasticity, is described as:

$$\sigma(x) = \sigma_{min} + \sigma_{lin}(x - R), \qquad (2.13)$$

where σ(x) is the stress at the certain point; σ_{min} is the minimal stress in the point $x = 0$. k_{lin} is the linear approximation factor. Corrosion velocity in accordance with [53], and depending on the mechanical stress, changes exponentially In general case the summarized below program enables to model any dependence of the corrosive velocity on

73

stresses. Taking into account that in the case of elastic deformation, stress dependences are linear on the coordinate x of SE cross-section and the corrosive velocity depends exponentially on tensile stresses, the dependence of the mentioned velocity on x can be approximated as:

$$v(x) = v_0 \text{ at } 0 \leq x < R$$

$$v(x) = v_0 + v_A e^{k(x-R)} \text{ at } R \leq x < 2R,$$

(2.14)

where k, v_A are the coefficients of the corrosive velocity approximation on the coordinate of SE cross-section that take into account k_{lin}, σ_{min} and other parameters included in this equation]; v_0 is the corrosive velocity at the absence of stress impact. For instance, at Fig. 2.11 x_1 is the coordinate of wire longitudinal section. It corresponds to the stress σ_1, at which the corrosion rate is equal to v_1.

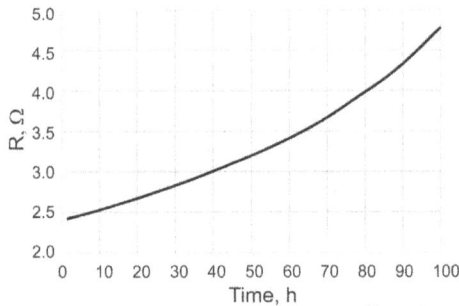

Fig. 2.9. SE resistance alters under the effect of corrosion: A=0.5 mm; B=0.1 mm; N=30; M=20; v=22.7·10⁻⁵ mm/h.

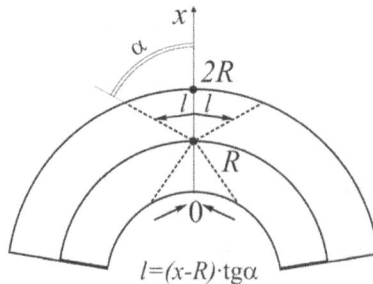

Fig. 2.10. Element of SE wire uniformly wound on the core and stress distribution in its cross-section.

It was conducted the modeling of time dependences of SE resistance under corrosion effect at the hostile environment for different stress gradients in the wire cross-section area. In results of modeling shown below, the corrosion rate gradient in the wire cross-section is approximated. For convenience of presented results it is necessary to normalize the corrosion rate increment along the wire cross-section on this rate at the point $x = 0$. Otherwise v_0 corresponds to 1, or 100 % value, and the corrosion rate in the wire opposite end; here corrosion velocity is maximal $(v_A \cdot \left(e^{k2R}/v_0\right)$, or in per cent $100 v_A \cdot \left(e^{k2R}/v_0\right))$ and is caused by the stress maximum (Fig. 2.11).

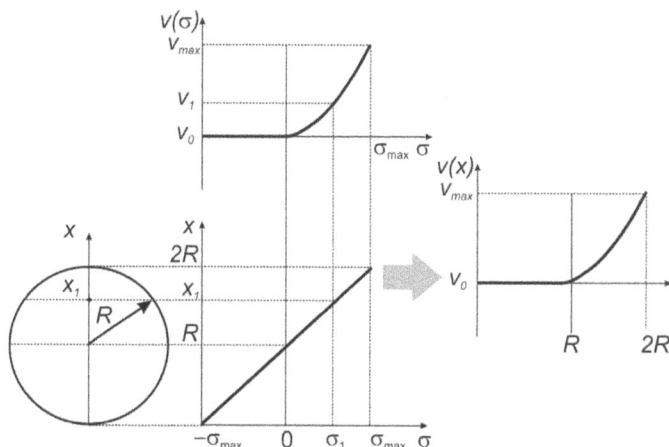

Fig. 2.11. Determination of the corrosion rate gradient of the sensitive element cross-section area.

Let's introduce the index K (Relative Corrosion Speed Difference) that indicates on the ratio of maximal corrosion velocity v_{max} (at $x = 2R$) to minimal corrosion velocity v_0 (at $x = 0$). Taking into account approximation, K is determined by the next formula:

$$K = \frac{v_A e^{k2R}}{v_0} \qquad (2.15)$$

or in per cent:

$$K = \frac{v_{max} - v_0}{v_0} \cdot 100 \ \% \qquad (2.16)$$

Next results of modeling are received at the fixed approximation coefficient k=12000, and factor v_A is determined basing on given K. In the Fig. 2.12 it is shown the corrosion rate change gradient under the stress effect. In this case corrosion has increased on 50 % (K=50 %). The column nearby the thermal map contains the gradient of colors to which K values (numbers from 1 to 1.5) correspond. At K=1 the corrosion rate is minimal, and when K=1.5 the corrosion rate is maximal. The results of SE corrosion modelling and SE resistance temporal dependence under the hostile environment impact for different K values of the corrosion rate gradient caused by stress are shown in Fig. 2.12. Here are shown the curves at K=0 %, 20 %, 50 % (that have been considered above) and at 100 %.

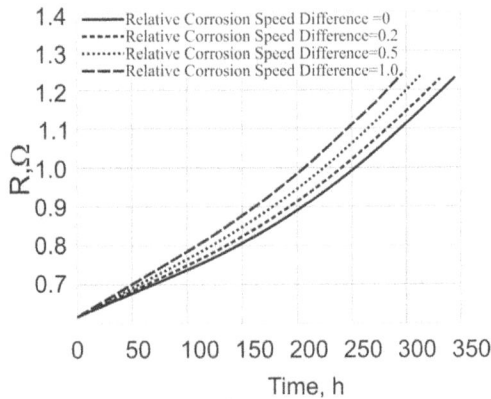

Fig. 2.12. Resistance drift under the hostile environment impact at different values of the corrosion rate gradient caused by stresses.

We defined the SE resistance alters ΔR due to the stress impact as the difference between resistance R at K=0 %, 20 %, 50 %, 100 % and resistance R_0 at $K = 0$ %. SE temperature changes ΔT (Fig. 2.13), caused by stress impact at the different corrosion rates, permit to determine the maximum operation time of thermotransducer, within which it ensures the specified metrological characteristics (resource of work for a given acceptable drift) are obtained from the dependences ΔR:

$$\Delta T = \frac{\Delta R}{\alpha R_0},\qquad(2.17)$$

where α is the temperature factor of electrical resistance of SE substance, R_0 is the resistance at $K = 0\,\%$. For instance, the permitted drift of temperature is 5 K. Then from Fig. 2.14 we can define and predict that SE of thermotransducer at K=20 % would be metrological reliable up to 200 hours of operation, at K=50 % the same can be argued for 100 hours of operation, and at K=100 % only for 50 hours.

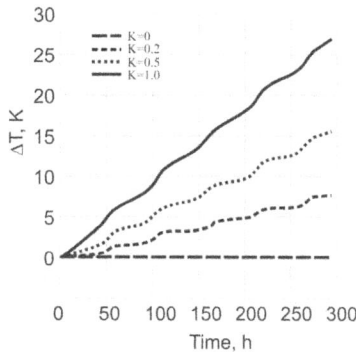

Fig. 2.13. Drift of SE temperature readout ΔT at the certain corrosion rate gradients due to the stresses impact.

2.4. Smart Sensors Grids Implementation

Modern life is dependent on the networks for the transmission of energy, data and so on, that became a part of civilization. So the essential problem of created CPS has to be provided by the more effective Smart Energetics, Smart Transport, and so on, equipped with the smart sensors grids. The maximum possible implementation by sensors for data acquisition and processing will enable to offer the best solutions that ensure economic efficiency, continuity of supply, environmental safety and even protection from terrorist attacks.

2.4.1. Smart Energetics and Information Support

The national energy system is a system consisting of set of the companies, each with their resources: from the power plants to the meters in houses of consumers. A unique feature of electric networks is their complete dependence on the consumer. Current consumption almost entirely determines the power output. Attempts to save and accumulate the generated electricity are few, since they are too complex and expensive. The existing systems of regulation can be called cyber-physical, as they provide dynamic control of power generation facilities in line with uncontrollable and alternating in time loads. Till now not all of the tasks of complex are automated and are resolved by operators, who are guided by their own experience of evaluating data obtained through feedback channels (see. Fig. 2.14).

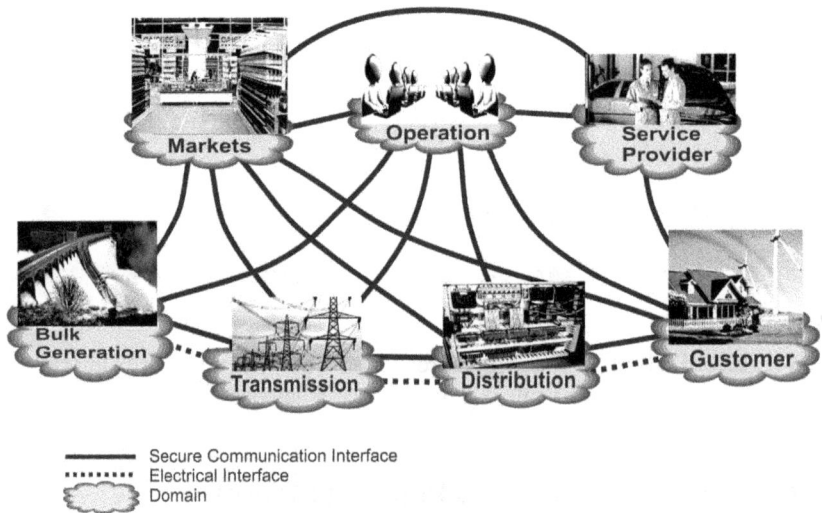

Fig. 2.14. Two components (information and energetic) of electrical networks.

2.4.2. Smart Energy-Efficient Buildings

Modern technologies can create smart buildings, construction with minimal or zero energy consumption. But they require the permanent monitoring, should be connected to smart grids and properly controlled by CPS tools so that the most expedient to use the world resources and services. Combining data from different sources, can achieve operating conditions close to optimal.

In addition, the main task for constructions is to monitor their own design, and environmental factors that affect them. For example, traditional control of the bridges comes down to periodic inspection of their state without taking into account the seismic, temperature, wind and other effects, so they have built with huge reserve, and even then periodically occur different kinds of troubles. Today we have the opportunity to provide construction by sensors and transmit telemetric data to checkpoints. Inclusion in the circuitry of CPS management would not only safely operate the existing facilities and extend their lifetime, but also would establish qualitatively new designs.

The considered below kind of CPS is a building in design, construction or operation which utilizes modern technology to manage the total lifecycle of an object and its subsystems as a whole, providing the modern level of assured operation of engineering subsystems, optimized utilization and economic consumption of energy and other resources. A distributed control system of building, such as LonWorks-based technology, combines sensors, controllers and actuators network by means of routing units.

Open architecture and the only communication protocol in network allow the use of system equipment and devices from different manufacturers, which significantly reduce the cost of system, its maintenance and operation. Controllers and sensors make up the basis of the network. The degree of system integration can be quite high and includes the controls of environment and energy supply, alarm and fire sensors subsystems, as well as security functions, access control features, check presence and residence time, control lighting, lifts and parking, video identification, and the monitoring of house engineering subsystems, its temperature and humidity.

An important factor in SEEH is engineering equipment providing comfortable accommodation. SEEH is inherent in two basic available units which are air handling unit with heat recovery and heat pump which minimize heat and energy costs; so, the inclusion of renewable energy sources in this system makes the building entirely independent.

In this case, we are interested in underpinning functions of SEEH development. Such a function emerges as a relationship between human comfort and energy consumption what is currently the most important. Therefore, we are to study below the different aspects of researching SEEHs in this area.

Smart house, as a component of CPS, is in many aspects of its construction and operation among a number of other CPS components, which interact at different structural levels. For example, there is a direct relationship between Smart Houses and Smart Energetics that consists in principally new possibilities of power structures of SEEHs to generate energy, including electricity, and transmit it to power supply. On the other hand, Smart house can be represented as any house in line: passive house → NZE-house → active house (Fig. 2.15). This is due to the use of modern information-based technologies, including the use of Smart Sensors and other achievements of metrology, capable within their intellectual ability of forming their own conduct, especially in the field of energy efficiency → energy saving → energy generation, and of forming independent structure of type of Smart Houses' Grid.

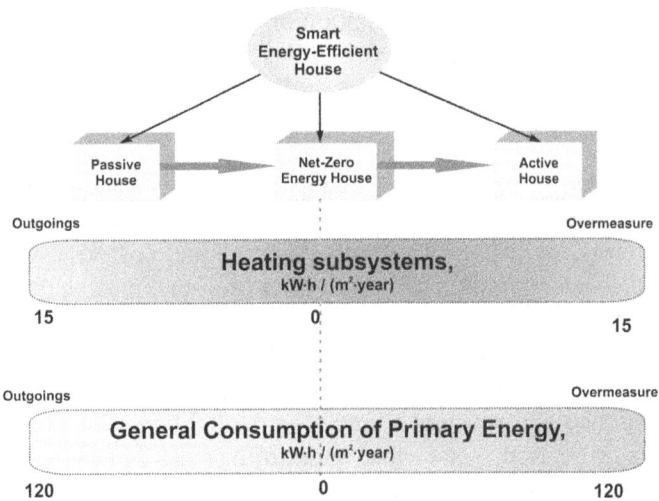

Fig. 2.15. Evolution of Smart Houses and level
of energy consumption / generation.

For this, Smart House as an element of the latter exchanges information with central management network and determines the need for the real-time implementation of the operation. It is known that electric power supply cannot exceed its consumption in each given moment of time. That is, SEEH, with its own energy-accumulating capabilities, acts as an additional power supply unit.

The relationship between Smart House and Smart Manufacturing manifests itself in another defining bond, since electrical power and the whole infrastructure of house can be frequently applied to organizing modern production, for instance, chip crystals, or, on the contrary, to creating the high-performance SEEH with building nanostructured materials [54] and special black coverage [55].

Appropriate attention is paid to the relationship between Smart House and Smart Life Safety. As for the proper safe operation of Smart House the latter is inevitably equipped with a number of security subsystems for electricity, gas, and extremely dangerous to humans carbon monoxide, and others. Finally, already today the emergence of Smart Transports makes it possible to discuss the aspects of the interaction of this CPS component with SEEHs.

In all cases, the effectiveness of relationships between CPS's different components is determined by their sensor equipment, management means, installed software, etc., particular attention to which is paid. However, the detailed systematic study of energy efficiency of Smart house, including the monitoring of its leading characteristics, is not sufficiently conducted, and obtained data are fragmented and unstructured resulting in the need of recurrent measurements at each object.

In this regard, Program [50] perfectly appropriates for CPS Objects and namely involves "Developing measurement science that will enable rapid design-to-product transformation; in-process sensing, monitoring, and model-based optimal control; performance qualification of materials, processes and parts; and end-to-end digital implementation of processes and systems". For SEEHs the similarly can be expressed as follows. Due to the deploy of measurement systems, energy monitoring, horizontal and vertical interlinks of CPS's various types, we have to achieve the real-time and efficient management of the Smart Houses on the basis of procedures for ongoing assessment of SEEH situation, power input/output regulation while considering energy state, reserve, environmental conditions, etc.

Primarily, we consider the passive house. Its main feature seems to be low power consumption, equal to ~10 % of the specific energy per unit volume that is consumed by ordinary buildings. Lower consumption is achieved especially by reducing the heat loss of the building.

Architectural concept of the passive house is based on the principles of compactness, high-quality and effective heat insulation, lack of cold bridges in materials and junctions sites, the correct geometry of the building, and orientation to the cardinal point. From among active methods in a passive house one can propose, for instance, the obligatory application of balanced ventilation with heat recovery. Ideally, such a passive house becomes energy independent (NZE House). The latter does not require energy expenditure on a comfortable temperature support. Recently two new Passive House Standards were introduced by Canadian Passive House Institute to the North American audience in the keynote address by Dr. Wolfgang Feist: Passive House Plus and Passive House Premium.

Heating a passive house should take place thanks to energy that has been released by appliances and people who live in it. If additional heating is necessary, it is desirable to use alternative energy sources. Hot water may also be supplied by the units of renewable energy, such as heat pumps or solar water heaters. To solve the problem of cooling (conditioning), the house is also provided with a corresponding architectural decision and, if necessary, with additional cooling at the expense of alternative energy sources, such as geothermal heat pump.

Energy criteria in the construction of passive houses are shown in Fig. 2.15. Here the specific heat energy consumption for heating determined by [56] should not exceed 15 kW·h/(m²) per year. And the total primary energy consumption for all the household needs (heating, hot water and electricity) must not exceed 120 kW·h/(m²) per year.

It is necessary to consider how much this state reflects the aims of the American development program in the field of CPS [50], which, in our opinion, is ideally suited to the development of construction and architecture industry of CEEHs. It may concerns here, in authors' opinion, the implementation of projects and built houses in the environment at the 1st stage and at the 2nd stage the studies of Smart houses' community with appreciation of their interdependent relations and impact on significantly greater areas of this very environment.

The characterization of construction materials, according to Program [50], means that they should be described by the same parameters during their operation as parts of the ready structures of buildings. However, at this stage of the formation of the modern construction industry it has not been done. Moreover, CPS metrological support principles, namely the real-time monitoring of buildings, have not been performed. The next item of Program is considered to be the main qualification process and the product which in the case of smart buildings ensures their energy efficiency at a certain level (≤ -15 kW·h/m² per year for passive house) through zero consumption for NZE House to positive level for an active house that transmits its produced and accumulated energy to the electrical network. The last item relating system integration presupposes not only the integration of multiple subsystems within one building, but also vertical integration, including energy, within the set of SEEHs of local community.

Technical Requirements for Smart Energy-Efficient House and Its Construction. The thermal insulation of exterior walls must be sufficient to ensure closing the indoor space by covering heat shell; then, winter indoor temperature should be comfortable and exceed 15 °C. The shell creates heat insulation in every place of the house; the minimum thickness of shell insulation makes at least 25 cm while thermal conductivity coefficient λ is equal to 0.04 W/(m·K). Passive House average heat flow U of wall is equal to 0.1 W/(m²·K), whereas for the best windows U=0.5 W/(m²·K).

The technology and method of performing heat insulation are essentially arbitrary. Beyond the correctly insulated ordinary walls and roof significant attention is paid to details trying to prevent the emergence of cold bridges. Even if high-quality insulation of shell is performed, there happen to be energetically unprofitable low temperature zones in the vicinity of cold bridges or other places of insulation shell discontinuity. Insufficient tightness of shell construction causes the flow of air passing into the building. In summer this means cumbersome, too high temperature in rooms, while in winter air becomes dry and, as a consequence, deteriorates the health of people. The shell is checked out with a special examination entitled the Blower Door test. By pumping air into the room and its spontaneous leakage for the given pressure difference inside and outside of the building, the total volumetric and mass flow rates of air are measured. Air exchange is considered to be optimal when all the air of the room is replaced within one hour at an excess pressure of 50 Pa. Then uncontrolled leakage through the gap in the shell should not exceed 60 % of the total volume of the space V (factor of 0.6) that is described by the expression: 50n = 0.6 V, m³/h., where n is the constant.

The heat loss through the walls, roof and windows of maximum-insulated passive house is effectively reduced. Additional energy saving is possible due to the reduction of energy losses at ventilation. Therefore, the ventilation equipment with recuperative function (the degree of recovery being not lower than 75 %) is embedded. The optimal energy savings are achieved with the help of two effects. On the one hand, by means of controlled system the so-called "comfort" ventilation provides the constant flow of optimum outdated air; on the other hand, entering fresh air is warmed up in a heat exchanger by the energy of the former.

The south direction of the main facade of the building (allowed deviation from the axis being 30 % in the west or east) ensures solar energy optimum utilization. Thus, windows operate as batteries that "collect" solar energy heating the space located behind them. Special windows with triple glazing and heat transfer factor of 0.75 W/(m²·K) become a source of significant solar heat savings. Window space filled with special gases, such as argon or xenon, substantially reduces heat transfer. Since the building during the heating season loses energy through its external enclosure surface, the passive house construction standards require the ratio of the building's shell to the total volume of its space as small as possible.

Methodological basis of consolidated heat engineering calculation consists in the necessary to measure the heat conductivity of building materials during their production. For builders and architects it is sufficient to examine in projects only the certified materials products and components (windows, etc.) with known thermal characteristics. Then, the previous assessment of buildings' thermal efficiency can be carried out at the design stage. This will make it impossible to yield the heat outdoor. Building SEEH is a bit more complicated than ordinary house, though the design and construction stages require increased attention to all the details.

For example, to reduce the heat loss, it is not enough to take possibly thicker heat insulation and to cover with it the outer surface of the building. This eliminates the so-called heat bridges, where, due to the violation of the integrity of insulation shell, the heat transfer is enhanced. Unfortunately, at the design stage, having some practical experience one can eliminate some of the "cold bridges" in the enclosing shell structures. The rest of the "cold bridges" appear during the house construction and can be detected only during its operation, i.e. at temperature drop existent between the indoors and outdoors.

Already today in the building design there should be set such characteristics of heat loss which ensure the possibility of human habitation as well as the operation of water supply and sewage in case of heating subsystem failure. Therefore, in the given work we focus on the working out thermal energy optimization principles for achieving the controlled parameters, including temperature. Understanding the loss origins is deemed to be helpful in simplified calculation. The latter requires the application of specific techniques based on the possibility to analyse each impact. First of all, the influence of relative size or relative parameters of thermal conductivity of different construction materials is important. Further calculation is performed according to the methodology that is more suitable for architectural design stage. The number of "effective restrictive surfaces" (walls) in each embodiment is predetermined for this aim. Hereinafter, 1-, 2-, 3-, 4- and 5-floor buildings with different length to width multiplicity are analyzed.

The number of "effective restrictive surfaces" N is the factor in the expression $Q_\Sigma = Q \cdot N$, which determines the total heat loss due to the heat conduction of restrictive surfaces that are further reduced to one apartment. In other words, N is the amount of "walls" which should be considered for assessing the total heat loss of separate apartment in a passive house; Q is the effective heat loss of averaged restrictive surface, accounted below for different values of surface area S_x and the specific effective thermal conductivity U_x of construction materials. Simultaneously the heat loss through the ceiling and the floor the area of which is larger than wall area is considered to be approximately equal to the loss through the wall. It is due to the higher heat conductivity of the latter. Then the heat loss of a single-storey passive house may be equal to 6Q, where "6" corresponds to the number of restrictive surfaces.

The heat loss of a two-storey building made of similar thermal insulation materials slightly increases, but is relatively smaller since the ceiling of ground floor serves as the basis of the next floor. Heat loss reduced to one floor is defined as $Q_{red} = Q_\Sigma/2 = 5Q$. The heat loss of a three-storey house is gradually growing, and being reduced to one floor constitutes 4.67Q. Here, for a four-storey building similar loss is equal to 4.5Q.

The equation for estimating heat loss due to the transfer through "effective wall" is:

$$Q = US(T_{int} - T_{ext})t, \qquad (2.18)$$

where the specific heat conductivity of wall U is normalized with the help of the coefficient of heat transfer resistance R. It is determined by the calculated thermal conductivity 1 and the thickness of insulation layer d:

$$R = \frac{d}{l} = \frac{l}{U} \frac{W}{m^2 K} \qquad (2.19)$$

A rather interesting factor of influence is deemed to be a relative value of windows total area in the efficient wall area of a projected house $X = S_{wind}/S_{\Sigma}$. With it increasing, light intensity and room comfort grow; however, simultaneously, heat losses caused by convection and radiation also increase. Then the relative total area of pure wall without windows is determined as 1–X and it becomes possible to replace the sum of products U·S of two constituents in the expression below with the smaller quantity of variables:

$$U_{\Sigma}S_{\Sigma} = U_{win}S_{win} + U_{wal}S_{wal}$$
$$= [U_{win}X + U_{wal}(1 - X)]S_{\Sigma} \qquad (2.20)$$

Thus, the general formula for calculating heat loss reduced to one house's apartment of a certain size and shape (according to the solution adopted by an architect of a designed house) is defined as:

$$Q_{\Sigma} = U_{\Sigma}S_{\Sigma}N(T_{int} - T_{ext})t$$
$$= [U_{win}X + U_{wal}(1 - X)]S_{\Sigma}N(T_{int} - T_{ext}) \qquad (2.21)$$
$$= aT_{int} - aT_{ext},$$

where a is the coefficient; T_{int}, T_{ext} are the temperatures inside and outside the house, respectively. Apparently, within studied linear model the extreme dependence on any of the set of variables is absent.

Development of optimization principles of temperature indoors mode subsequently concentrates attention on parameter describing comfort conditions in a passive house with its small energy consumption. This parameter is temperature which should be established in inner rooms individually, for instance, at the level equal to 20±1 °C (293±1 K). Basing on the consideration of heat energy balance, as the amount of heat expenditure energy flows, including radiation flow, one may obtain the nonlinear algebraic equation of the 4th degree concerning the indoor temperature. It is suitable for searching the optimal solution, which would be the desired temperature mode:

$$A - aT_{int} + aT_{ext} - bT_{int}^4 + bT_{ext}^4 = 0, \qquad (2.22)$$

where A is the parameter bound with interior spaces additional heating units that include ground heat accumulator, heat exchanger, etc. The time t is included as a factor in constants A, a, b. When all the parameters are averaged over a certain period, for example, over one year, t is eliminated. So, energy balance is transformed into the power one.

For studying the operation efficiency of an auxiliary heating system the mentioned approach is ineffective. Hereinafter in the equation (2.22) we consider t as an additionally selected variable. The dimension of the rest of products corresponds to the sources power:

$$At_H - at_C T_{int} + at_C T_{ext}(t) - bt_R T_{int}^4 + bt_R T_{ext}^4(t) = 0, \quad (2.23)$$

where t_H is the duration of turned on additional heat sources (electric and gas heating, ground accumulator heating, etc.); is the duration of conductive heat removal; t_R is the duration of radiant heat withdrawal (through the windows). Of course, these variables are of different magnitude.

When determining diurnal period, the main tasks are to evaluate the impact of periodic (night-time) turned on heating, for example, at $t_H = 8$ hours duration; to accept the averaged duration t_C of cooling by heat conduction through the shell enclosing; to take into account the acquisition of solar energy during the day-time (additional component in $A \cdot t_H$), and lowered heat loss due to curtained radiant cooling during the night within duration t_R. The indoor temperature T_{int} is determined from the following equation:

$$A(t)t_H - at_C T_{int} + at_C T_{ext}(t) - bt_R T_{int}^4$$
$$+ bt_R T_{ext}^4(t) = 0$$

$$\quad (2.24)$$

$$at_C T_{int} + bt_R T_{int}^4$$
$$= A(t)t_H + at_C T_{ext}(t) + bt_R T_{ext}^4(t)$$

It retained a permanent by automatics intervention, in particular, the precise level as a result of additional heating system temperature sensors operation. Temperature $T_{int} = F[T_{ext}(t); a; b; t_C; t_R; t_H; A(t)]$ depends on a number of impacts. Most magnitudes of the given above variables excluding parameter A are considered known values that are bound with interior heating by additional energy sources. The latter becomes possible to be set by the thermometer indications in apartments automatically adjusting electricity, gas or other heating power.

To compensate heat loss the required energy has to be replenished by the heat accumulator energy, since there is no restriction on its reception. The amount of accumulated energy is of primary importance in this process. For this purpose, we estimated the energy of heat accumulator located in the house's 100 m³ basement. The accumulator could be charged during the summer season (for known specific heat capacity and mass of basalt filler) by $3200\ kW \cdot h$. It can deliver each day of wintertime up to $4.4\ kW \cdot h$. of heat energy within 8-hour operation.

The polynomial roots of algebraic equation of the 4th degree are defined at its numerical solution if the totals of the mentioned equation constituents tend to approach zero. Thus, for the best solution this equation should be recorded in a specified form. Within the energy concept, the following equation the constituents of which include a temporal factor was obtained:

$$bt_R T_{int}^4 + at_C T_{int} - at_C T_{ext}(t) - bt_R T_{ext}^4(t) - A(t)t_H = 0$$

$$bt_R T_{int}^4 + at_C T_{int} - C = 0$$
(2.25)

The equation (2.25) can be simplified in power concept while carrying out further settlements:

$$bT_{int}^4 + aT_{int} - A(t) - aT_{ext}(t) - bT_{ext}^4(t) = 0$$

$$bT_{int}^4 + aT_{int} - C = 0$$
(2.26)

The numerical values of the coefficients in equations (2.25) - (2.26) are determined for diurnal calculation in winter conditions: $b = C_0 K_{tr} X S_{\Sigma} P$, where the constant $C_0 = 5.67 \cdot 10^{-8}\ W/(m^2 K^4)$; K_{tr} is the transmittance of infrared radiation by window panes accepted as 0.5; X is the ratio of windows area to wall area, its value is chosen 0.2; 0.3; 0.4; S_{Σ} is the area of mean wall (set as 36 m² that corresponds to the input data of the project, based on the model standard cubic area with 12 m sides at height of 3 m); N is the number of walls reduced to one storey (N=3.5 for a 4-storey building at 4-fold length). As a result, for $X = 0.2$ the coefficient b can be assessed: $b = 5.67 \cdot 10^{-8} \cdot 0.5 \cdot 0.2 \cdot 36 \cdot 3.5 = 72.7 \cdot 10^{-8}\ W/K^4$. For $X = 0.3$; 0.4 factor b is equal to $109.0 \cdot 10^{-8}\ W/K^4$ and $145.4 \cdot 10^{-8}\ W/K^4$ respectively.

The constant a in the expressions dependent on heat conductivity is determined for $X = 0.2$; $U_{wall} = 0.1\ W/(m^2 K)$; $U_{win} = 0.7\ W/(m^2 K)$, which together adjust the thermal conductivity of the wall to the value of $U_{red} = 0.7 \cdot 0.2 + 0.1 \cdot 0.8 = 0.22\ W/(m^2 K)$. The constant $a = U_{red} \cdot S_{\Sigma} \cdot N = 0.22 \cdot 36 \cdot 3.5 = 27.7\ W/K$ is defined for accepted 36 m² area of the wall at $N = 3.5$. For $X = 0.3$; 0.4 coefficient b is equal to 35.28 and 42.84 W/K respectively.

Also, t_C is considered to be equal to 24 hours, since conductive cooling takes place continuously. It is assumed that during the winter night of 8 hours closing window blinds completely overlaps inner radiant heat exchange with the environment; so t_R is equal to 16 hours. The factor C in (2.25) has 3 main components. Two of them are determined by cooling due to the thermal conductivity and radiant heat exchange with pre-determined factors a, b and the mentioned time durations. And the 3rd component relates to auxiliary heating systems the value of which is being searched. To determine them, it was previously adopted that the outdoor temperature in winter night is $-3°C = 270\ K$. Then aT_{ext} was defined as 27.7 $W/K \cdot 270\ K = 7479\ W$, and the reverse power radiation from the outside into apartments through windows was specified as $bT_{ext}^4 = 72.7 \cdot 10^{-8} \cdot 2704 = 3863\ W$.

The solved equation of the 4th degree with pre-defined factors at unknown indoor temperature, where $A(t)$ is the power of additional energy sources (for the outside temperature of 270 K and the relative total area of windows $X = 0.2$), is as follows:

$$\sum A_t(t)t_H = 72.7 \cdot 10^{-8} t_R T_{int}^4 + 27.7 t_C T_{int} \\ - 479 t_C - 3863 t_R \tag{2.27}$$

For the outdoor temperature of 270 K and the relative total windows area $X = 0.3$, the last equation changes into:

$$\sum A_t(t)t_H = 109.1 \cdot 10^{-8} t_R T_{int}^4 + 35.3 t_C T_{int} \\ - 9527 t_C - 5794 t_R \tag{2.28}$$

At the relative total area $X = 0.4$ and the same temperature outside, similar equation is:

$$\sum A_t(t)t_H = 145.4 \cdot 10^{-8} t_R T_{int}^4 + 42.8 t_C T_{int} \\ - 11567 t_C - 7726 t_R \tag{2.29}$$

Determined power of sources $\Sigma A_i(t)$ that are responsible for heating at the given duration must compensate the heat loss of one apartment area of $12^2\ m^2$ (external dimensions) or of $11^2\ m^2$ (internal dimensions).

Energy sources should be divided into special types (heat accumulator with solar collector, insolation through windows, air/water heat exchanger, geothermal heat pump, solar batteries, etc.) and reserve types (boilers with gas and electric heating and others similar devices). The latter are not considered below believing that for the proper execution of the project one can get rid of them.

Determination of indoor temperature in winter To we begin by assuming that $A = 0$ and by applying software: we define the temperature inside T_{int}, at which the value of the left side of (2.29) tends to 0. Then the increment of temperature is set and the necessary heating power is determined. Calculation results are shown in Table 2.1. As one can see, at the relative total windows area of $X = 0.2$ and power of $A = 1.63\ kW$ the acceptable temperature of 293 K is reached.

Table 2.1. The required additional heating power A of a 4-storey passive house (at the outside temperature of 270 K) depending on the preset indoor temperature T_{int}, K for different values of factor X.

T_{int}, K	280	285	290	293	295	300
0,2	0.68	1.04	1.41	1.63	1.79	2.18
0,3	0.98	1.48	2	2.33	2.55	3.11
0,4	1.24	1.89	2.56	2.98	3.29	3.99

The values of power A include the heat power radiated by electric lights, refrigerator, etc.

The calculations given above relate to one embodiment of the architectural and construction project that combines the use of specific building materials and of certain architectural methods to reduce heat loss to a tolerable level. If to use other building materials, techniques and approaches, it becomes necessary to change the values of factors in calculations. To accelerate and reduce the cost of designing, it should be automated with advanced computer technology application. For its creation we used the following expression that summarizes the formulas above:

$$\sum K_i$$
$$= aXS_\Sigma N_{real}T_{int}^4 + [U_b X + U_w(1-X)]S_\Sigma PT_{int} \quad (2.30)$$
$$- aXS_\Sigma N_{real}T_{ext}^4 - [U_b X + U_w(1-X)]S_\Sigma PT_{ext},$$

where, the total value of power minimized for different energy sources is defined as:

$$\sum K_i(t)$$
$$= A_{H-ac}(t) + B_{Insol}(t) + C_{S\_bat}(t) + D_{Geoterm}(t) \quad (2.31)$$
$$+ E_{rek}(t) + F_{Re\,s\_source}(t),$$

where the 1st constituent concerns the heat power of accumulator, the 2nd one is the insolation power, the 3rd one concerns an electric solar cell power, the 4th one is the geothermal heat pipe power, the 5th constituent is equal to the recuperation power, and the 6th constituent is the non-renewable backup.

The expressions (2.24)-(2.26) determine the mathematical foundations and with the help of software package Excel form the methodology for estimating the $energy/power$ efficiency of designed SEEHs. The latter is not based on article by article balance of heat loss and heat income, as it is recommended by [56], but is based upon the evaluation of maintaining the optimal temperature conditions in interior rooms [57]. Here the values of thermal characteristics of used building materials and structures serve as variables specified with a certain increments at a particular indoor and smoothly changing outdoor temperature ranges.

The latter corresponds to the ambient temperature and even temperature of the outer surface of walls, which do not always coincide with each other (in particular, while coating the exterior walls with a special black coverage [55]). However, the method enables us to specify certain values of transmission coefficients of window structures in infrared spectrum, to alter discretely the interior insolation during the day-time, and to reverse radiation at night-time.

As a result, at the predetermined temperature of SEEH interior (in the given case it is equal to 293 K) and gauged outdoor temperature (environment or/and wall surface), on the basis of the solution of equation (2.30) with specific the additional sources of energy supply described by (2.31) we can obtain the results of the examination for every considered architectural and engineering project at the design stage.

These results could be expressed in a graphic dependency due to input data.

2.4.3. Metrological Study of Heat Energy Balance

It is worth noting that according to EU Norms the buildings after reconstruction (at the area over 100 m²) are subjects to compulsory testing on heat insulation loss not exceeding the normalized values, which are different in the certain countries.

We have developed 3 methods for the study of heat loss which are applied at various stages of research. The 1st one is based on the thermal imaging (by Fluke TI25), which gives momentary picture of the building. The 2nd method envisages the examination of heat loss over a long time, but at one point of the building shell through readings of 2 Chipset types ATMEL with built-in temperature sensors. Circuits are connected by conductor or wirelessly, and the heat loss of the given element can be determined for the known thickness of shell and its specific thermal conductivity. The 3rd method in time-spatial sense represents the superposition of the first two. According to it all the points of a gauged surface are consistently surveyed by a narrowly directed photodiode, and the direction to a particular point on the surface is given by 2 stepping motors with mutually perpendicular directions of rotation and displacement of enshrined photodiode.

By applying the thermal imager we investigated the initial state of Inter-district hospital heat insulation which preceded the EU grant reconstruction. It was carried out by the outside and inside tomography in winter conditions in the presence of the operating heating system. A certain amount of cold bridges was fixed, such as construction defects, quasi-closed openings of water and gas supply, as well as points of poor quality operation of electrical equipment (extra heating of sockets), and significant heat emission through windows (Fig. 2.16).

Relying on studies performed by the Fourier law: $Q = \lambda \frac{\Delta T}{d}$, where λ is the thermal conductivity of wall; ΔT is the measured temperature drop on the wall; d is the thickness of wall; we calculated the heat losses under known λ, d. The calculation of the impact of cold bridge (Fig. 2.17) seems to more complicated, since there is a priori unknown velocity of air through the given bridge.

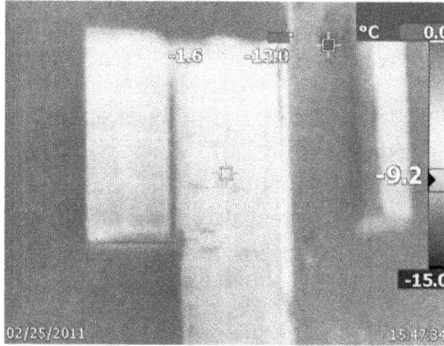

Fig. 2.16. Thermogram of studied building (view from outside).

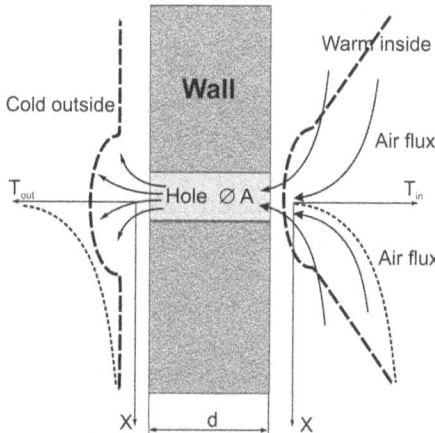

Fig. 2.17. Distribution of temperature outside and inside the walls of thickness d with an existing cold bridge of diameter A.

However, setting the boundary conditions for this task, basing on the obtained temperature distribution on the outer and inner surfaces of the walls, and developed mathematical apparatus [58] of solving such problems in applied heat engineering, it becomes possible to define the heat loss in our case. Thus, the error of heat flux determination depends on the error of temperature field gauging, the relative velocity of convective flux that is the function of diameter of hole.

Energy exchange devices as technical equipment are considered to be an important factor in SEEHs. Optimal living conditions can be achieved only with modern engineering. The combination of at least two units enables to reach comfortable living conditions with minimum costs and makes the building energy-sufficient (Net-Zero Energy House).

Here can be used three or more basic units of energy exchange devices. These are the heliosystem with basalt filler in SEEH basement; the solar cells on the roof or around the house as the source of additional power; balanced ventilation with heat recovery; heat pump, including a series of exchange chambers, and a device for geothermal energy usage. The energy-efficient devices that apply the heat of substance phase transition are still being developed.

However, intensive radiation through the windows (Fig. 2.16) worsens the energy balance of a house. Therefore, we can recommend applying electric-driven automatics for screening the windows at night-time by curtains. This decreases the heat loss through windows threefold that corresponds to the diurnal heat loss of 30...40 %.

As shown above, additional energy efficiency can be achieved if air enters the building through underground duct. Then, it becomes easier to maintain the comfortable environment in SEEHs. Taking into account mutual influence of closely located similar houses through air and ground, the grid of houses in contemporary suburbs can be regarded as a unified ecosystem with gauged and controlled reproducible characteristics. So problem of full compensation of energy losses and restoration of resources and ecological reserves should already be considered in such grid's design, construction and operation.

Practical measuring the heat loss of studied surface by using the pyrometric methods consists in measurements of intensity of radiant heat exchange between this surface and the heat receiver that would be the sensitive element(s) of MI. Thus, in the measurement result is introduced significant error caused by an unknown emissivity of specified surface.

Known method for determining the emissivity of materials is based on the Stefan-Boltzmann law and realized on the equation: $\varepsilon = \left(T_P / T \right)^4$, where T_P is the radiative surface temperature; T is the thermodynamic temperature [59]. Thus, the measurement result is inherent in a significant error caused by significant heat withdrawal through electrodes of thermocouples, and the error value increases with temperature.

Therefore, pyrometer provides the gauging the intensity of radiant heat intensity (between the studied surface and the sensitive element of the certain temperature), expressed as a radiation temperature that is attributed to its scale marks during calibration [60].

Thus, the calibration of radiation pyrometer is carried out on "blackbody" for which emissivity is taken a close to 1. As the real samples are characterized by values of emissivity lower than 1, the considerable errors ($\geq$10 %) emerge in pyrometer readings. As a result, for one group of radiation pyrometers is accepted beforehand that they operate at the certain factor of emissivity (0.95) of studied sample, or rather reduced factor of system "pyrometer - sample" which is introduced below. For more complicated radiation pyrometers this factor can be adjusted manually within the limits 0.1 (0.3)...1.0. It is clear that for both groups the significant methodical error arises due to ignorance of real values of the emissivity.

Even in greater extent the above factors concerns thermal imagers, that, as a result, makes it possible to get only qualitative picture, for example, of heat loss of buildings. To prevent significant errors, the special tables are enjoyed, where emissivity of various materials and surface finish degrees are given. To prevent significant errors, the special tables are enjoyed, where emissivity of various materials with different surface finish grades are specified. F.i., presence of oxide film on the surface significantly affects its emissivity. So, the emissivity factor of oxidized steel is equal to 0.85 and of polished steel 0.075).

Prerequisite for the implementation of correct temperature measurement of object could be considered the determination of surface emissivity factor in the time immediately preceding the stage of temperature measurement. To understand the essence of the proposed method, first of all, consider the flow of energy E_0 transmitted from the studied surface of object in the radiant way to pyrometer, or rather to its blackened plate on which the receiver of heat energy is located (thermopile, bolometer, etc.).

The Stefan–Boltzmann law states that the total energy radiated per unit surface area of a black body across all wavelengths per unit time (emissive power), E_0 is directly proportional to the fourth power of the black body's thermodynamic temperature T:

$$E_0 = \sigma T^4; \quad E_0 = C_0\left(T/100\right)^4,$$

where $C_0 = 10^8 \cdot \sigma = 5.7 \left[\frac{W}{m^2 K^4}\right]$ is the blackbody's emissivity factor. The grey body is described by the emissivity $\varepsilon < 1$ indicating how much the object radiation is less intense against blackbody radiation. Radiant heat transfer between 2 bodies (tested surface area S with its ε_{surf} and sensitive element of pyrometer with its ε_{SE}) is defined by the difference of 2 fluxes of efficient radiation:

$$q = \varepsilon_{red} C_0 S \left[\left(\frac{T_{surf}}{100}\right)^4 - \left(\frac{T_{SE}}{100}\right)^4\right], \qquad (2.32)$$

where

$$\varepsilon_{red} = \left(\frac{1}{\varepsilon_{surf}} + \frac{1}{\varepsilon_{SE}} = 1\right)^{-1}, \qquad (2.33)$$

It is considered for the system of 2 bodies and depends on ratio of the areas of mutually radiating objects. At pyrometer calibration, to each value of q is attributed the certain value of radiation temperature having regard to particular construction with its own factor A of transformation of energy flow into heat of sensitive element and with further elaboration of received signal. Note that the coefficient A should also consider the value $C_0 S$. Thus, calibration is carried out by blackbody, trying to reach $\varepsilon_{red} \to 1$. As a result, you can obtain equations of pyrometer transformation function, expressed through radiant heat exchange between the sensitive element and tested surface: $T_r = T_r(T_{surf}; T_{se}; \varepsilon_{red})$. There are two unknown quantities (T_{surf} and ε_{red}) here. This equation can be solved only taking that $\varepsilon_{red} \to 1$. The solution of real equation with emissivity $\varepsilon_{red} < 1$ results in emergence of significant methodic error.

Therefore, conditions for preliminary determination of reduced emissivity are created in proposed method aiming the obtainment of unambiguous transfer function of pyrometer in the form of $T_r = T_r(T_{surf})\big|_{T_{SE}=Const; \varepsilon_{red}=Const}$ at known emissivity and temperature of sensitive element. Let us give following system of 2 equations with 2 above-mentioned unknowns, and the 2nd equation describes the state of energy exchange between the studied surface and heated by the few degrees (ΔT) sensitive element of MI:

$$T_{r1} = \varepsilon_{SE} A \left[\left({}^{T_{surf}}\!/_{100} \right)^4 - \left({}^{T_{SE}}\!/_{100} \right)^4 \right], \qquad (2.34)$$

$$T_{r2} = \varepsilon_{red} A \left[\left(\frac{T_{surf}}{100} \right)^4 - \left(\frac{T_{SE} + \Delta T}{100} \right)^4 \right], \qquad (2.35)$$

Subtracting (2.35) from (2.34), we obtain:

$$\Delta T_P = \varepsilon_{SE} A \left[\left(\frac{T_{SE} + \Delta T}{100} \right)^4 - \left(\frac{T_{SE}}{100} \right)^4 \right], \qquad (2.36)$$

As the hot junctions of pyrometer thermopile are mainly located on Pt petal, coated with Pt niello, the initial condition ($\varepsilon_{se} \to 1$) is achieved due to varied equation (2.33) at $\varepsilon_{red} \approx \varepsilon_{surf}$. Then (2.36) is changed to:

$$10^8 \Delta T_P = \varepsilon_{surf} A [4 T_{SE}^3 \Delta T + 6 T_{SE}^2 (\Delta T)^2 + 4 T_{SE} (\Delta T)^3 + (\Delta T)^4] \quad (2.37)$$

It is sufficient for calculation to make use of first two components of the polynomial in square brackets because each following component is in 100-fold lesser than previous component (at $\Delta T = 0.01 T_{SE}$):

$$10^8 \Delta T_r = \varepsilon_{surf} C_0 S [4 T_{SE}^3 \Delta T + 6 T_{SE}^2 (\Delta T)^2] \qquad (2.38)$$

Hence the expression for calculating the emissivity factor of studied surface is defined:

$$\varepsilon_{surf} = \frac{10^8 \Delta T_r}{A [4 T_{SE}^3 \Delta T + 6 T_{SE}^2 (\Delta T)^2]} \qquad (2.39)$$

Implementation of method is the following. Let us assume that the way to ensure efficiency and ease of further calculations is established temperature gain of sensitive element equal to 1 % of T_{SE}. Then expression (2.39) to be simplified to:

$$\varepsilon_{surf} = \frac{10^8 \Delta T_r}{0.0406 \cdot A T_{SE}^4} = \frac{\Delta T_r}{0.0406 \cdot A \left(\frac{T_{SE}}{100} \right)^4} \qquad (2.40)$$

When the measuring device with sensing element is at 293 K, then expression for computing can be simplified to:

$$\varepsilon_{surf} = \frac{2.9924}{K^4} \cdot \frac{\Delta T_r}{A},$$

(2.41)

where A [1/K] is the factor of conversion of radiant energy flux into the pyrometer readout under conditions resulted in device specifications.

Determination of emissivity factor of the object surface consists in measurement of intensity of radiant heat transfer by radiation temperature with help of MI. Herewith, measuring the intensity is carried out twice (at different temperatures of SE) by readouts of additional built-in sensor. Results of measurements of intensity are subtracted from each other, getting change in units of radiation temperature (ΔT) which is inserted in (2.41). The emissivity factor od studied object is computed in such a way.

Note the following. We have to know in advance the factor A for calibrated pyrometer or thermal imager. MI should be equipped with block of sensitive element heating and should be provided with electric power supply. In addition, closely to SE has to be the additional sensor, for example thermistor, switched into the bridge scheme. It makes possible to set the certain temperature rise on the SE.

So, the accuracy of radiant flow measurement is improved. It can be considered a significant factor while establishing the state of energy efficiency of Housing through energy audit of newly constructed and reconstructed buildings.

2.5. Conclusions

Implementation of Cyber-Physical Systems is impossible without easily deployed Smart Measuring Instruments that suppose an opportunity for a wide spectrum of various applications. Numerous advantages include the large-scale flexible architecture, high-resolution data, and application-adaptive mechanisms as well as the row of metrologically specific functions (self-check, self-validation, self-verification, self-calibration, self-adjustment, etc.), which unprecedentedly improve performance.

Predictable ways of Smart Instruments development consist in further studying the Smart Sensors Grids with improved parameters that are reached not only by software & middleware enhancing and metrix introduction, but also by ensuring their scalability, flexibility and number of other characteristics as well as by providing the specific metrological checks, for instance of software, or by model-based diagnosing of Smart Instruments reliability.

Particularly the similar challenges for smart metrology are manifested on the example of AM metrology, which operates on a double transformation, especially on building a computer model of created object, and then on its reproduction by this model. In such a way, the method of segmental polynomial approximation based on invariance of signal intensity of white-light interferogram from optical path difference, is applicable for surface reconstruction at complicated and non-stationary surface topology. Here, metrological properties ascertained the boundary conventional error of reconstruction of tilted surface ≤ 0.2 % and the RMS error ~ 0.05 %.

The model-based diagnostics and prognostics are especially important for Cyber-Physical Systems with powerful measuring equipment. Particular attention has to be paid the manufacturing systems with inwardly controlled hostile medium. Within MatLab environment there has been developed the CorrSim2014 program and the model on its basis of sensors performance drift at the certain chemical and mechanical impacts; it has been proved the possibility of prediction the resistance thermometer readouts in hostile environment throughout prolonged operation period.

Smart Energetics and Smart Energy-Efficient Houses design, considered in this chapter, have envisaged the necessity of further development of heat energy characteristics assessment. The requirement of full compensation of house's energy consumption and resources' reduction results in the formation of Smart Houses Set with its impact on the environment and in strengthening the information and energy exchange within Smart Energetics Grid, improving the operation of region critical infrastructure. Next stage of smart house development should be establishing the structures with adaptive properties of enclosing shells susceptible to environmental alterations.

References

[1]. A. Platzer, Logical Foundations of Cyber-Physical Systems, *Carnegie Mellon University* http://symbolaris.com/logic/lfcps.html
[2]. C. Mathas, Smart Sensors – Not Only Intelligent, but Adaptable, Contributed by Electronic Products, 2011-09-29.

[3]. S. Y. Yurish, Sensors: Smart vs. Intelligent, *Sensors & Transducers*, Vol. 114, Issue 3, March 2010, pp. 1-6.

[4]. B. Deb, S. Bhatnagar, B. Nath, A Topology Discovery Algorithm for Sensor Networks with Applications to Network Management, Technical Report DCS-TR-441, May 2001, http://descanso.jpl.nasa.gov/symposia/ieee_cas2002/short/deb.pdf

[5]. D. Sharma, S. Verma, K. Sharma, Network Topologies in Wireless Sensors Networks: A Review, *International Journal of Electronics & Communication Technology,* Vol. 4, Issue Spl-3, April – June 2013, pp. 93-97.

[6]. O. Younis, S. Fahmy, HEED: A Hybrid, Energy-Efficient, Distributed Clustering Approach for Ad-hoc Sensor Networks, Department of Computer Sciences, Purdue University, http://citeseerx.ist.psu.edu/viewdoc/download?doi=10.1.1.81.2738&rep=rep1&type=pdf

[7]. Sensor technology handbook, Jon S. Wilson (Ed.), *Elsevier*, SA, 2005.

[8]. Je. Ruban, Review of Wireless Technologies for the Smart Grids, Tallinn University of Technology, *in Proceedings of 8-th International Symposium on Topical Problems in the Field of Electrical and Power Engineering*, Pärnu, Estonia, January 11-16, 2010, pp. 50-53.

[9]. A. K. Singh, Smart Grid Sensor, *International Journal of Computational Engineering Research*, Vol. 2, Issue 7, November 2012, pp. 930-963.

[10]. Tekpea Solutions, Automating Demand Response, http://tekpea.com/solutions/demand-response/

[11]. H. Salem, M. Nader, Middleware: Middleware Challenges and Approaches for Wireless Sensor Networks, *IEEE Distributed Systems Online,* 1541-4922, Published by IEEE Computer Society, Vol. 7, No. 3, March 2006, pp. 1-23.

[12]. 'What is Middleware?', Middleware.org, Defining Technology, 2008, Retrieved 2013-08-11.

[13]. S. Hadim, N. Mohamed, Middleware challenges and approaches for wireless sensor networks, *IEEE Distributed Systems Online,* Vol. 7, Issue 3, 2006, Retrieved March 4, 2009.

[14]. Yu. Bobalo, Z. Kolodiy, B. Stadnyk, S. Yatsyshyn, Development of Noise Measurements. Part 3. Passive Method of Electronic Elements Quality Characterization, *Sensors & Transducers,* Vol. 152, Issue 5, May 2013, pp. 164-168.

[15]. M. Kazahaya, A Mathematical Model and Error Analysis of Coriolis Mass Flowmeters, *IEEE Transactions on Instrum. and Measurement,* Vol. 60, Issue 4, 2011, pp. 1163 – 1174.

[16]. D. Spitzer, The Consumer Guide to Coriolis Mass Flowmeters, Seminar, *Spitzer and Boyes, LLC,* 2004, http://www.spitzerandboyes.com/wp-content/uploads/2014/03/Slides-Flow-Coriolis.pdf

[17]. The procedure for certification of software of measuring instruments, *State Committee of Ukraine for Technical Regulation and Consumer Policy,* 2006 (in Ukrainian): http://www.uazakon.com/documents

[18]. WELMEC 7.1, Issue 2, Information document, Development of software requirements, *WELMEC*, Vien, 48 p., 2005, http://www.vniims.ru/009lab/docs/welmec_7_1.pdf

[19]. MI 3286 – 2010, Testing of software protection, with the determination of its level at the measuring instruments tests in order to type approval, Moscow, 33 pp., 2010 (in Russian).

[20]. B. Selic, The Pragmatics of Model-Driven Development, *IEEE Software 20.5*, Sep./Oct. 2003, pp. 19-25.

[21]. TechTarget, What is CSSLP, Posted by M. Rouse, http://searchsecurity.techtarget.com/definition/CSSLP-certified-secure-software-lifecycle-professional

[22]. J. Jarzombek, PMP, CSSLP, Software Assurance: Enabling Security and Resilience throughout the Software Lifecycle, Software Assurance Forum, *US Department of Homeland Security*, October 2012, 70 p.

[23]. G. V. Zlygosteva, S. I. Muraviev, The generalized model of test procedure of measuring software, *Bulletin of Tomsk Polytechnic. Univ.*, Vol. 318, No. 4, 2011, pp. 62-67. (in Russian).

[24]. E. V. Kovalevskaya, Metrology, software quality and certification, *Moscow State University of Economy, Statistics and Informatics*, Moscow, 2004. 96 p. (in Russian).

[25]. O. Oleskiv, I. Kunets, I. Mykytyn, Review of Techniques and Methods of Software Verification of Metrological Means, *Publishing House of Lviv Polytechnic*, Lviv, No. 75, 2014, http://vlp.com.ua/node/12707 (in Ukrainian).

[26]. H. Sheng, Ya. Chen, T. Qiu, Fractional Processes and Fractional-Order Signal Processing. Techniques and Applications, *Springer*, 2012.

[27]. Yu. Bobalo et al., Design and Hardware Implementation of Fracktal Comb-Structured Signals, *Smart Computing Review*, Vol. 4, No. 6, Dec. 2014, pp. 459-469.

[28]. S. Yatsyshyn, B. Stadnyk, Ya. Lutsyk, L. Buniak, Handbook of Thermometry and Nanothermometry, *IFSA Publishing*, Spain, 2015.

[29]. Measurement Science Roadmap for Metal-Based Additive Manufacturing, *National Institute of Standards and Technology*, May 2013.

[30]. S. Ko, H. Pan, C. Grigoropoulos, et al., All-inkjet-printed flexible electronics fabrication on a polymer substrate by low-temperature high-resolution selective laser sintering of metal nanoparticles, *Nanotechnology*, Vol. 18, 2007, 345202, 8 p.

[31]. Concept Laser Introduces QA Tool for In Process AM: QMmeltpool 3D Available Next Year, July 2, 2015, http://disruptivemagazine.com/concept-laser-introduces-qmmeltpool-3d-qa-tool-available-next-year/

[32]. K. Totsu, Y. Haga, et al., 125 mm diameter fiber-optic pressure sensor system using spectrometer-based white light interferometry with high-speed wavelength tracking, in *Proceedings of the Conference on Microtechnologies in Medicine and Biology*, Kahuku, Oahu, Hawaii, 12 - 15 May 2005, pp. 170-173.

[33]. K. Schwenzer-Zimmerer, J. Haberstok, 3D Surface Measurement for Medical Application—Technical Comparison of Two Established Industrial Surface Scanning Systems, *Journal of Medical Systems,* Vol. 32, Issue 1, 2008, pp. 59-64.

[34]. W. Cong-Fei, W. Guang-Long, et al., The signal interrogation technology of MEMS optical fiber pressure sensor, in *Proceedings of the International Conference on Information and Automation,* China, Zhuhai/Macau, 22 -25 June 2009, pp. 1285-1288.

[35]. K. Kitagawa, 3D Profiling of a Transparent Film using White-Light Interferometry, in *Proceedings of the SICE Annual Conference,* Japan, Sapporo, 4-6 August 2004, pp. 585-590.

[36]. W. J. Bock, W. Urbanczyk, Coherence multiplexing of fiber-optic pressure and temperature sensors based on highly birefringent fibers, *IEEE Transactions on Instrumentation and Measurement,* Vol. 49, No. 2, April 2000, pp. 392-397.

[37]. R. Leach, Optical Measurement of Surface Topography, *Springer-Verlag,* 2011.

[38]. P. Hariharan, Basics of Interferometry, 2nd ed., *Elsevier,* 2007.

[39]. J. C. Wyant, White light interferometry, *Proceedings of SPIE,* Vol. 4737, 2002, pp. 98-107.

[40]. D. Apostol, V. Damian, P. C. Logofatu, Nanometrology of Microsystems: Interferometry, *Romanian Reports in Physics,* Vol. 60, No. 3, 2008, pp. 815–828.

[41]. T. Guo, S. Wang, D. J. Dorantes-Gonzalez, J. Chen, X. Fu and X. Hu, Development of a Hybrid Atomic Force Microscopic Measurement System Combined with White Light Scanning Interferometry, *Sensors,* 12, 1, 2012, pp. 175-188.

[42]. R. Cincio, W. Kacalak, C. Łukianowicz, System Talysurf CCI 6000 – methodic of analysis surface feature with using TalyMap Platinium, *Pomiary, Automatyka, Kontrola,* 54, 4, 2008, pp.187-191 (in Polish).

[43]. The Online Industrial Exhibition, Taylor Hobson Precision, Talysurf CCI 6000. The world's highest resolution automated optical 3D profiler, 2005. http://pdf.directindustry.com/pdf/taylor-hobson/talysurf-cci-6000/7159-132100.html

[44]. Th. Seiffert, Schnelle Signalvorverarbeitung in der Weißlichtinterferometrie durch nichtlineare Signalaufnahme, *DGaO-Proceedings,* 2004.

[45]. L. Mingzhou, Development of fringe analysis techniques in white light interferometry for micro-component measurement, Ph. D. Thesis, *National University of Singapore,* 2008.

[46]. H. Abdul-Rahman, Three-dimensional Fourier fringe analysis and phase unwrapping, Ph. D. Thesis, *Liverpool John Moores University,* 2007.

[47]. H. M. Muhamedsalih, Investigation of wavelength scanning interferometry for embedded metrology, Ph. D. Thesis, *University of Huddersfield,* 2013.

[48]. B. Stadnyk, E. Manske, A. Khoma, State and prospects of computerized systems monitoring the topology of surfaces, based on white light interferometry, *Computational Problems of Electrical Engineering,* Vol. 4, No. 1, 2014, pp. 75-80.

[49]. V. Heikkinen, R. Kurppa, et al., Quality control of ultrasonic bonding tools using a scanning white light interferometer, in *Proceedings of the IEEE International Ultrasonics Symposium,* 2010, pp. 1428-1430.

[50]. NIST National Technical Information Service, FY 2014, Budget Submission to Congress, 2014.

[51]. Michael Guckes, Measurement Technology for both Development and Production, Industry 4.0, *HBM,* http://www.hbm.com/en/4112/industry-4/

[52]. E. Mankowska, B. Stadnyk, I. Mykytyn, P. Skoropad, Analiza zmian rezystancji elementów czułych termoprzetworników i metod określania rodzaju ich korozji w środowiskach agresywnych, *Zeszyty Naukowe Wydziału Elektrotechniki i Automatyki Politechniki Gdańskiej: XIX Międzynarodowe Seminarium Metrologów MSM'2014,* Gdańsk – Stokholm, Issue 38, 2014, p.41-43 (in Polish).

[53]. E. M. Hutman, Mechanochemistry of metals and protection against corrosion, *Metallurgy,* Moscow, 1981 (in Russian).

[54]. M. Mahdavinejad et al., Challenges Regarding to Usage of Nanostructured Materials in Contemporary Building Construction, *Advanced Materials Research,* Vol. 829, 2013, pp. 426-430.

[55]. Build It Solar, The Renewable Energy site for Do-It-Yourselfers, Cristian's Earth Sheltered Passive Solar Home in Romania, March 2011. http://www.builditsolar.com/Projects/SolarHomes/Romania/CristianHous

[56]. W. Feist, Forschungs projekt passive Häuser, *IWU,* Darmstadt, 1988.

[57]. M. Yatsyshyn, Yu. Kryvenchuk, Thermal imager estimation of quality of heat insulation of buildings, in *Proceedings of the International Scient. Conf. on Infrared Thermography and Thermometry, Metrological Assurance of Measurement and Testing,* Lviv, Ukraine, 23-27 Sept. 2013, pp. 137-138 (in Ukrainian).

[58]. L. Ingersoll, O. Zobel, A. Ingersoll, Heat conduction with engineering, geological and other applications, *McGraw-Hill,* New York, 1954.

[59]. B. Stadnyk et al., Peculiarities of determining the emissivity of materials at low temperatures, *Measuring Equipment and Metrology,* No 68, 2008, pp. 165-168 (in Ukrainian).

[60]. SIMVOLT.UA, Catalog. Modern infrared pyrometers - temperature measurement without contact, http://simvolt.ua/suchasn-nfrachervon-prometri-vimryuvannya-temperaturi-bez-dotiku.html, (O. Hnatiuk, in Ukrainian), 2016.

3.

Embedded Measures as the Measuring Instruments

S. Yatsyshyn, B. Stadnyk and M. Mykyychuk

Major problem of CPSs operation is determined mostly by credibility of obtained information which depends as on sensors metrological reliability as well as on actuators precision. The latter has to be gauged and control by a set of different sensors whose participation in the management is determined in the design phase or changes automatically by adjusting. Unfortunately these units become obsolete, and more importantly metrological characteristics drift up to mechanical failure. Possible consequence of running processes affect in lowering the QoS or QoP.

According to current practice of standardization the traceability of measurements is provided by periodic calibrations (graduations, verifications, etc.). Then duration of the intercalibration interval defines the period of operation of the mentioned unit with a certain, previously accepted probability of mainly metrological or total failure.

The state of the measuring instrument is usually verified by comparing with measure or standard, or by supplying electrical signal of reference value to its input, or by verifying the installed metrological software versus the checked one. Since two from three failures of measuring instruments are caused by metrological failures and they usually precede major failures, increases the need for cost calibration procedures of every sensor within calibration period (~2-3 years) or its substitution. The latter may be unrealized for instance for temperature and pressure sensors of nuclear power plants. The special issue seems to be a necessity to suspend the production cycle aiming to provide the calibration of sensor(s). Unreliable information received from the measuring instruments with the considerable drift of characteristics degrades the quality of the final product.

3.1. Checked Instrument Based on the Inverse of Conductance Quantum

CPS technologies companies have to utilize the sophisticated metrology equipment for production lines. This involves the estimation of the comparability of CPS component measuring instrument by verification. Development of portable, highly-precise devices is able to provide in-place precision measurements. Chip-scale devices could be directly integrated into equipment to provide continuous quality control and assurance, freeing manufacturers and customers from complex measurement traceability chains and lengthy calibration procedures [1].

New outcoming aim has emerged – metrological assurance embracing the adjustment of necessary measurement precision by means of checked measuring instruments and obtaining the required metrological traceability.

For metrological calibration of measuring instruments usually one applies the direct measurement by the verified measuring instrument of outgoing signal of multivalued measure with determination of the error as a difference of its readout and the mentioned signal. Correction methods of systematic error constituent are realized by operator impact or automatically in offline mode when, for example, self-calibration is carried out [2].

The unique and newly created CPSs often require checking and verification of metrology facilities to ensure their quality work. Unfortunately, existing measures lose their values (accuracy characteristics) by several orders while transferring them to the end user and that is actually considered a normal metrological practice. Such practice cannot be deemed adequate for development of CPSs.

Currently the standards of SI units replace the standards of old generation in the State Laboratories of Metrology and Standardization practice. They are significantly more precise since are built on the usage of fundamental constants of matter just as it was regarded [3] by participants of Royal Society Discussion Meeting. We discuss below the possibility of implementing the high-precision measure of electrical resistance on the basis of inverse of conductance quantum in checking the measuring instruments for CPS purposes.

The aforementioned measure is proposed to be developed by applying the latest scientific achievements in the area of electrical conductivity.

For example, it was established [4] for graphene nanopatterns that the quantum of conductance quantum is equal 12906.34 ± 0.20 Ω at quantization error $(5 \pm 15) \cdot 10^{-6}$. Similar works have been conducted during the last 20 years. There were studied some metrological problems of proposed standards of electrical resistance made of different kinds of substance and samples, estimated their values and uncertainties etc. [5-7]. As a result, in [8] there were compared the quantum Hall Effect resistance standards of National Physical Laboratory and the Bureau International de Poids et Mesures. Obtained results have been agreed for the 100 Ω resistance standard as well as for the 10000 Ω / 100 Ω and 100 Ω / 1 Ω ratios within the few parts in 10^9, to a value consistent with the estimated uncertainties. Moreover, it was shown that to limit the impact of the Peltier effect on 100 Ω / 1 Ω ratios measurement, it was recommended to withstand the measures for long time after reversing the DC currents.

This type of work is performed for purposes of CPS Metrological Services whose responsibilities include transferring the researched physical quantity unit (R) to the CPS measuring instrument by its calibration and sequentially checking the CPS electrical circuits or studying the products' performance. We are trying to engage in research of proposed measure and its implementation in dispersed CPS-components that enables to provide precise operation with high metrological characteristics.

We accept not only the above experimental studies of establishing the inverse of conductance quantum, but its metrological substantiation as a physical quantity directly related to the Boltzmann constant and the charge of electron with particular uncertainties. Then basing on the results for these quantities received by means of different experimental methods and applying the theory of metrology, we study the correlation between $R_0 = \dfrac{h}{2e^2} = 12906.4037217$ Ω uncertainty and the other two known uncertainties: the Planck constant $h=6.62606957(29) \cdot 10^{-34}$ J/s and the charge of electron $e=1.602176565 \cdot 10^{-19}$ C. So we try to establish the value of inverse of conductance quantum and its uncertainty. Therefore, we are absolutely not interested in with which uncertainty the studied quantity was measured before in a certain metrological center and in one or another scientific work. For instance, in [9] there was submitted a value of relative standard uncertainty of von Klitzing constant (25812.8074434 Ω) equal to $3.2 \cdot 10^{-10}$, or it is much more precise than the obtained value. Apparently, it refers to the coverage interval of the results of repeated measurements of resistance standard and ignores the systematic constituent of error because the true value of the quantity is unknown at least in uncertainty model approach.

So it is sufficiently to know the current values of fundamental physical constants h and e. They were defined with high precision by means of a set of different physical methods. According to the data of the method of Watt balance, installations of studies, such as of X-rays crystal density, Magnetic resonance, Faraday constant, and Josephson constant, CODATA 2010 recommended value of weighted mean Planck constant relative uncertainty to be equal to $u_h = 4.4 \cdot 10^{-8}$ [9]. The same concerns the charge of electron, which uses the results of different research and its evaluation methods (Millikan's oil-drop experiments, E.Rutherford and other investigations). Relative standard uncertainty of electron charge determination is estimated as $u_e = 2.2 \cdot 10^{-8}$ [10].

Below we consider the appropriate prototype of resistance measure 12906 Ω applied for calibration of high-precise measuring instrument. Then we are able to practically obtain a reference point of its scale that is the main in checking the measuring instrument and so raising the accuracy class. This way, the calibration of measuring instruments and sequentially validation of gauging data can be realized. The advantage of the similar methods of metrological checking is evident. It was demonstrated [11] on examples of checking the temperature, pressure and so on. By continuous controlling the reliability of metrological data and basing on the checked results for previous duration period, forecasting the instrument's metrological state is developed.

Being at superconductive state, graphene is inherent in the resistance value which corresponds to inverse of conductance quantum that is equal to (12906.4037±0.0020) Ω due to transient resistance of contacts. Four-wire circuit is sufficient to carry out the measurement of this resistance [12], and the Wheatstone bridge together with the Hamon network [13] would be sufficient to transfer this value to the current standards (1.0; 100.0 Ω), and then to adjust precise values of the working resistors to the required denominations.

Due to the development of nanotechnology there is an opportunity to implement the superconductive CNTs [14-15] as ideal resistive elements in the State standards of electrical resistance. Valid State standards are regulated by current verification schemes. For example, in Ukraine this is the decisive document [16]. It establishes the destination of State standard primary unit of electrical resistance – Ω, set of basic measuring instruments which are the part of it, the basic metrological specifications of the Standard, and transferring procedure for the magnitude of electrical resistance unit from State primary standard via Secondary standards and Working standards to Working measuring instruments herewith indicating uncertainties and basic methods of verification.

For transferring the value of unit to reference standards in the range $3 \cdot 10^{-3} \dots 1 \cdot 10^{9}$ Ω at DC the State primary standard consists of a set of measuring devices that includes the group of 20 electrical resistance measures of the nominal quantities of 1 and 100 Ω, comparator, groups of three 1Ω measures, set of transitional electrical resistance measures. Ohmmeters, DC/AC bridges, unambiguous and ambiguous measures of electrical resistance are usually applied as working measures.

The range of values of a physical quantity that is realized by the mentioned standard is equal to 1 Ω or 100 Ω. State primary standard provides storage, verification and supervision of the mentioned unit with standard deviation measurement result S_m not exceeding $3 \cdot 10^{-8}$ Ω (for 1 Ω standard the relative deviation is $3 \cdot 10^{-8}$) at 10 independent observations (Fig. 3.1).

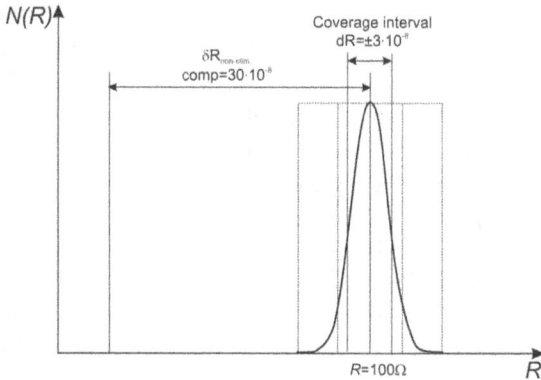

Fig. 3.1. Instrumental error of State electrical resistance standard.

Non-eliminated component of systematic constituent of relative error is $3 \cdot 10^{-7}$. Reference standards are used for transferring with standard deviation $1 \cdot 10^{-7} \dots 2 \cdot 10^{-5}$ Ω by means of verification method (DC comparator) of the unit magnitude to electrical resistance measures of the 1st, 2nd categories and so on. The relative instability (drift) of reference standard's value for one year operation does not exceed $2\cdot10^{-6}$.

Transferring scheme for the certain physical quantity is rather complex. At each stage of transfer accuracy is lost significantly; almost an order of magnitude. Main attention is paid to the final facility which steers transferring the unit size of electrical resistance. It would be an ambiguous measure of electrical resistance, or the precision measures.

If such measure is a precision resistor, whose operation is based on the reproduction of inverse of conductance quantum, a few metrological problems arise. Namely: a) How to fit a resistor in the State scheme of verification; b) Whether there are additional problems related to instability of structure and the drift of determining properties of CNT's substance as a result of its transition from the superconductive to a semiconductor state [14]; c) Number of other special metrological problems [13].

However, the benefits of its implementation are obvious: a) The performance drift while using fresh specimens of superconducting nanotubes as working elements of the precision resistor becomes negligible; at the same time the relative instability of measures that make up the reference standard for one year duration, determined at 20 °C and at DC, is defined at nominal reference value: from 10 to $1\cdot10^3\Omega$ – up to $5\cdot10^{-6}$; $1\cdot10^4\,\Omega$ – up to $6\cdot10^{-6}$; $1\cdot10^5 \ldots 1\cdot10^7\,\Omega$ – up to $8\cdot10^{-6}$ [16]; b) The error due to leakage current of the reference measure is quite essential and in contrast the same error of CNT working element becomes negligible owing to its superconductivity.

Uncertainty analysis involves proper attention to further assuring the sufficient level of metrological maintenance. Core of it lies in the triangle "embedded metrological hardware – installed metrological software – implemented metrological firmware". In particular, it facilitates the implementation of industry measuring methods that try to verify the performance of proposed embedded measure of electrical resistance. Aforesaid enables to realize qualitatively new model of secondary SI unit transfer from the Reference standard to Measuring instruments for CPS purposes by applying the embedded unambiguous measure (of electrical resistance on the basis of inverse of conductance quantum). The proposed model allows to get rid of intermediaries in transferring scheme.

Basing on the mentioned data we have estimated [17] the relative uncertainty of CNT resistor's value: $u_R = \pm(u_h + 2u_e) = \pm 8.8 \cdot 10^{-8}$. So, the uncertainty of the studied resistivity is determined by Planck constant and the electron charge uncertainties. Transition to the mean square error is made by the equation: $\sigma = u\sqrt{3} = \pm 15.242 \cdot 10^{-8}$ (Fig. 3.2). The absolute value of proposed standard tolerance limit is defined as $\pm 0.0019672\ \Omega$.

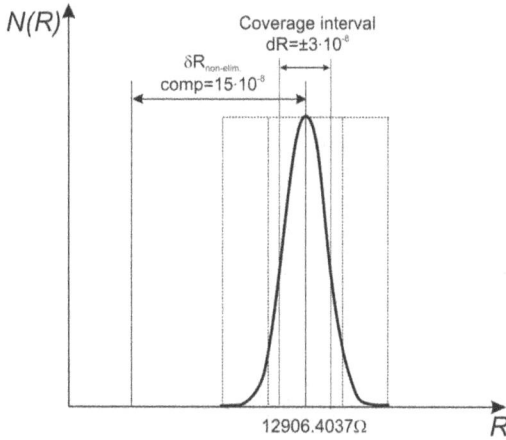

Fig. 3.2. Instrumental error of electrical resistance measure, based on the inverse of conductance quantum.

So, performed studies permit to consider the relative non-eliminated component of systematic constituent of the proposed measure (Fig. 3.2). It proved to be twice smaller, in comparison with the current State primary standard value (Fig. 3.1). At the same time, the relative standard instability, as such, cannot exist since the Planck constant and electron charge are basically unchangeable. Resistance value will only be determined more exactly when the methods of h and e studies are further improved.

Ordinary 4-wire scheme is sufficient to measure the resistance of CNTs. Current is fed by one pair of subminiature electrodes. Another pair of electrodes is used to connect a voltmeter. The scheme of measuring the inverse of conductance quantum may be similar to the described one in [15]. By switching two CNTs in series we can get the value of $25812.8\ \Omega$, and by switching in parallel – $6453.2\ \Omega$ (Fig. 3.3).

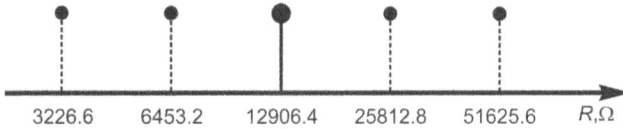

Fig. 3.3. Measurement range of Ohmmeter within which calibration is provided.

For 4 CNTs two more denominations of electric resistance – 51625.6 Ω by sequential switching, and 3226.6 Ω by parallel switching can be obtained. Otherwise, the easiest way that is simultaneously the most accurate one seems to overlap possible values of range from 3226.6 Ω to 51625.6 Ω with the increment of 3226.6 Ω. The transfer of electrical resistance measure value to smaller values of working resistors without losing precision can be carried out as shown below.

For standardization service purpose the mentioned scheme could be metrologically expedient and a real resistance reference measure could be composed of two rows - (7+6) of superconductive CNTs that form together a resistance value $R_{\Sigma} = \dfrac{h}{26e^2} = 992.8002863\ \Omega$. Then the non-eliminated component of systematic constituent of relative error is unchangeable and equal to $15.24 \cdot 10^{-8}$.

The transfer size scheme of electrical resistance unit on the basis of fundamental constants is realized as follows. At first stage the unit is transferred to the higher value – 1000 Ω resistance that has to be formed by 13-parallel-connected carbon nanotubes (992.8 Ω) with 7 connected in series resistance of 1.0 Ω value (Fig. 3.4).

The transfer is realized by the substitution method while equalizing values of 2 compared resistances – the standard one (999.8 Ω) and 10 working measures with a nominal value of 100 Ω each (the Hamon network is made up of ten resistors with a switch). Then the high precision of measurement is reached because the error does not depend on the measuring instrument; it is determined only by quantization error. The latter can be reduced to the required value by selecting the minimal resistance of bridge circuit branch. Hamon network is placed inside cylindrical box which contains also the guarding resistors and the servo control to perform parallel connection. The device is directly connected to the input of pico-ammeter under calibration in order to minimize the leakage current and electromagnetic noise.

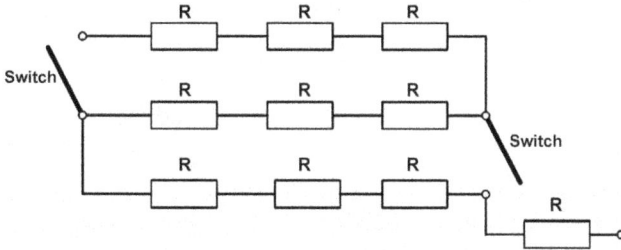

(a) The switch contacts are open, and 10 resistors are connected in series;

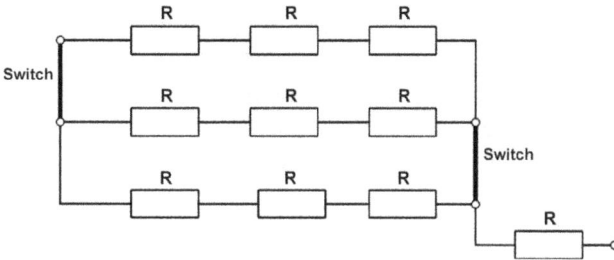

(b) The switch contacts are closed, and 10 resistors are connected
in a complicated manner: 9 – in parallel (3 branches of 3 resistors in each),
and 1resistor – sequentially with them.

Fig. 3.4. Hamon network of resistance measure.

At second stage, with previously determined numerical value of 10 mentioned working standards, we transfer an electrical resistance unit from proposed resistance measure to the certain 100 Ω working standard of electrical resistance. For this aim we again apply the method of measuring the ratio of two similar values resistors. The mentioned 9 working standards are switched by closing the contacts (Fig. 3.4, b), aiming at forming 100 Ω resistance value. Then, similarly, the specified on the 1st stage measurement method is carried out to gauge the ratio of the composed 100 Ω resistance and the 10th working standard of 100 Ω-denomination.

As a result, the non-eliminated component of systematic error of inverse of conductance quantum is transferred successfully to 100 Ω current working standard while maintaining the sustainable adjusted value of the latter – the current numerical value of the State standard.

3.2. Study of Quantum Unit of Temperature and Temperature Measuring Standard

CPS technologies have to utilize the sophisticated metrology equipment for production lines. Similar to nanotechnology, try to ensure the final measurement result by the next generation of Standards [3] and by introducing a number of metrological procedures such as self-verification, self-validation and self-calibration. Development of portable, highly-precise devices is able to provide in-place precision measurements. The studied quantum standard may be recommended firstly to apply as intrinsic standard; such a standard does not need permanently recurring measurements against the realization of the SI unit in order to validate its accuracy. Intrinsic standards are important instruments in disseminating accurate measurements in an efficient way for instance in CPS operation. The developed standard may be considered as intrinsic standards of temperature that could be embedded into CPSs ensuring their precision operation.

At the end of last century there were successfully implemented the atomic standards of SI units. As result of quantum discreteness qualities extend in nanosphere, for example, the ability appears to realize not only measuring instruments but to create also the standards of measurands. Currently temperature remains the last only value among seven main units of International unit system that is still not regulated at the atomic and hence much higher level in terms of accuracy.

3.2.1. Primary Thermometry and Quantum Units of Temperature

In the chain of leading metrological centers (USA, GB and other countries), through several years the intensive endeavors of elaborating and assuring the unit of temperature scale in the form of a quantum energetic unit (minimal by size a discrete value of energy or heat energy that can be defined, established and fixed by the experimenter) are carried out at the high methodological level [18]. The recommended by Ia. Mills et al [3] new format of unit with new definition is the next. The kelvin, K, is the unit of thermodynamic temperature; its magnitude is set by fixing the numerical value of the Boltzmann constant to be equal to exactly $1.38065 \ 10^{-23}$ when it is expressed in the unit $s^{-2}m^2kgK^{-1}$, which is equal to $J \cdot K^{-1}$. The effect of proposed definition is that the kelvin is equal to the change of thermodynamic temperature that results in a change of thermal energy $k_B T$ by 1.38065 10^{-23} J/K. Then using k_B rather than T_{TPW} to define kelvin better reflects modern practice in determining thermodynamic temperature directly by primary methods, particularly at very high and low temperatures.

The unit of thermodynamic temperature, the kelvin, will be redefined in 2018 by fixing the value of the Boltzmann constant, k_B. The present CODATA recommended value of k_B is determined predominantly by acoustic gas-thermometry results. To provide a value of k_B based on different physical principles, purely electronic measurements were performed by using a Johnson noise thermometer to compare the thermal noise power of a 200 Ω sensing resistor immersed in a triple-point-of-water cell to the noise power of a quantum-accurate pseudo-random noise waveform of nominally equal noise power. Measurements integrated over a bandwidth of 550 kHz and total integration time of 33 days gave a measured value of $k_B = 1.3806514(48) \cdot 10^{-23} \, {}^J\!/_K$, for which the relative standard uncertainty is 3.5·10^{-6} and the relative offset from the CODATA 2010 value is +1.9·10^{-6} [19].

Considering the mentioned primary methods of temperature determination and having determined Boltzmann constant with a very small error, scientists could develop the unit of a temperature scale due to the energetic/power unit, endued with a certain determination error. Unfortunately, these hard works, details of which were descript earlier in [3, Table 1 (Summary uncertainty budget for a determination of Boltzmann constant by QVNS thermometry)], practically not eliminate the major principal shortcoming. It consists in the necessity to calibrate thermometer at TPW temperature. Othergates, researchers are unable to get rid of traditional calibration and only replace the outdated method by the modern one. Nevertheless the replacement of the temperature measuring instruments for the energy ones will raise especially severe difficulties precisely in the area of ultralow energies gauging [20-21] that can be associated with minimal energy (temperature) unit.

Primary thermometry envisages that particular measuring instrument concerns the concrete measurand (T) that can be defined by the calculating the gained results excluding other unknown quantities and applying only fundamental constants of matter as proportionality factors. Secondary thermometry develops the methods of measurement of another kind temperature than thermodynamic one or the methods using any dependence of properties on temperature, and then some points of received dependence are ascribed the certain values of thermodynamic temperature.

The classic set of primary methods includes five methods of thermometry: noise, gas, acoustic, optical, and magnetic. Those methods are based on the fundamental physical laws whose mathematical descriptions comprise the thermodynamic temperature. Among them the gas and optical thermometry have gained widest application in the reproducibility of thermodynamic temperature. The last problem is particularly inherent in Nanothermometry where methodical errors rise when the thermal capacities or sizes of the thermometry-processed body and the thermometer become comparable [22].

Noise method of thermometry still remains in a state of metrological elaboration. In the conditions of durable development of nanotechnology with its unrepeatable measurements the uncertainty accumulation causes a substantial decrease in authenticity of information extracted from the received experimental results. From Johnson-Nyquist equation we could state that the concrete and precise determination of root-mean-square voltage and hence electric power implies the usage of a studied object with a priori known value of electric resistance. At precisely defined Boltzmann constant a relative instrumental error of the measured voltage is formed as the sum of notified values: the relative errors of determining object resistance and temperature, and also the relative error of specifying frequency bandwidth of measurement. It indicates the undoubted advantages (resistance and temperature are known with high accuracy) of employing the substance of its sensitive element in systematic noise studies. Recently the considered method is implemented as second one in addition to acoustic gas-thermometry method aiming the determination of the Boltzmann constant and trying to introduce the metrologically obvious step in redefining the notion "Temperature" by CODATA.

The noise method is increasingly applied in Nanothermometry [23]. Particular attention is paid to the investigation of $1/_{f\gamma}$ electrical noise and especially to their connection with changes of entropy of studied thermodynamic system [24], including the nanodimensional system [25]. Similar results were obtained by us in [26]. The deduced equation of noise thermometer transformation function relates the power of electric noise P_{el} with the thermodynamic temperature T through the dissipation rate of entropy dS/dt:

$$T\frac{dS}{dt} = -e\phi\frac{dN}{dT} = P_{el}, \tag{3.1}$$

where e is the electron charge; N is the amount of charge carriers. Inside the researched substance the latter could be estimated by following the thermodynamic approach. Earlier we have followed from the most probable physical processes in sensitive substance under the given conditions that: $\frac{dS}{dt} \approx -\frac{\Delta S}{\tau}$, where ΔS is the entropy changes taking place as consequence of a relaxation process with constant τ .

Raman method of thermometry is a contactless temperature-measuring method. The measurement of the solid body surface temperature with taking advantage of a Raman phenomenon is related to one of few methods of primary thermometry being realized with the help of a thermometer whose state equation could be written in an explicit form, avoiding the involvement of unknown constants dependable on temperature. The given method helps to measure the temperature for objects ranged from 100 nM to 100 µM as well as within this from cryogen till mid-high temperatures, which in addition does not demand calibration before measurement, could be distinguished. This method is considered in detail in chapter 6.

Magnetic method of thermometry is the method of measuring temperatures lower than 1 K. It is based on the temperature dependence of magnetic susceptibility χ of a paramagnetic substance. Gas method of thermometry operates basing on volume expansion effect of substance. Here the measuring temperature changes ΔT are proportional to the changes of sensitive substance volume. This method seems to be a separate issue that requires a special approach. Acoustic method of thermometry was studied by us earlier in [22] in conjunction with research of thermosensitive materials including nanostructured ones.

3.2.2. Promising Methods of Thermometry

Coulomb blockade thermometer is the primary thermometer based on electric conductance characteristics of tunnel junction arrays. The parameter $U^{1/2} = 5.439\,{}^{Nk_BT}\!/_e$ (k_B is the Boltzmann constant), the full width at half minimum of the measured differential conductance dip over an array of N junctions together with the physical constants, provide the absolute temperature [27]. So, half width $U^{1/2}$ depends only on the constants of matter and known parameter N that seems to be quite close to design of primary thermometer based on fundamental physical constants.

A typical Coulomb blockade thermometer is made from an array of metallic islands, connected to each other through a thin insulating layer. A tunnel junction forms between the islands, and as voltage is applied, electrons may tunnel across this junction. The tunnelling rates and hence the conductance vary according to the charging energy of the islands as well as the thermal energy of the system. In order for the Coulomb blockade to be observable, the temperature has to be low enough so that the characteristic charging energy (the energy that is required to charge the junction with one elementary charge) is larger than the thermal energy of the charge carriers. For capacitances above 1 fF (10^{-15} farad), this implied that temperature has to be below about 1 kelvin. This temperature range is routinely reached for example by ^{3}He refrigerators. Thanks to small sized quantum dots of only few nanometers, Coulomb blockade has been observed currently above liquid helium temperature, up to room temperature.

Free electron gas primary thermometer is based on bipolar transistor, temperature of which is extracted by probing its carrier energy distribution through its collector current, obtained under appropriate polarization conditions, following a rigorous mathematical method. The obtained temperature is independent of the transistor physical properties as current gain, structure (homo-junction or hetero-junction), and geometrical parameters, resulting to be a primary thermometer. This assumption has been tested using off the row of silicon transistors at thermal equilibrium with water at its triple point. The obtained transistor temperature values involve an uncertainty of a few mK. Further free electron gas primary thermometer has been successfully tested in the temperature range of 77...450 K [28].

Rather complicated Josephson junction noise thermometer is a thermometer, operation principle of which is built on the Josephson Effect and readouts are proportional to thermodynamic temperature. And here we want to underline that emerging the oscillations at frequency $f_0 = \frac{2e}{h} u_0$, where h is the Planck constant, while the DC voltage u_0 is applied to Josephson element, makes it possible to perform a generator of sinusoidal current. Its main advantage is the next. Half-width of the spectral line of thermal noise in Josephson element, measured by radio spectrometer and given by the equation $\Delta f_0 = 4\pi k_B T r \left(\frac{2e}{h}\right)^2 \left(1 + \frac{Ir}{u_0}\right)$, where r is the resistance, which shunts the element in measuring scheme, I is the current that passes through element, enables [29] to link thermodynamic temperature T.

3.2.3. Investigation in Creating the Quantum Unit of Temperature

Temperature in nanothermometry is the statistically formed value of quantity, determined by the inner energy of a body of sufficient sizes for purpose of applying the thermodynamic consideration to this body. It seems to be one of the fittest terms among the considerable number of temperature definitions which try to identify temperature in nanothermometry. A thermodynamical notion of temperature is related to heat exchange between two systems. The quality of supplying or not to the balance among themselves under some predetermined conditions pertains to all macroscopic systems. The necessity to characterize a state of thermodynamic systems by some specific quantity becomes obvious. So, a notion "Thermodynamic temperature" has been introduced for this purpose. The objective measurement of temperature is possible due to the transitivity of a thermodynamic equilibrium. Therefore there is a possibility to compare the object temperatures among themselves without the objects' per se contact. Current definition of the unit of thermodynamic temperature, kelvin, is based on a material artefact, namely, the triple-point-of-water temperature [30, p. 175] which can be realized with uncertainty about 10^{-7}.

Triple-point-of-water temperature is the temperature of phase equilibrium between vapor, solid and liquid phases of water. It is the basic reference point of Thermodynamic temperature scale, as well as one of the basic reference points of ITS-90. Triple point of water implementation scheme is shown in Figure below. Glass flask with tightly inserted tube (for thermometer) is filled with water of isotopic composition closed to the composition of ocean water, vacuumed and sealed. In middle of tube firstly pour the liquefied air or fill with crushed ice, as a result around test tube the ice layer appears, then pour warm water to form between ice and walls of the tube a thin water layer. In this state, the flask is immersed in Dewar vessel with mixture of water and crushed ice.

After a certain time above the water surface in the flask is set the equilibrium vapor pressure of water and the installation can be used to reproduce a reference point 273.16 K. Measurements in the triple point of water are recommended to start in 2…3 days after frosting of icy shell. Temperature 273.16 K value can be maintained inside the test tube for a long time, i.e. up to hundreds of hours (duration depends on triple point of water intensity of usage). Tube for thermometer filled with water, improves a heat transfer between it and the medium. Triple point of water reproducibility is $2 \cdot 10^{-5}$ K [22] (Fig. 3.5).

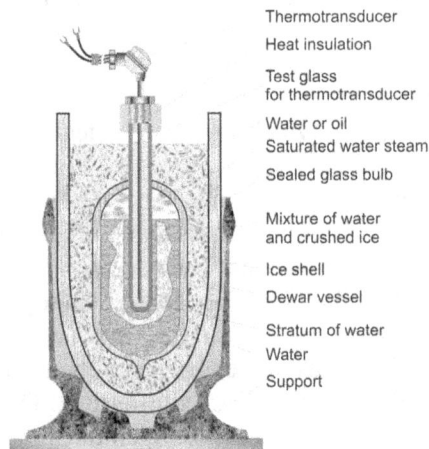

Thermotransducer
Heat insulation
Test glass
for thermotransducer
Water or oil
Saturated water steam
Sealed glass bulb
Mixture of water
and crushed ice
Ice shell
Dewar vessel
Stratum of water
Water
Support

Fig. 3.5. Scheme of realization of Triple point of water.

It is discussed below the possibility of researching the most contemporary measure of temperature on the basis of fundamental constants of matter with involvement of the Standard of electrical resistance on the basis of Inverse of Conductance Quantum [31] as well as the Standard of voltage based on the Josephson junctions [27] that can produce voltage pulses with time-integrated areas perfectly quantized in integer values of h/2e. The synthesized voltage is intrinsically accurate because it is exactly determined from the known sequence of pulses, the clock frequency, and fundamental physical constants.

Thus, we consider the investigation of the electrical resistance value of which is based on Klitzing constant, and of the electrical voltage standard on the Josephson Effect for exact frequency-to-voltage conversion, combined with the clock. As the mentioned resistance we propose to study one of widespread FET constructions, namely the CNTFET with built-in CNT [32] which has to be superconductive. Source and drain have to be manufactured from dissimilar metals that form the thermoelectric pair via CNT. The latter, being in superconductive state, is inherent in resistance which corresponds to 25812.807557±0.0040 Ω, due to transient resistance of contacts. While studying the dissipation of electric power $I^2 R = U^2/R$ on such an electric resistance in temperature measurement area:

$$E = \frac{U^2 \Delta t}{R_{Kl}} = I^2 R_{Kl} \Delta t = N \frac{3}{2} k_B T a, \qquad (3.2)$$

was noted that we are able to estimate the change of thermodynamic temperature T, or substituting this equation by $I = \frac{\Delta Q}{\Delta t} = \frac{Ne}{\Delta t}$ (Δt is the time period), we clarify it to:

$$\frac{(Ne)^2 h}{(\Delta t)^2 e^2} \Delta t = N \frac{3}{2} k_B T, \qquad (3.3)$$

when the electrical current is formed per unit time by N conduction electrons that transfer the energy $\frac{3}{2} k_B T$ to the atoms of matter. From here the TT jump ΔT at current transmission I through superconductive CNT (cooling is considered to be negligible), is defined as:

$$\Delta T = \frac{2hN}{3k_B \Delta t} = \frac{2hI}{3k_B e}, K \qquad (3.4)$$

On condition of power supply from Johnston junctions array it appears an opportunity to pass a discrete particular number of electrons through nanotube of FET. Then the resulting value of the temperature increase of atom reduced to one electron that was scattered on its phonon at the unit time is identified as Reduced Quantum Unit of Temperature:

$$\Delta T|_{\substack{\Delta t \to 1s. \\ N \to 1}} = T_{UR} = \frac{2h}{3k_B} \left[\frac{K}{s.}\right] \cdot 1[s.]$$
$$= 3.19949342 \cdot 10^{-11} \approx 3.2 \cdot 10^{-11}[K] \qquad (3.5)$$

Otherwise, the temperature increment reduced due to single-electron dissipation on phonon of the superconductive CNT junction with source/drain and due to unit time application is defined only by fundamental physical constants (h and k_B); it is equal to $2h \cdot 1s./3k_B = 3.2 \cdot 10^{-11}\ K$.

Hence, the proposed in [3] figures regarding interrelation and inter-definition of basic SI units and the principles of study of the mentioned units via the fundamental constants of matter are modified by results of the performed study (see Figs. 3.6 and 3.7).

In such a way the Reduced Quantum Unit of Temperature that is independent of kind of matter and recommended in the creation of Temperature standard, can be regarded. It would be the Standard based on a 2 quantum effects (von Klitzing Effect and Josephson Effect) and, having been measured against the SI system of units, has a certain value with uncertainty determined by sum of 2 uncertainties: of Planck constant and of Boltzmann constant [19] which together make its total relative uncertainty value that equals to $59.2 \cdot 10^{-8}$. The last value also includes the relative standard uncertainty of atomic unit of time that is 5 orders of magnitude smaller ($5.9 \cdot 10^{-12}$ [9]) and therefore is neglected at this stage of study.

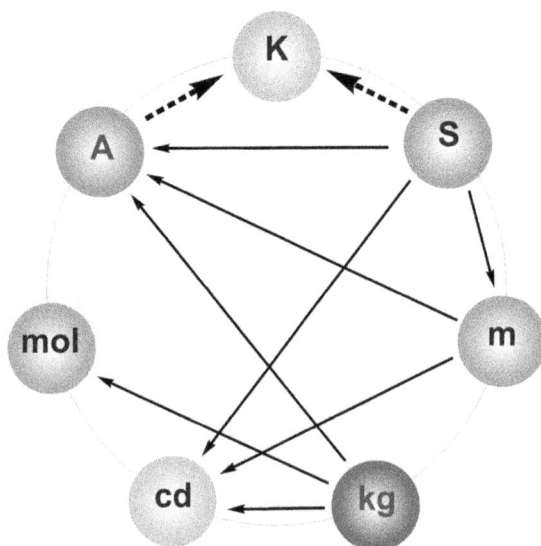

Fig. 3.6. Interrelation and inter-definition of basic SI units: dashed arrows show the revealed relationship of the studied unit T (K) with unit I, A (by unit V and unit R) and with unit t, s.

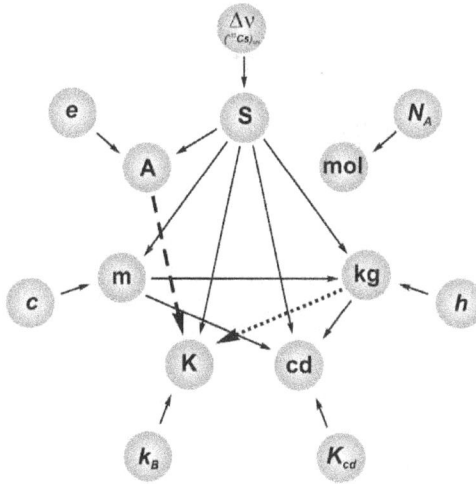

Fig. 3.7. Principles of mentioned units study via the fundamental physical constants of matter: elimination of interrelation between unit m and unit T (dotted arrow) as well as the emergence (dashed arrow) of interrelation between unit I, A and unit T, K.

Operating mode is as follows. The studied appliance is propose to supply by short (~10^{-3} s) pulse voltage consequences, effect of which is measured at the 2nd stage at power absence. Measuring temperature with minimal methodical error is easiest with the built-in thermocouple. It's enough at manufacturing the CNTFET to make source and drain from two dissimilar conductive metals (f.i., Ni and Cu). Superconductive CNT as the 3rd intermediate body forms a quasi junction of produced thermocouple.

3.3. Mass Measures with Coded Remote Access

Realization of existing CPS development programs is impossible without taking into account the metrological aspects of their designing, constructing, and operating. CPS technologies companies have to utilize the sophisticated metrology equipment for material and geometric characterization [1]. One of the main types of characterization seems to be the weight determination, and with the lowest possible error. The leading production centers based on modern machinery are not able to be tested by virtue of their own complexity and problems of delivery to certified laboratories. Moreover, unique and newly created machinery often requires self-verification and standardization of metrology facilities to ensure the quality operation in a continuous loop. As well, the existing measures lose their accuracy characteristics by several orders while transferring them to the end users, or to CPSs.

Smart Manufacturing, namely in metal production with application the precision smart balances provided by the remote calibration, in-place executable, is considered below. On this way the creation of remote horizontal-vertical structure of the metrological surveillance of smart instrumentation as the structural units of CPSs, envisages the remote control and verification with using code access to previously installed software and hardware. Especially it concerns the strain gauge balances applicable in aviation [33], in water control [34] etc. If complex CPS equipment, in our case strain gauge balance, is located at great distance from the control device or an information terminal, then it becomes necessary to monitor the state of these units and to manage their characteristics. There are crucially important for balances their timely calibration, compliance their readouts to passport data, because false impressions usually entail significant financial loss. Therefore, it becomes urgent the problem of balance's calibration at great distances via the industrial Internet [35]. The main requirements for such networks are: the stability of their work in time, rapid transfer of information from system elements, easy network expansion without loss of performance, compatibility with other systems standard protocols, communication over long distance etc.

So we consider below the aforementioned production cycle equipped with balances connected with data networks based on TCP/IP protocol. For proper functioning the balance has to be ensured calibration that is performed in its operating place. Due to man-in-loop technology the remotely accreditation of precision balance is accomplished, improving their performance in certain operating ranges up to working standards values. To implement this technology the multi-level networks and Ethernet are widely applied.

Calibration of balance is a set of technical arrangements that establish the mass ratio obtained by balance and standards (or kettlebell) in order to bring metrological characteristics of the first one in compliance with its technical parameters specified in balance passport details. There exist several types of balance calibration. Initial calibration is performed at balance manufacturing or after repair. Periodic calibration is carried out during the operation when substantial changing of metrological characteristics was fixed after the diagnosis. The same is fulfilled within the terms defined by adjusting calibration intervals which is approved by the results of metrological attestation as it is realized for instance by [36]. Extraordinary calibration is performed while abrupt drifting the balance metrological characteristics was noticed, moreover faster than calibration interval ended. Producers in their operational documents indicate the recommended periodicity of calibration.

By type of calibration the balances are divided into the following groups: a) With the external calibration (operator performs the calibration with use of kettlebells of appropriate accuracy; b) With the built-in automatic calibration (the need for calibration is determined by the operator, and procedure is carried out by embedded unit; c) With the built-in self-calibration (the need for calibration is determined automatically and performed without operator) [36].

There are several types of networks. In this work the multi-level network for remote calibration of balance with built-in working measure is studied. The information terminal performs role of server. It saves information about the state of system elements and manages them. Each of the elements of primary network management may have its subnet in which it is able to act as a server. In each of subnets the access to their elements can receive only the network server that enables to split load data processing. Thus, the generalized information on system status is transmitted to central control device that sends common control commands. Until recently the most spread data interface of management information systems was RS-485 imposing several restrictions: number of connected elements not exceed 32, low speed of data transmission, need of creating the separate network for system etc.

Built-in Weighting Subsystem with Remote Access. Applying the data interface Ethernet in management of information systems of CPSs units allowed to circumvent these restrictions. So, can be transferred large quantity of information at a greater rate, and also may be used the existing networks (Fig. 3.8).

To send information in Ethernet networks there were worked out a few protocols operating at different levels: application one, transport, gateway level as well as network access layer. The basic network layer protocol of technology TCP/IP is internetwork protocol IP and its subsidiary protocols. This protocol is used to identify the device address of the recipient (in this case particular balance) and the route by which information will be transferred to it.

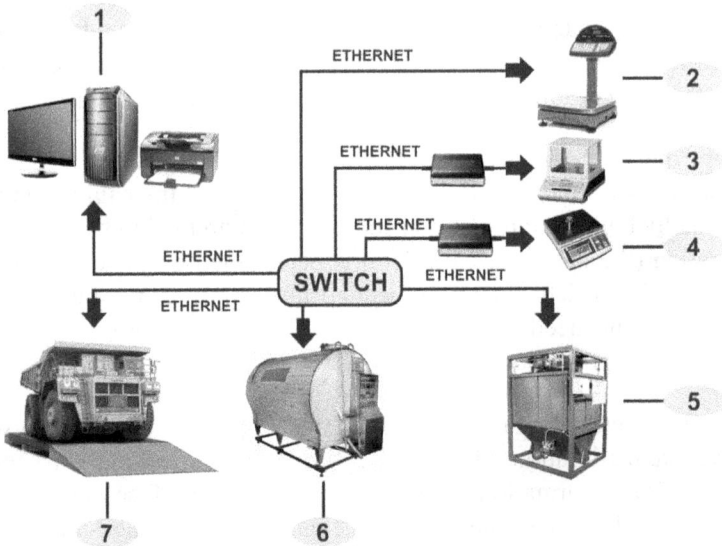

Fig. 3.8. Block diagram of balances communicating over Ethernet:
1 is the workstation; 2 is the balance-operating microprocessor;
3, 4, 5, 6, 7 are the strain gauge weights of different types.

Having written appropriate software for mentioned system application, it can be greatly facilitate the work of operators of weights staff, since weights calibration procedures at remote more than 1000 km enterprise such as "Avdiivka Coke Plant" can be conducted by specialist located in Lviv Scientific and Production Enterprise "Technobalances", Ukraine. The above system is designed to work with car weights. Weighing takes place in two stages that are the gross weighing and the container weighing. In addition to basic calibration function, this system ensures the transfer of information on vehicles weighing (gross and Tara), multi-level user authentication, prevents unauthorized use of information on carried weighing. It creates directories with the data: customers, drivers, cargo, warehouses, cars, and vehicle numbers. Also it becomes possible the revision status of weighting, filtering data by specific criteria. In addition, is searching information on previous weightings, formation of common reports for a given period, printing consignment notes, and data archiving and applying. An important component of metrological testing procedures can be considered the weight testing, settings periodic transfer of data weighing by E-mail. By continuous controlling the reliability of metrological data and basing on the self-checking results for previous time duration, the forecasting of device's metrological state is developed. It gives opportunity to correct measurands and to introduce the amendments in production.

Evaluation of measurement error is conducted in the next way. Information from output of verified instruments is sent to computing facilities for calculating of errors and creating signal for proper recording of the verification results. It forms the basis for the subsequent execution of self-adaptation and self-correction of the performance. So, let us consider more consistently the methodological basis of errors' structure of projected means intended for weighing in proclaimed goal. Fig. 3.9 depicts some metrological terms, for defining the basic error, or error in the normal operation mode.

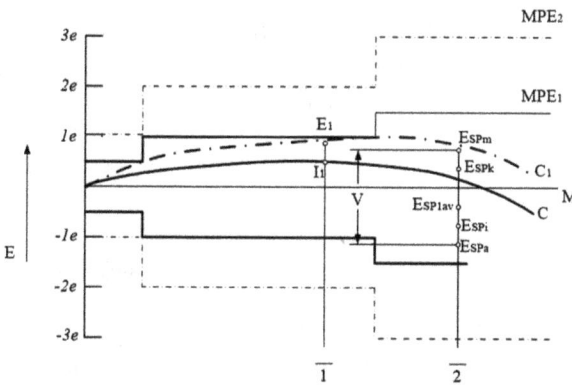

Fig. 3.9. Terms in usage: M is the value of measured mass, E is the readout's error, MRE1 and MRE2 are the maximum permissible error at initial verification and at operation period respectively, C is the characterization under normal conditions, C1 is the characteristic caused by impact factor (here is envisaged that impact factor acts on unchangeable characteristic); ESP is the readouts' error defined by the long-term instability testing results; I is the basic error; V is the variation of readout's error during the long-term instability testing.

Cut 1 reflects an error E1 of MI caused by impact factor, I1 is the basic error. Cut 2 displays the errors average value ESPlav for the first measurement during the long-term instability testing as well as some other errors ESPi and ESPk, and errors and boundary values of errors ESPm and ESPn. All these errors should be assessed at different moments of long-term instability testing. At the same time the errors span V of readouts during the latter is equal to ESPm - ESPn.

Within this method of remote calibration the considered MI consists of individual units; each of them is inherent in its own error. While evaluating aforementioned errors it needs to apply the following requirements. Maximum permissible error M_i, of module that is tested separately, are equal to share pi of mentioned error, or permitted variations of readouts of complete unit in which this module is included. Shares for any module must be chosen according to the same class of exactness and the same number of verification divisions that has a particular measuring instrument that includes this module. These shares p_i must meet the following condition:

$$p_1^2 + p_2^2 + p_3^2 + \cdots \leq 1 \tag{3.6}$$

Wherein module's producer should choose the share p_i, and its value must be confirmed by appropriate tests. However, the share shall not exceed 0.8 and not be less than 0.3 at effect more than one module.

For mechanical structures such as balance's platforms, weighing transmitters, and mechanical or electrical connecting elements that are designed and manufactured according to the conventional engineering practice, generally apply the share $p_i=0.5$ without any trial. For instance, similar occurs if whole lever is made from the same material and if leverage has two planes of symmetry (longitudinal and transverse), and if characteristics stability of electrical connecting elements corresponds to transmitted signal, such as a sensor output signal, total resistance etc. For devices that include typical modules given in Table 3.1, the shares p_i can match values specified below.

3.4. Conclusions

Promising way to improve the efficiency of Cyber-Physical Systems is equipping them with embedded metrological subsystems that include the working measures of physical units built on the basis of fundamental physical constants.

Table 3.1. Shares p_i for typical modules.

Implementation criteria	Sensor	Electronic indicator	Fasteners etc.
The combined impact*	0.7	0.5	0.5
Temperature impact on readouts without loading	0.7	0.7	0.5
Change of supply voltage	-	1	-
Impact of drift	1	-	-
Impact of damp heat	0.7	0.5	0.5

*Combined effects of: non-linearity, hysteresis, short-term effect of temperature. After warming up, the duration of which is set by the manufacturer, constituents of combined impacts error are applicable to modules.
Sign "-" means "does not apply".

The first such intrinsic measure seems to be the measure of electrical resistance (12906.4037 Ω), based on the superconductive carbon nanotube with its particular resistivity, that is equal to an inverse of conductance quantum. It enables to improve significantly the precision of electrical subsystems and, as a result, to raise the quality of manufactured products. To achieve sufficient precision in CPS production cycle, is enough to hold in-place ohmmeter calibration at the points that correspond to values, multiple to the inverse of conductance quantum or its certain part accomplished by series-parallel connection of several identical measures.

Advance in Cyber-Physical Systems is impossible without the temperature gauging that demands the continuous development of experimental techniques due to progress in Thermometry, namely in creation of Temperature Standard. Expanding the set of Quantum Standards of the SI units, towards the study of major pillars of the Temperature Standard on the basis of fundamental physical constants becomes possible as a result of emerged opportunities of unique electronic devices, in particular Resistance Standard (on the basis of Inverse of Conductance Quantum) and Voltage Standard (on the basis of Josephson junctions array) combined in addition with the Cesium Frequency Standard. Researching the foundations of creation of the Quantum Unit of Temperature envisages the minimum value of temperature jump caused by electron-phonon dissipation. It is proved that the Reduced Quantum Unit of Temperature is determined by the electric energy dissipated on CNTFET contacts at passing a current, via ratio of h and k_B and is equal to $3.19949342 \cdot 10^{-11}$ K with relative standard uncertainty $59.2 \cdot 10^{-8}$ (at single electron-phonon relaxation per unit time).

To ensure the quality and reproducibility of CPS, a remote access on the basis of TCP/IP protocol to such a precision measuring instrument as balances has been realized. Moreover, installation was equipped with the built-in mass measures that enabled to perform in-place operations of metrological checking, verification, calibration etc. without interrupting the production cycle. Consequently the measures, preset in mentioned information-measuring subsystems, could be activated distantly by qualified metrologists with help of start-coded signal. Possible errors of measurement results have been estimated considering that weighting subsystem consists of individual modules, every of which shares its own intrinsic error.

References

[1]. NIST Three-Year Programmatic Plan, FY 2014-2016, 2014.
[2]. R. Taymanov, K. Sapozhnikova, I. Druzhinin, Sensor devices with metrological self-check, *Sensors & Transducers,* Vol. 10, Special Issue, February 2012, pp. 30-45.
[3]. Ia. Mills, T. Quinn, P. Mohr, B. Taylor, and E. Williams, The New SI: units and fundamental constants, *Royal Society Discussing Meeting,* Jan. 2011.
[4]. A. J. Giesbers, G. Rietveld, E. Houtzageretal, Quantum resistance metrology in graphene, *Applied Physics Letters,* Vol. 93, 2008, pp. 222109-13.
[5]. B. Jeckelmann, B. Jeanneret, The Quantum Hall Effect as an Electrical Resistance Standard, *Meas. Sci. Technol.,* Vol. 14, 2003, pp. 1229-1236.
[6]. B. Jeckelmann, B. Jeanneret, The Quantum Hall Effect as an Electrical Resistance Standard, *Rep. Prog. Phys.,* Vol. 64, 2001, pp. 1603-1655.
[7]. A. Tzalenchuk, S. Lara-Avila, A. Kalaboukhovetal, Towards a quantum resistance standard based on epitaxial graphene, *Nature Nanotechnology,* Issue 5, 17 January 2010, pp. 186-189.
[8]. F. Delahaye, T. Witt, Comparison of quantum Hall Effect resistance standards of the NPL and the BIMP, *Rapport BIMP-99/18,* Dec. 1999.
[9]. The NIST Reference on Constants, Units, and Uncertainty, CODATA Internationally Recommended 2014 Values on Fundamental Physical Constants, http://physics.nist.gov/cuu/Constants/index.html
[10]. P. J. Mohr, B. N. Taylor, D. B. Newell, The 2010 CODATA Recommended Values of the Fundamental Physical Constants, 2011.

[11]. The Total Calibration Solution, *CPS. Instrumentation & Calibration Experts,* 24 Sep 2014 (http://www.cps.co.nz/blog-display.aspx? ArticleId=98&categoryId=63).

[12]. Physical Property Measurement System. Resistivity Option User's Manual, Part Number 1076-100A, *Quantum Design,* August 1999.

[13]. M. Lisowski, K. Krawczyk, Insulation resistance influence on high resistance Hamon transfer accuracy, PAN, *Metrology and Measurement Systems,* Vol. 16, No. 1, 2009, pp. 33-45.

[14]. M. Dresselhaus, G. Dresselhaus, R. Saitoc, A. Joriod, Raman spectroscopy of carbon nanotubes, *Physics Reports,* 2004, pp. 1-53.

[15]. A. J. Giesbers, G. Rietveld, E. Houtzager et al, Quantum resistance metrology in graphene, *Applied Physics Letters,* Vol. 93, 2008, pp. 222109-1 … 3.

[16]. Metrology. State verification scheme for means of measuring the electrical resistance, *State Standard of Ukraine,* No. 3712-98, 1998, Ukraine (in Ukrainian).

[17]. B. Stadnyk, S. Yatsyshyn, State standard on the basis of von Klitzing constant, in *Proceedings of 58rd Internationales Wissenschaftliches Kolloquium,* Technische Universität- Ilmenau, Germany, 08–12 September 2014, pp. 36-39.

[18]. Consultative Committee for Thermometry, Mise en Pratique for the definition of the Kelvin, *Bureau International des Poids et Measures,* S'evres, France, 2006.

[19]. S. P. Benz, A. Pollarolo, J. Qu, H. Rogalla, C. Urano, W. L. Tew, P. D. Dresselhaus, D. R. White, An Electronic Measurement of the Boltzmann Constant, *Metrologia,* 48142, 2011, 23 p.

[20]. M. Hohmann, P. Breitkreutz, M. Schalles, T. Fröhlich, Calibration of heat flux sensors with small heat fluxes, in *Proceedings of the Internationales Wissenschaftliches Kolloquium: 'In Shaping the Future by Engineering',* Technische Universität, 58, Ilmenau, Germany, 08-12 Sept. 2014, p. 29.

[21]. M. Lindeman, Microcalorimetry and transition-edge sensor, Thesis UCRL-LR-142199, *US Department of Energy, Lawrence Livermore National Laboratory,* April 2000.

[22]. S. Yatsyshyn, B. Stadnyk, Ya. Lutsyk, L. Buniak, Handbook of Thermometry and Nanothermometry, *IFSA Publishing,* 2015.

[23]. B. Stadnyk, S. Yatsyshyn, Ya. Lutsyk, Research in Nanothermometry. Part 1. Temperature of Micro- and Nanosized Objects, *Sensors & Transducers,* Vol. 140, Issue 5, May 2012, pp. 1-7.

[24]. V. P. Koverda, V. N. Skokov, Stability of a random process with 1/f spectrum at the determined impact, *Journal of Technical Physics,* Vol. 83, Issue 3, 2013, pp. 1-5 (in Russian).

[25]. F. Gasparyan, Excess Noises in (Bio-) Chemical Nanoscale Sensors, *Sensors & Transducers,* Vol. 122, Issue 11, November 2010, pp. 72-84.

[26]. S. Yatsyshyn, B. Stadnyk, Z. Kolodiy, Development of Noise Measurements. Part 1. Fluctuations and Thermodynamics, Proper Noise

and Thermometry, *Sensors & Transducers,* Vol. 150, Issue 3, March 2013, pp. 59-65.

[27]. J. Schindler, Coulomb Blockade and Josephson Noise Thermometry, Physics of Noise, *Karlsruher Institut für Technologie,* http://www.phi.kit.edu/noise/abbildungen/Noise-06_Josephson_Noise_and_Coulomb_Blockad_Thermometry.pdf

[28]. J. Mimila-Arroyo, Free electron gas primary thermometer: The bipolar junction transistor, *Appl. Phys. Lett.,* 103, 2013, 193509.

[29]. P. Joyez, D. Vion, M. Götz, M. Devoret and D. Esteve, The Josephson effect in nanoscale tunnel junctions, *Journ. of Superconductivity,* Vol. 12, No. 6, 1999, pp. 757-766.

[30]. E. O.Göbel, U. Siegner, Quantum Metrology. Foundations of Units and Measurements, *Wiley-VCH Verlag GmbH &Co. KGaA, 2015.*

[31]. A. J. Giesbers, G. Rietveld, E. Houtzager et al., Quantum resistance metrology in graphene, *Applied Physics Letters,* Vol. 93, 2008, pp. 222109-13.

[32]. R. Sahoo, R. Mishra, Simulations of Carbon Nanotube Field Effect Transistors, *Internat. Journ. of Electronic Engineering Research,* Vol. 1, Issue 2, 2009, pp. 117-125.

[33]. Strain Gauge Balances, *RUAG Aviation,* Switzerland, 2016, online, http://www.ruag.com/aviation/subsystems-products/engineering/ aerodynamics/instrumentation/strain-gauge-balance/

[34]. L. Erm, Development of Two-Component Strain-Gauge-Balance Load-Measurement System for the DSTO Water tunnel, Technical Report DSTO-TR-1835, *Defence Science and Technology Organization, Australia Government, Department of Defence,* 2006. http://dspace.dsto.defence.gov.au/dspace/handle/1947/4422

[35]. Industrial Internet Insights Report for 2015, General Electric Company, 2014, https://www.accenture.com/us-en/_acnmedia/Accenture/next-gen/ reassembling-industry/pdf/Accenture-Industrial-Internet-Changing-Competitive-Landscape-Industries.pdf

[36]. Setting and Adjusting Instrument Calibration Intervals. Application Note, Agilent Technologies, 2013. http://cp.literature.agilent.com/litweb/pdf/ 5991-1220EN.pdf

[37]. ČSN EN 45501+AC Metrological aspects of non-automatic weighing instruments, *Česká technická norma,* Publ. 1. 8. 1995, https://www.nlfnorm.cz/en/ehn/3964.

4.

Code-Controllable Measures for Correction of Measuring Channels

V. Yatsuk, M. Mykyychuk and S. Yatsyshyn

Built-in MIs are the integral part of CPSs. Since such MIs contain ADCs and DACs and they have to operate with predetermined metrological performance throughout the duration of operation. Tasks of their regular and irregular metrological checking, validation, verification, calibration, and errors adjustment emerge, complicated by heterogeneous nature of the measured values.

4.1. Quality Assurance in Measuring Instrument Design

Realizations of metrological verifications is linked with chain of technical and technological problems because these built-in units are dispersed and their output signals can be used to verify the state of CPS (Fig. 4.1) [1-2]. Objective information about technological processes running is obtained by the way of measuring that is considered as entire process from the perception and transforming of the object measurement information to its processing, saving, transmission and usage with the aim to elaborate retroaction at controlled objects. Therefore it is reasonable to study possibilities of smart MI realization to fulfil the function of the CPS measuring channels control while operating.

Nowadays the number of built-in MIs is sufficiently large with tendency to considerable increase in the future [1-2]. Given this and taking into account their dispersion the issue of metrological assurance optimization becomes an urgent task. Considering tendencies of CPS design as distant dispersed applied infrastructure it is practically impossible to use traditional procedures of the metrological verification [3-8]. Requirements for measurement precision can significantly differ for every concrete measurement task but all these tasks must be performed with ensuring of tracing [9-14]. The large number of types and mass of used MI and labor-intensity of metrological procedures in modern measurement systems, tracing of physical quantities measurement is not always ensured as normative documents require [12-14]. Another important aspect of CPS metrological providing is an achievement of necessary precision and metrological reliability of measurements performed at all stages of producing (for instance, microelectronic production) [7-8, 15-16].

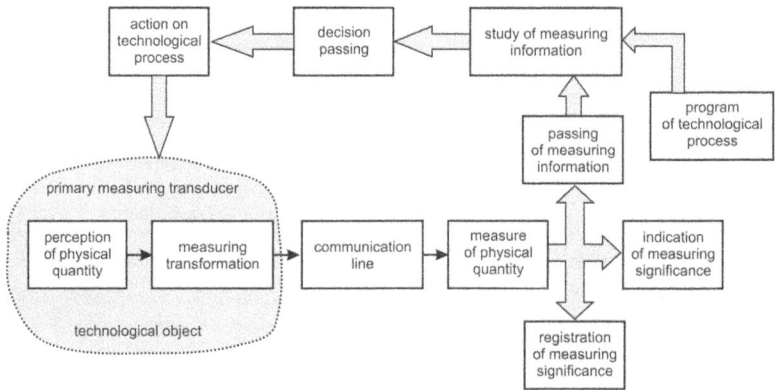

Fig. 4.1. Functional scheme of measuring instruments.

Significant quantities of cheap MIs are available in the market that substantially complicates the problem of the CPS designing [1, 2, 17-18]. Simultaneously one of the central problems is quality estimation of measurements performed by means of MIs. It predicts the ascertainment of MIs capabilities, correct choice of the indexes list and determination of ponderability coefficient values during the measurement quality estimation. Therefore the problem of the performed measurements quality estimation is not solved in theory and practice [11-16].

Furthermore, at the stage of MI design are absent the theoretical and practical recommendations related ensuring the necessary level of MI quality in conditions of the operation. Today international metrological organizations recommend using the conception of the uncertainty of received results as the main standard of performed measurements quality [19-21]. By this way it was attempted to eliminate the basic practical defect of measurement errors approach as difference between measured and true value of measurand.

However, during designing it is sufficiently difficult (or impossible in general) to use the uncertainty approach because the measurement results are still absent. But in this case the measurement error approach can be successfully applied taking as true the nominal value of MI transformation coefficient (function). Logically to suppose that it is the value which the designed MI measurement result has to tend to.

Traceability, accuracy/trueness, precision, systematic bias, evaluation of measurement uncertainty are critical parts of quality management system. Basic provisions concerning metrological support of the development, production, testing and operation of MIs used during goods producing, science researches conducting and other kinds of activity are regulated by the number of international standards. These standards establish general requirements for measurement processes and equipment, the order of its usage and competence of testing and calibrating laboratories [12-14]. The technical basis of metrological assurance involves the systems of national and working standards of physical units, standards, comparators for transferring dimensions of physical units, certified reference materials and substances standards component pattern and characteristics of substances, and working MIs. In many countries metrological verification and calibration of MIs are performed on the ground of special testing schemes which documentarily establish the tools, techniques and exactness of transmission of unit dimension of particular value from national to working standards.

For MI verification, the methods of direct comparison of tested MI with the standard mean of the same type, and also direct verification by the tested MI of output signal of multi-valued measure and determination of the uncertainty as difference of measurement results are applied. Furthermore, sufficiently precise portable multivalued code-controllable measures can be used too and it provides the possibility to perform operative metrological control of MIs in the operation conditions. In the accordance with normative documents the MIs verification is performed in the laboratory where there are special (normal) conditions including the absence of hazardous radiation, interferences, impact loads, vibrations which can affect MI parameters. Metrological verification fulfilment demands the dismantling of instruments from technological objects and transportation them to laboratories. In addition to only technical, organizational inconveniences and financial expenditures the MI verification in such "hothouse" conditions doesn't meet a lot of metrological aspects of their operation. Besides, other units of CPS's measuring circuit are not verified, and verification of whole measuring systems in laboratory environment is meaningless because it demands the dismantling of system.

It is reasonable to revise traditional verification methods and techniques for virtual systems of gathering and processing of measurement data in the direction of automation. Economically expedient seems to conduct verification in place. To accomplish it portable code-controllable measures of physical values and development of proper software are required. This enables to enhance significantly labor productivity [22-23].

Metrological reliability seems to be one of the main important MI parameters. For its providing on proper level, there is a need of permanent operation processes control [12]. Reliable measurement information of necessary exactness can be received only by the way of technically well-grounded choice of MI that includes next data [22-23]. Firstly this is the list of measurement parameters of the object with permissible measurement errors. Considering the stochasticity of the analyzed processes it is necessary to know acceptable probability of false and unidentified failures for every controlled parameter, values of confidence probabilities, distributing laws of the measured parameters and their measurement errors. MI operation conditions affect significantly their metrological properties or properties of their elements. These are mechanical loading, climatic effect (temperature, humidity, pressure etc.), and presence/absence of the environment (corrosive gases and liquids, high temperatures or voltage, fungi, mold, electromagnetic fields, radioactive and other emissions).

Then after determining the restrictions, the precision characteristics (permissible limits of basic and additional errors) of MI are estimated. Reliability of measuring parameter is indicated at probability that measurement error value does not exceed permitted values with given probability $P(\Delta_\Sigma \leq \Delta_{allow}) \geq P_{conf}$, where P is the actual value of the probability, Δ_Σ is the total error of result obtained by selected MI, Δ_{allow} is the boundary value of measurement allowable error, P_{conf} is the set value confidence probability. Reliability is ensured if the probability of false and unidentified failures on results of monitoring parameters by selected MIs does not exceed a specified level: $P_f \leq P_{fal}$; $P_{ui} \leq P_{uial}$, where $P_f, P_{fal}, P_{ui}, P_{uial}$ are the actual and allowable probabilities of false and unidentified failures respectively. They as well as measurement error and tolerance for the controlled parameters are theoretically examined in [22-23] and described by complicated expressions.

136

Method of simplified calculations leads to overstated demands regarding precision of chosen MI. For practical realization of mentioned method there is a need of quite significant amount of a priori information about stochastic properties of measured parameters X_i, and also about MI transformation function. Mainly such information is absent. To receive it, becomes necessary the conduction of additional studies. Moreover, under drastic changes of any measurement conditions (for example, emergency conditions and electromagnetic induction effects) crude errors appear that abruptly worsens the measurement reliability. Incremental degradation changes of MI components and units lead over time to the noticeable uncertainty or even to MI metrological failure within intercalibration interval. Today there is no theory of the determination of this interval duration which mainly is determined on the experience basis of similar MI operation and on "engineering intuition". For this reason the standard ISO 10012:2003 recommends to implement methods of the control of measurement processes based on regular verification and gradual analysis of measurement results. The mentioned approach should be applied at all measurement stages: from standards calibration by state metrological laboratory to own regular verifications [12]. In practice the measurement process control is applied with the aim to enhance the safety of technological process (f.i., of energy plants) or providing of guaranteed quality of final product. The metrological management system ISO 10012:2003 has to guarantee that measurements (performed with usage of measurement equipment within intercalibration interval) is quiet precise for given problem. But MIs are checked metrologically for ultimate consumer of single models or small samplings of their general totality.

The distribution law of the measured parameter for totality of MIs, which are given on periodic verification, differs considerably from the distribution law of the same parameter at the moment of MI release from producing and practically it cannot be studied [22-23].

Wide MI operating conditions and different nomenclature of their parameters complicate the quality analysis. Its ultimate aim is close to optimal (in coordinates "functionalities - cost") choice of necessary for current measurement MI from the variety of modern devices. To ensure required quality level of goods, production and services produced with MI application the normative documents recommend to perform control of measurement processes as separate long procedure [12-14]. Traditionally to estimate any production quality of the same kind differential, complex and mixed-mode methods are applied. The differential method includes such shortcomings as some subjectivity and insufficient precision of quality coefficient determination [16]. While using of complex method it is necessary to choose natural generalized quality index for this kind of production and to determine its functional dependence from main single indexes that partly affects the quality. During estimation of such complex technical production as MI mainly mixed method is applied. This method is based on the exact calculating the number of separate groups which include MI significant properties. For this MI, the generalized quality complex indexes are determined with following their comparison with the help of the differential method concerning appropriate indexes of basic MI [16].

General peculiarity of such product as MI is that technical and economic indexes can be differently characterized depending on the type of measured value and these indexes can be changed at all stages along life cycle that includes designing, producing and operation. Now operation expenditures that mainly connected with expenses on periodic verifications of MIs, their maintenance and calibration [11, 20-24], determine MIs operating time. MIs cost continuously decreases, even for devices of high exactness (0.01...0.05 %) because of wider involvement of integrated electronics [1, 17-18].

With regarding to spectra of MI opportunities it is proposed to consider the informational energy efficiency h_3 as functional dependence which connects their major metrological parameters and reveals interaction with the object of measurement [24]. But definition of MI mentioned efficiency is partly subjective. Really the main function of any MI is receiving measurement information. It is naturally to consider the object of the measurement as source, and MI as a receiver of measurement information. This analog information is transmitted by the communication channel in which reduced to the input MI noises operate.

It is known that equivalent dispersion of the random input signals is determined as sum of different sources dispersions. During its definition it is necessary to take into account the sources of thermal and flicker noise. So we propose to determine the amount of MI received information M_{U_x} during voltage measurement of direct current on the basis of C.-E.Shannon theorem [25]:

$$M_x = Ct_x = t_x B \, log_2\left(1 + \frac{S_x}{N_n}\right) = t_x B \, log_2\left(1 + \frac{U_x^2}{D_{nU}}\right), \qquad (4.1)$$

where t_x is the duration of measurement, $B = f_H - f_L$ is the bandwidth of the measuring channel, $S_x = \left(U_x^2 / R_{in}\right) t_{mx}$ is the useful signal energy, $N_n = \left(U_n^2 / R_{in}\right) t_{mx} = \left(D_n / R_{in}\right) t_{mx}$ is an energy of equivalent input noise, U_x is the measured voltage, R_{in} is the MI input resistance, U_n and $D_{nU} = D_{1U} + R_x D_{Iin} + 4k_B T R_x B$ are the voltage and equivalent input noise variance respectively, D_{n1U}, D_{Iin} are the variances of the voltage and the current equivalent noise respectively, k_B is the Boltzmann constant, T is the temperature of internal resistance of measured voltage source.

During electric current measurement I_x (with appropriate voltage drop on the R_{sh}) dispersion D_{nI} of equivalent noise is determined by the expression $D_{nI} = D_{1U} + R_{sh} D_{Iin} + 4k_B T R_{sh} B$, where R_{sh} is the shunt resistance. Analysis of the dispersions D_{nU} and D_{nI} of equivalent input noise at the measuring voltage or current reveals that they are formal similar and additive regarding active measurement values.

In the case of measurement of the electrical passive value resistance the expression $D_{nR} = D_{1U} + R_x(D_{Iin} + D_{Ics} + 4k_B T B)$ for equivalent noise dispersion contains multiplicative component besides additive one.

On the other hand, the main goal of measurement process fulfilment is receiving the numerical value, which by the entropy (Shannon measure) or by the logarithmic measure of the uncertainty (Hartley measure) is usually determined. It is naturally to accept that output uncertainty is equal to width $X_2 - X_1$ of MI measurement range, and uncertainty after measurement performing has to be determined by resolution error Δ_p of particular measurement results. Then on conditions of equiprobable dispersion laws of measured values and of their resolution error $\Delta_{r.e.}$, the amount of received information after measurement is estimated by relation [26]:

$$H_{mx} = log(X_2 - X_1) - log \Delta_{r.e.} = log \frac{X_2 - X_1}{\Delta_{r.e.}}, \qquad (4.2)$$

where $X_2 - X_1$ is the measurement bandwidth, $\Delta_{r.e.}$ is the resolution error of separate measurement results for given MI.

Quality factor K_x or efficiency of MI can be represented as the product of the two quantities of received measurement information:

$$K_x = M_x H_{mx} = t_x B \, log \left(1 + \frac{X^2}{D_n} \right) log \frac{X_2 - X_1}{\Delta_{r.e.}}, \qquad (4.3)$$

where $X = U_x$; $I_x R_{sh}$; $I_{sc} R_x$ are the input electrical signals at voltage, current and resistance measurements respectively. From the analysis (4.3) we conclude that under all another similar conditions the amount of received measurement information with the help of MI is determined by the errors of measurement circuit which are caused by characteristics of the MI and object, measurement conditions, operator's qualification, duration etc. [11]. In practice, under applying of polynomial model the expression for the resolution error can be submitted by the quadratic function [22-23]. At this moment necessary information on laws of probability distribution of elementary errors for estimating the based on their composition the distribution resultant error is needed. By this distribution the confidence coefficient $k(P)$ can be determined correctly. Its value depends on confidence probability P. As the result, the resolution error estimation is difficult to assess. Besides with time t as a result of MI aging and wear its error alters (degrades often). To ensure metrological reliability MI is produced with some technological reserve factor of which is equal to $k_{MH} = 0.4...0.8$ concerning the acceptable error value [11, 21-24]. In most practical cases, the expression for the MI error $\Delta_{r.e.}(x, P, \vec{Q}, \vec{\xi}, k_H, t)$ in operation condition at moment t can be submitted as:

$$
\begin{aligned}
&\Delta_{r.e.}\left(x, P, \vec{Q}, \vec{\xi}, k_H, t\right) \\
&= \bar{\Delta}\left(x, \vec{Q}, \vec{\xi}, k_H, t\right) \pm k\left(P, \vec{Q}, \vec{\xi}, t\right) \cdot \sigma\left(x, P, \vec{Q}, \vec{\xi}, k_H, t\right) \\
&= \Delta_{0pe}\left(P, \vec{Q}, \vec{\xi}, k_H, t\right) + \delta_s\left(P, \vec{Q}, \vec{\xi}, k_H, t\right) x \\
&+ \varepsilon\left(P, \vec{Q}, \vec{\xi}, k_H, t\right) x^2,
\end{aligned}
\qquad (4.4)
$$

where $\Delta_0\left(P, \vec{Q}, \vec{\xi}, k_H, t\right) = \bar{\Delta}_0\left(\vec{Q}, \vec{\xi}, k_H, t\right) \pm k\left(P, \vec{Q}, \vec{\xi}, t\right)\sigma_0$ is the additive CoE, $\delta_s\left(P, \vec{Q}, \vec{\xi}, k_H, t\right) = \bar{\delta}_s\left(\vec{Q}, \vec{\xi}, k_H, t\right) \pm k\left(P, \vec{Q}, \vec{\xi}, t\right)\sigma_\delta$ is the multiplicative CoE, $\varepsilon\left(P, \vec{Q}, \vec{\xi}, k_H, t\right) = \bar{\varepsilon}\left(\vec{Q}, \vec{\xi}, k_H, t\right) \pm k\left(P, \vec{Q}, \vec{\xi}, t\right)\sigma_\varepsilon$ is the quadratic CoE, $\vec{Q}$ is the MI parameter vector, $\vec{\xi}$ is the vector of elementary errors, k_H is the MI nominal transfer factor, P is the confidence probability.

Random component of additive CoE is determined by MI noise components. Simultaneously it is necessary to take into account thermal noise and $1/f$ noise. With $1/f$ noise spectrum divergence in low-frequency limit a question about its stationary raises. However if MI low limit of frequency range is accepted, the frequency f_{kl} of its calibration (determination of "zero" indexes) and its upper limit of transmission frequency f_{hf}, then $1/f$ noise is limited by frequency. These processes are stationary in first approximation and their amplitudes (in the broad sense) are normally distributed [22-23].

Noise signal dispersions D_{nU} and D_{nI} reduced to MI input in the frequency range from $\omega_{kl} = 2\pi f_{kl}$ to $\omega_{hf} = 2\pi f_{hf}$ can be determined by means of Wiener-Khinchin theorem as [22-23, 27-28]:

$$D_n = \lim_{\tau \to 0} D_n(\tau) = A_{0e}(f_{hf} - f_{kl}) + A_{fe}f_{fe}\,ln\frac{f_{hf}}{f_{kl}}, \quad (4.5)$$

where $A_{0e}, A_{fe}, \omega_{fe}$ are the spectral densities of thermal noise, $1/f$ noise and circular frequency conjugation of noise equivalent densities respectively; $D_n = D_{nU}$; D_{nI}. So, generally the equivalent noise signal dispersions D_{nU} and D_{nI} reduced to MI input are inherent in the additive and multiplicative components:

$$D_{nR} = D_{nU} + R_x\left[C_{0e}(f_{hf} - f_{kl}) + C_{fe}f_{fe}\,ln\frac{f_{hf}}{f_{kl}}\right], \quad (4.6)$$

where C_{0e}, C_{fe}, f_{1fe} are the equivalent spectral densities of thermal noise current, $1/f$ noise and circular frequency conjugation of these noise respectively. Except parameters $A_{0e}, C_{0e}, A_{fe}, C_{fe}$ and f_{fe}, f_{1fe} which describe respectively thermal and $1/f$ noise, the dispersion also depends on MI parameters (upper frequency f_{hf} of passband) and applied measurement algorithm including considering the frequency f_{kl} of its calibration (zero indexes determination). At commensurate values A_{0e} and A_{0e} and dimension of f_{fe} at vicinity of hundreds Hz [17-18, 27-28], equivalent noise signal dispersion D_{nU} reduced to MI input increases with upper frequency f_{hf} of passband, and it decreases hyperbolically with increase of the frequency f_{kl} of the determination of MI zero indexes. Excluding additive component under the same conditions the dispersion D_{nR} power of equivalent noise signal reduced to MI input depends multiplicatively on measured resistance R_x value.

Correlation of (4.5) and (4.6) underlines that systematic CoE leads to shifted estimation of the measurement result uncertainty; moreover it also depends on change of ambient conditions and MI parameters. Among the variety of methods of errors automatic correction, the method of input channel inversion should be preferred. Such MI contains the compulsory polarity switch. Then measurement result N_x is determined as algebraic sum of results N_1, N_2 obtained at the opposite polarities of input signal. It enables to correct not only additive CoE, and also the even powers of polynomial approximation of MI transformation function [22-23]. High degree of correction and simplicity of technical realization facilitate the implementation of input signal inversion in Δ_Σ ADC micro circuits which inherent in quite satisfactory performance. F.i., when the offset error specification of the additive CoE is ±0.5 µV typically, the drift of these ADCs is specified as ±5 nV/°C that actually is immeasurable [22-23, 29].

4.2. Additive Error Correction for Measuring Instrument

Certain amount of values is compulsory measured during operation of any CPS. Hereby, the measuring circuit traditionally consists of sensors, CLs, CCs, scale devices, and ADCs. Development trends of multiplex smart measuring systems is the unification based on converting whichever signals into electrical signals [1-2] that are characterized by the prominent advantages in transferring, transformation, storing, registration etc. The direct current voltage is mainly used for precise ADC conversion among all variety of electric signals [17-18]. Most output signals of sensors are low-levelled that is coupled with additive CoE significant impact for measuring circuits (Fig. 4.2). Here additive CoE is determined by residual parameters of CLs, CCs, equivalent offset voltages, input currents, and drift of SABs and ADCs [1-2, 11].

The task of multichannel measuring the low-level signals in the interchannel interference affect conditions and necessity to assurance of spark and explosion safety demands is especially important in oil and gas, energy and chemical industries [3-8, 30-34]. When IAs are used for this purpose it results in the growing costs of the measuring circuit, and also needs the errors correction of every measuring channel. Generally IAs are not insulated sufficiently. It often demands to apply the mentioned barriers in every measuring channel to provide the safety. Necessary to emphasize that multichannel thermoelectric thermometers of complicated design with their prolonged lines performed from expensive materials have to be provided by auto correcting the cold junction temperature of every measuring channel [22-24, 30-34].

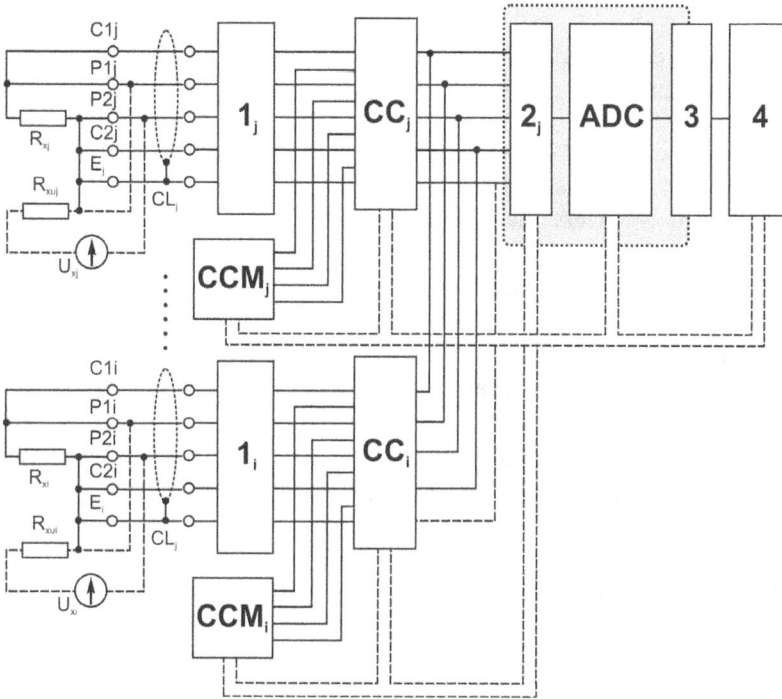

Fig. 4.2. Equivalent scheme of multichannel instruments for both self-generating and modulating sensors: 1 is the antispark and explosion damage barrier; 2 is a scale analog block; 3 is an isolation block; 4 is an instrumental controller.

Currently, the measurement is considered as integral process of measuring data acquirement from object, further conversion to its processing, storing, transferring and simultaneously trying to ensure the opposite effect to controlled object [1, 5, 24]. Information measuring systems based on integrated micro electronics are distributed and smart enough to perform functions of automatic checking, verification, validation and so on in CPS measuring circuits.

There are a number of low-cost variable MIs and transducers in market. It significantly complicates the problem of their choice for the concrete measuring tasks at CPS design stage [11, 17-18]. Simultaneously, an additive CoE automatic correction becomes the principal problem that makes the greatest contribution to Errors Budget. Errors value depends considerably on the sensor type and measuring circuit configuration (Fig. 4.2). It can change substantially with MI operation conditions. To decrease error, various correction methods are used [1, 35-42]. Apparently, parameters of smart ADCs and DACs largely determine the CPS performance.

To measure the parameters of spark and explosive dangerous objects, the cheap isolated devices of a few per cent exactitude and the isolated power supply units for every measuring channel are generally applied. These channels are ensured by additive CoE auto adjustment means resulting in complexity. Spark dangerous or explosion safety is provided by special block 2 installed in every channel (see Fig. 4.2). It may cause the additive CoE due to residual parameters of this block and SAB input currents impacts. The best electric isolation is attained while applying the relays with sealed contacts. However during operation, the commutators are inherent in unstable residual voltage which determines the additive CoE of measuring circuit [17, 18, 30-33].

Fundamental drawback of known methods and MIs for designing the smart measuring circuits consists in difficulty of auto calibration in the workplace through impact of residual parameters of CLs, CCs, and antispark and explosion damage barrier.

Additive CoE of multi-channel instruments is determined by residual voltages of input measuring circuit and by leakage currents of all input electronic keys. So, urgent purpose of MIs is developing the valid self-calibrating circuits. It enables to provide the MIs reliability, to carry out the current test of running processes and to ensure in-place verification of MIs in CPS operation.

Due to dissemination of the sensors array as well as MTs [1, 2], an urgent problem of metrological perfection of distributed measuring systems emerges. Only considering trends of quality improvement, cost reduction and chips rapid obsolescence, we may to reach the preset metrological reliability of MTs for the certain duration of operation. At first we try to minimize the additive CoE to value lesser than one of least significant digit of MI's reading. Regardless of their precision, an adequate traceability must be achieved [30-42]. In practice, the traceabilities of measuring systems required by normative documents are not principally ensured due to substantial labor intensity of metrological procedures.

Main principles of metrological assurance of MT development, production, testing, and operation are regulated by number of normative documents. They determine general demands for measuring processes, devices, application procedure and competence of testing and calibrating laboratories [12-14]. The basis of foregoing assurance is mathematical model of MI's transfer functions and errors analysis at operating conditions. General and special measuring techniques are permanently developed [11, 43] aiming the error adjustment.

Mainly, the error value $\Delta(X, \vec{Q}, \vec{\xi}, t)$ depends on quantity X of input informative signal. For its analysis, it is convenient to develop multinomial model [5, 28-29]:

$$\Delta_x(X, \vec{Q}, \vec{\xi}, t)$$
$$= \Delta_{0x}(\vec{Q}, \vec{\xi}, t) + \delta_{Sx}(\vec{Q}, \vec{\xi}, t)X + \varepsilon_x(\vec{Q}, \vec{\xi}, t)X^2 + \cdots \quad (4.7)$$
$$= \overline{\Delta_x}(\vec{Q}, \vec{\xi}, t, X) + \ddot{\Delta}_x(\vec{Q}, \vec{\xi}, t, X),$$

where $\Delta_{0x}(\dots), \delta_{Sx}(\dots)X, \varepsilon_x(\dots)X^2$ are the additive, multiplicative and nonlinear CoEs respectively, and $\overline{\Delta_x}(\dots)$, $\ddot{\Delta}_x(\dots) = \Delta_x(\dots) - \overline{\Delta_x}(\dots)$ are the systematic and random CoEs respectively.

Random values dependent on vectors of $\vec{Q}$ parameters of measuring circuit and $\vec{\xi}$ errors factors are included in coefficients $\Delta_{0x}, \delta_{Sx}, \varepsilon_x$. They are independent of informative parameter X. Additive CoE always distorts the measuring results, and introduction of amendments on its basis seems to be an important metrological problem. Unfortunately, universal method of systematic CoE adjustment is not elaborated sufficiently due to complexity of methods, MIs and operating conditions [11, 19-21]. Systematic errors are eliminated if they (or sum of their non-eliminated residual values) don't exceed the half of the decimal digit unit which has one least significant unit of permittable error Δ_{al} of measuring result [11, 20-21]. Forecasting of errors drift demands conducting the complex research of particular MIs in operation conditions and therefore it is difficult to implement.

On the basis of (4.4) an additive CoE leads to biased estimation of data precision. To avoid undesirable consequences, the set of general methods of error minimization are applied. The accuracy rises since the errors value decreases. Another set of errors diminishing is a set of special methods conjugated with the certain measuring algorithm. Their joint action can be implemented particularly (see Chapter 5) depending on measuring realization [11, 43]. As methods of errors decreasing are widely applicable in metrology, it is worth to consider the ways of MIs' quality ensuring. First of them is based on statistic minimization, consists in random errors decreasing, and is utilized in transducer design [11, 19-21].

Errors auto correction with spatial separation of measuring channels and correction channels is based on the measurement invariance principle and has to cover as many additional correction channels as the number of destabilizing factors exists [11, 43]. However, this method is effective only at impact decreasing of mentioned factors. Moreover, it is compulsory to know the dependence of MT errors on these factors and, ultimately, to appreciate of their impact actions on the certain units.

Let's consider limiting capabilities of error auto correction methods if correction and measuring channels are inherent in distribution over time that is simplest in realization. In practice, MT's conversion function is submitted as:

$$y = KX + \Delta_{0x} = K_H(1 + \delta_{Sx})X + \Delta_{0x}, \qquad (4.8)$$

where δ_{Sx}, Δ_{0x} are the multiplicative and additive CoEs of MTs respectively; X is the measuring value; K, K_H are MT's conversion coefficient and nominal value respectively. Iterative method is apparently attractive due to its realization simplicity. Correction algorithm consists in step-by-step specification of the MT output signal y_i with help of precise inverse transducer [11, 43]. The first corrected value of the conversion result y_i is given with consideration of members of second infinitesimal order:

$$\begin{aligned} y_1 \\ = K_H X[1 - \delta_{SM} + (\delta_{Sx} + \delta_{SM})^2] \qquad (4.9) \\ + K_H(\Delta_{12M} - \Delta_1 \delta_{SM}), \end{aligned}$$

where R_X, R_{K1}, R_{LZ} are the measuring source internal resistance, closed key S_1 resistance of input switch, and CL resistance respectively, I_{B1}, I_{C1} are source and drain electrodes inverse current of S_1 key, I_{B2} is source electrode inverse current of input S_2 key, I_{XZ} is MT input current, $I_{B123} = I_{B1} + I_{B2} + I_{B3}$; R_{K2}, R_{SM} are closed key S_2 resistance of input switch and DAC output resistance respectively, I_{C2} is drain electrode inverse current of input switch S_2 key, X_M, y_i are the DAC and MT output signal respectively (y_i is the calculating block output signal for step i iteration); $\Delta_{12M} = \Delta_1 - \Delta_2 - \Delta_M$, $\Delta_1 = \Delta_{0x} + I_{C1}R_X + I_{B123}(R_{K1} + R_{SM})$, $\Delta_2 = \Delta_{0x} + I_{C2}R_{SM} + (I_{B12} + I_{SZ})(R_{SM} + R_{K2})$; δ_{SM}, Δ_{1M} are the DAC's multiplicative CoE coefficient and additive CoE respectively.

Analysis of (4.9) after the first iteration envisages that the measuring result error is practically determined by DAC errors. Furthermore, it is noticed the impact of additive CoE difference Δ_{12}, which is specified by configuration change of MT input measuring circuit and this configuration's value, can exceed additive CoE value of the DAC accuracy. If the DAC error is significantly less than MT error, then at 1st iteration the measuring result code doesn't depend on MT multiplicative CoE:

$$y_2 = K_H[X(1 - \delta_{SM} - \delta_{SM}^2)] + K_H\Delta_{12M}(1 - \delta_{SM}).$$

Practically it eliminates the necessity for following specification of measuring results, and additive CoE remains unchanged comparing with 1st iteration result.

Simplicity in implementation of reference measures method promotes to its dissemination for ADC and electric MIs auto calibration. If the transform function is described by (4.87) and dimension of the certain measures is equal to zero, then the result appears successively in 3 cycles at alternate measuring of: signal X, zero signal and standard measure X_0. It is defined from the system of 3 equations with 3 unknown quantities X, Δ_a, δ_{Sx} [11, 43]:

$$\begin{cases} y_1 = K_H X(1 + \delta_{Sx}) + \Delta_1 \\ \quad\quad y_2 = \Delta_2 \\ y_3 = K_H X_0(1 + \delta_{Sx}) + \Delta_3 \end{cases}, \quad\quad (4.10)$$

where $\Delta_1 = I_{C1}R_X + (I_{B123} + I_{SZ})(R_x + R_{K1}) + \Delta_{0x}$ is the additive CoE of signal X; $I_{C1}, I_{B1}, I_{C2}, I_{B2}, I_{C3}, I_{B3}$ are the drain and sourse inverse current of electrodes of S_1, S_2, S_3 keys of input switch respectively, $\Delta_2 = \Delta_{0x} + I_{C2}R_{01} + (I_{B123} + I_{SZ})(R_{01} + R_{K2})$ is the additive CoE for measuring the zero signal, $I_{B123} = I_{B1} + I_{B2} + I_{B3}$, R_{K1}, R_{K2}, R_{K3} are the resistance of closed key $S_1, S_2, S_3, R_x, R_{01}, R_0$ are the inner resistances of gauged signal X, zero signal and signal X_0 respectively, $\Delta_3 = \Delta_{0x} + I_{C3}R_0 + (I_{B123} + I_{SZ})(R_0 + R_{K3})$ is the additive CoE for measurement of the signal of reference measure X_0.

The value of measured quantity X is determined from (4.10):

$$X = X_0 \frac{y_1 - y_2}{y_3 - y_2}\left(1 + \frac{\Delta_{32}}{X_0}\right) - \frac{\Delta_{12}}{K}, \qquad (4.11)$$

where $R_{K12} = R_{K1} - R_{K2}$; $R_{K32} = R_{K3} - R_{K2}$; $\Delta_{12} = (I_{C1}R_x - I_{C2}R_{01}) + [(R_x - R_{01}) + R_{K12}](I_{B123} + I_{SZ})$ is the additive CoE difference for 1st and the 2nd measuring cycles, $\Delta_{32} = (I_{C3}R_0 - I_{C2}R_{01}) + [(R_0 - R_{01}) + R_{K32}](I_{B123} + I_{SZ})$ is the difference for 3rd and 2nd measuring cycles.

Expression (4.11) analysis presents that uncorrected additive CoE Δ_{12} and multiplicative CoE ${(\Delta_{32}X)}/{X_0}$ values at similar conditions, are determined by the difference of internal resistances of measured R_x and reference signals R_{01}, R_0, and by residual parameters of input commutator. In practice, the resistance R_x of measuring signal source can change from 0 to several hundred Ω. For multipurpose commutative elements of parameters $I_{Ci} \approx I_{Bi} = 10\ nA$, $\Delta I_{Ci} \approx \Delta I_{Bi} = 1\ nA$, $R_{Ki} \le 50\ \Omega$; $\Delta R_{Ki} \le 5\ \Omega$, caused by the keys residual parameters can reach $\pm 10\ \mu V$ at normal operating conditions and at $R_x \le 200\ \Omega$, $R_0 \le 18\ \Omega, R_{01} \le 100\ \Omega$.

For electronic commutators at temperatures close to high operation limit, an error value drops down to our practical threshold that is a few tens μV. So even when MT parameters and reference measures method are invariable within correction process, then uncorrected error depends significantly on difference of resistances of measuring and correction channels, and it sufficiently changes with temperature while operating.

If test method implements, it needs several measuring cycles, which signals are formed without disconnecting the measured quantity. It enables to conduct measurement without the obtained information loss [11, 43]. While measuring electrical quantities and realizing the additive and multiplicative tests (at considering the residual parameters of commutative elements and additive CoE of multiplicative test formation block), the measuring result is received in 4 steps. The received results of each step (y_1, y_2, y_3, y_4) are saved:

$$\begin{cases} y_1 = KX + \Delta_1 \\ y_2 = K(X + X_0) + \Delta_2 \\ y_3 = KmX + \Delta_3 \\ y_4 = K[m(X + X_0)] + \Delta_4, \end{cases} \qquad (4.12)$$

where $\Delta_1, \Delta_2, \Delta_3, \Delta_4$ are the MT additive CoEs summarized to their input values for every of 4 cycles. After data processing of these conversions the measuring result is described by:

$$X = X_0 \frac{(y_3 - y_2)\left(1 - \frac{\Delta_{4231}}{X_{0E}}\right)}{y_4 - y_2 - y_3 + y_1} + \frac{\Delta_{31}}{X_{0E}}, \qquad (4.13)$$

where $\Delta_{4231} = (\Delta_4 - \Delta_2) - (\Delta_3 - \Delta_1)$, $X_{0E} = X_0 K(m - 1)$, $\Delta_{31} = \Delta_3 - \Delta_1$; $I_{C11}, I_{C12}, I_{C2}, I_{C3}, I_{B11}, I_{B12}, I_{B2}, I_{B3}$ are the drain and source electrodes inverse currents of input keys respectively; Δ_m is the additive CoE of multiplicative test formation block; I_{sz} is the MT inverse current; $R_x, R_0, R_{K11}, R_{K12}, R_{K2}, R_{K3}$ are the resistance of measuring source signal, additive test formation block, and input switch closed keys respectively.

As analysis of (4.13) clarifies the non-corrected value of additive CoE is mainly determined by the source resistances of measured signal and connecting wires R_x, and by the reverse currents of input commutator. Furthermore, the mentioned test signals method, compared to the reference measures method, rules out the reduction of errors. Then it needs to expand the dynamic measuring range for concluding about limits of test methods.

Since we consider the digital micro schemes, the algorithmic methods of accuracy become attractive. These methods consist in multi-stage conversion of algebraic sum of measured and one or more standard quantities. While linear approximating the MT transform function for ideal commutative elements, the nominal measuring equation is given as the certain conversion code relation. This relation is equal to ratio of differences of conversion codes in 1st and 3rd $N_1 - N_3$ steps and in 1st and 2nd $N_1 - N_2$ steps, where N_1, N_2, N_3 are the results conversion codes obtained correspondingly for such algebraic values of the input quantity $X + X_0, X - X_0, X_0 - X$ [11, 43]. The measuring result error is determined only by changing the reference voltage values and residual parameters of commutative elements:

$$\begin{cases} N_1 = K(X + X_0) + \Delta_1 \\ N_2 = K(-X + X_0) + \Delta_2 \\ N_3 = K(X + X_0) + \Delta_3, \end{cases} \qquad (4.14)$$

where $\Delta_1, \Delta_2, \Delta_3$ are the additive CoEs which are added up at MT input on the 1st, 2nd and 3rd measuring conversion cycles respectively, $\Delta_1 = \Delta_{11} + \Delta_{12} + \Delta_{13} + \Delta_{14} + \Delta_{15};$ $\Delta_{11} = (I_{C13} - I_{B13})R_{k1};$ $\Delta_{12} = (I_{C24} - I_{B24})R_{k2};$ $\Delta_{13} = (I_{C57} - I_{B57})R_{k5};$ $\Delta_{14} = (I_{C68} - I_{B68})R_{k6};$ $\Delta_{15} = I_{SZ}(R_x + R_{k1} + R_{k2} + R_{k5} + R_{k6} + R_0);$ $I_{C13} = I_{C1} + I_{C3};$ $\Delta_{21} = (I_{C68} - I_{B57})R_{k7};$ $\Delta_{22} = (I_{C57} - I_{B68})R_{k8};$ $I_{B13} = I_{B1} + I_{B3};$ $I_{B24} = I_{B2} + I_{B4};$ $I_{C24} = I_{C2} + I_{C4};$ $\Delta_2 = \Delta_{11} + \Delta_{12} + \Delta_{21} + \Delta_{22} + \Delta_{23};$ $\Delta_{23} = I_{SZ}(R_x + R_{k1} + R_{k2} + R_{k7} + R_{k8} + R_0);$ $\Delta_3 = \Delta_{31} + \Delta_{32} + \Delta_{13} + \Delta_{14} + \Delta_{33};$ $\Delta_{31} = (I_{C13} - I_{B24})R_{k3};$ $\Delta_{32} = (I_{C24} - I_{B13})R_{k4};$ $\Delta_{33} = I_{SZ}(R_x + R_{k3} + R_{k4} + R_{k5} + R_{k6} + R_0).$ $I_{C1}, ..., I_{C3},$ $I_{B1}, ..., I_{B3}$ are the drain and source electrodes inverse current of input keys $S1, ..., S8$ respectively; R_{SZ}, R_x, R_0 are the resistance respectively of MT input, measuring signal X and CL, reference signal X_0; $R_{k1}, ..., R_{k8}$ are the resistance of closed key $S1, ..., S8$ respectively; I_{SZ} is the MT input current.

Solving and simplifying (4.14), we receive:

$$X = X_0 \frac{N_1 - N_2}{N_1 - N_3}\left(1 + \frac{\Delta_1 - \Delta_3}{2KX_0}\right) - \frac{\Delta_1 - \Delta_2}{2KX_0} \qquad (4.15)$$

This equation analysis envisages that uncorrected additive and multiplicative CoEs are determined only by dispersion of closed resistances of keys and their reverse currents. If group of keys $S1, ..., S4$ and $S5, ..., S8$ are located within chip's scheme, the experimentally determined dispersion of parameters does not exceed few per cent [11, 19]. For mentioned keys parameters and their dispersion no more than ± 2 % the uncorrected additive CoE is not more than a few tenths of microvolt even at maximal operating temperatures, and uncorrected multiplicative CoE factor is $\leq \pm 10^{-5}$ % at $X_0 = 1.0 \, V$.

Advantages of algorithmic methods include the invariance provision to MT conversion function without complicating the measuring circuit and conducting the single additional transformation into algebraic code $(-X - X_0)$. Simultaneously ME value including the uncorrected additive CoE has not exceed high-mentioned values [43]. There exists sufficient impact of the error due to discreteness of transformation codes presentation $N1, ..., N4$. Therefore it seems necessary to choose $X < X_0$ that results in narrowing the MT measuring range.

General algorithmic methods drawbacks, that restrict the implementation, include both measuring duration and range expansion of aforementioned values. Special correction methods of systematic CoE include such ones as substitution, error compensation by sign, input quantity inverting, transposition and symmetrical observation [11, 19]. The peculiarity of these methods consists in changing the measuring scheme configuration performed with the key usage. Their residual parameters ultimately determine the quality of systematic CoE correction. The transposition method is applied rarely and is considered as efficient only to correction of the conversion coefficients of double-channel MT balancing. Symmetrical method is occasionally available for correcting the linear changes of systematic CoE.

To realize substitution method in digital form, the multivalued CCMs and MTs are required. In the 1st cycle, the quantity X is measured by MT and is obtained the transformation result code $N_x = k_{ADC}(X + \Delta_{0x1})$, where $\Delta_{0x1} = I_{B1}R_x + (I_{SZ} + I_{C12})(R_{k1} + R_x) + e_{SZ} + I_{B1234}R_{k2}$ is equivalent to ADC value at X-th cycle. At the 2nd cycle the reference quantity $X_0 \left({}^{N_x}/_{N_m} \right)$ is submitted to MT input and the code $N_{xn} = k_{ADC} \left[X_0 \left({}^{N_x}/_{N_m} \right) + \Delta_{0x2} \right]$ is obtained, where k_{ADC} is the ADC MT conversion coefficient; $\Delta_{0x2} = e_{SZ} + e_M + I_{B2}R_M + (I_{SZ} + I_{C12})(R_{k2} + R_M)$, where R_{k1}, R_{k2} are the resistance of closed keys pair of both measuring and reference values respectively, $I_{B1}, I_{C1}, I_{B2}, I_{C2}$ are the drain and source electrodes inverse currents of every closed pair of input keys accordingly, e_{SZ}, I_{SZ} are the bias voltage and input current, e_M is the additive CoE which transforms into CCM input value, $I_{C1} = I_{C1} + I_{C2}$, R_M is the CCM output resistance, $I_{B123} = I_{B1} + I_{B2} + I_{B3}$.

Assuming that 2 codes are equal each other, the measured quantity value X is determined as:

$$X = X_0 \frac{N_0}{N_m} [e_M + I_{SZ12}R_{MX21} + I_{B2}R_M - I_{B1}R_x], \qquad (4.16)$$

where $R_{MX} = R_M - R_x$, $R_{MX21} = R_{MX} + R_{k21}$, $R_{k21} = R_{k2} - R_{k1}$, $I_{SZ12} = I_{SZ} - I_{C12}$. Analysis of equation (4.16) by means of transposition method demonstrates that corrected additive CoE depends on the accuracy of applied CCM, source's signal resistance R_x and keys residual parameters of both input commutators.

If carry out digital measurement of electrical quantities in DC mode, the alteration of measuring circuit is not usually succeeded, aiming the maintenance the permanent systematic error at both signs of measured value. Therefore, the error compensation by the sign method is applied quite rarely. However, it is easy to fulfil inversion of input signal. Structure of such MT includes compulsory PS, and the measuring result N_x is computed as algebraic sum of conversion codes N_1 and N_2 at opposite polarity of input signal. It enables to correct not only additive CoE but also some members of MT polynomial approximation of conversion function [11, 19-21]. Indeed, the measuring results codes at positive $N_1 = (X + \Delta_{0x1})k_{ADC}$ and negative $N_2 = (-X + \Delta_{0x2})k_{ADC}$ polarities of measured quantity are derived:

$$N_x = 2k_{ADC}X + k_{ADC}\Delta_{0x12}, \qquad (4.17)$$

where k_{ADC} is the additive CoE MT conversion coefficient, e_{SZ}, I_{SZ} are the bias voltage and input current; $R_{k1}, R_{k2}, R_{k3}, R_{k4}$ are the resistances of closed keys accordingly $S1, ..., S4$; $I_{B1}, I_{C1}, I_{B2}, I_{C2}, I_{B3}, I_{C3}, I_{B4}, I_{C4}$ are the drain and source electrodes inverse currents of every keys pair respectively; R_x is the inner resistance of measuring signal source, $I_{B14} = I_{B1} + I_{B2} + I_{B3} + I_{B4}$; $I_{C14} = I_{C1} + I_{C4}$; $\Delta_{0x1} = e_{SZ} + (I_{SZ} + I_{C14})(R_{k12} + R_x) + I_{B14}R_{k2} + I_{B13}R_x$ is the additive CoE for the certain polarity of measuring voltage, $R_{k34} = R_{k3} + R_{k4}$; $\Delta_{0x2} = e_{SZ} + (I_{SZ} + I_{C14})(R_{k34} + R_x) + I_{B14}R_{k3} + I_{B24}R_x$ is the additive CoE for the opposite polarity of measuring voltage, $\Delta_{0x12} = \Delta_{0x1} - \Delta_{0x2} = (I_{SZ} + I_{C14})\Delta R_{1234} + (I_{B13} + I_{B24})\Delta R_{12} + (I_{B13} - I_{B24})R_x$ is the corrected the additive CoE value, $\Delta R_1 = R_{k1} - R_{k2}$; $R_{1234} = R_{k12} - R_{k34}$; $I_{B24} = I_{B2} + I_{B4}$; $I_{B13} = I_{B1} + I_{B3}$; $R_{k12} = R_{k1} + R_{k2}$.

Analysis suggests that the corrected additive CoE Δ_{0x12} is determined only by keys parameters dispersion. At usage of integrated microcircuits with ordinary dispersion parameters that does not exceed tenths of per cent [17-18, 35-36], additive CoE value can drop down more than two orders of magnitude in comparison with results obtained by other relevant methods [11, 19-21]. Input signal inverting method permits decreasing the MT nonlinearity error due to elimination of few members of transform function.

Small value of uncorrected error is promoted the auto implementation of the input signal inverting method in the nested-chopper technique amplifiers and chopping (Δ_Σ) ADC circuits. In these nested-chopper amplifiers both the main and additional pairs of choppers have been applied. Additional pair of choppers operates at considerably lower frequency (several tens Hz) than the other original pairs (several tens kHz). This technique allows to reduce the impacts of low-frequency interference and noise, including $1/_f$ noise, offset and offset drift as also cross interferences of electric scheme, to equivalent voltage level ~100 nV [35]. Then, the additive error can be reduced by means of nested-chopper amplifier exclusively.

Chopping (Δ_Σ) ADC are inherent in adequate performance: their offset error specification is ±0.5 µV at drift ±5 nV/°C that is practically immeasurable [29]. But it has not decrease additive CoE of whole measuring circuit including CLs, CCs, and antispark and explosion damage barrier.

Decentralized MTs are applied for data transmission in networks. To ensure reliability of MT for cheap systems, the major efforts have to be concentrated on optimal MT design. Widespread MTs, used in control desks and panels of manufacturing processes supervision, require temperature stability and low drift values. Attention should be paid to the duration of stable operation excluding labor-intensive adjustment. Moreover, to decrease costs and to raise production efficiency, the maximal simplifying of structure should be advised.

Considered method of A/D conversion is realized by choosing type and range of measured quantity that is especially inherent in direct current signals at the input. These include voltage or current, especially of small values (f.i., output electrical signals of temperature sensors). To convert exactly the current signals into code, the method of input signal inversion is applied [44].

Conducted errors analysis of the operation of serial devices A565 and CR7701 evidences the next. At MOS transistor application as polarity switch and at converting direct current signals from sources (of inner resistance not higher than 1 kΩ) connected to MT input, the corrected value of additive CoE does not exceed 0,1 µA. In operating conditions the MT errors are a few-fold higher than their limits of permitted values established for normal conditions. So, methodical error, specified by ratio of MT input resistance to input resistance of signal source and CLs, can affect the data quality.

Peculiarities of digital measuring devices structure, designed to operate with industrial MTs, are given below. They are the low level of input signals, high power of normal and common types of noise, linearization of transfer function, necessity of cold junctions' compensation, significant errors due to measurement current overheating the resistive transducers, providing the invariance of measuring result to values of currents and resistances of 3- and 4-wired CLs. The same concerns high-temperature action and enhanced drift, especially at MT operation in energy facilities.

Therefore MIs are manufactured with the additive CoE auto adjustment and digital linearization of transfer function. The sufficient interference impact reduction is achieved in ADCs and galvanic isolation of analog and digital units by averaging methods.

Company "MuckachivPrylad", Ukraine, produces universal MT control desk of CR7701 type intended to gauge the direct current, EMF and direct current voltage, temperature (by means of uniform primary transducers), and a number of physical quantities transformed in aforementioned signals. In particular, the desk is characterized by: permitted coverage interval of relative error not higher $\pm(0.1...0.2)$ %; operating temperatures from 5 to 50 °C; permitted coverage interval of additional relative error not higher than a third part of basic measurement error; operation time without maintenance less than 5 000 hours [44].

Intelligent data acquisition system of error correction in changeable environment is studied below. Nevertheless, both the devices of CR7701 type and the devices based on ADC chip make it possible to correct the additive CoE at their inputs. Here we deal only with additive CoE of processing device. In practice, quite long CLs connect the industrial devices with sensors, where contact EMFs may appear due to temperature gradient. This CoE is also emerged in the case of voltage drops caused by the leakage currents.

In operating conditions, the secondary devices form the segments of the measuring circuit which includes additionally the primary MT and CLs. Their considerable length can bring quite significant uncontrollable error into obtained result. Similar errors are invoked by contact EMF in wires junctions, or by CL leakage effect for inducted normal and common type interferences from the nearby electronic devices, etc. The analysis proves that the additive CoEs of CLs are of low frequency and slowly vary over time. Therefore, the measurement impact can be considered in first approximation as the additive; it utilizes the input signal inverting method for the additive CoEs adjusting and proposes to locate the polarity switch as close as possible to the sensor output (Fig. 4.3).

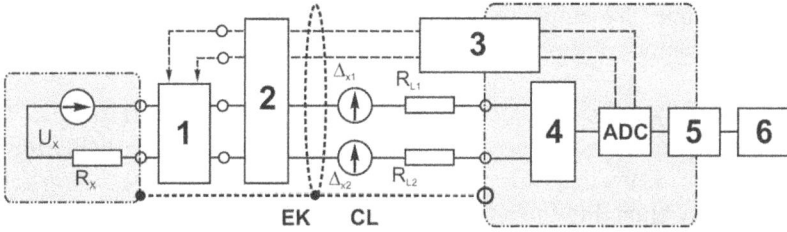

Fig. 4.3. Structure of digital voltmeter with auto adjustment of additive CoE of measuring circuit: 1 is the reverse polarity switch; 2 is the antispark and explosive damage block; 3 is the isolation block; 4 is the scale analog block; 5 is the A/D isolation block; 6 is the device controller.

Number of CL wires is raised to 5. To receive the correct result, it should be considered some additive CoEs caused by the leakage currents of the reverse polarity switch control electrodes through the insulation resistances R_{is1}, R_{is2} and the resistance of one of the R_{LZ} CL wires. Then the measurement result is determined as subtraction of two codes $N_{11} = k_{ADC}[(U_X + \Delta_{XL1})k_1 + \Delta_{0x}]$ and $N_{12} = k_{ADC}[(k_1\Delta_{XL2} + \Delta_{0x}) - k_{ADC}U_X]$:

$$N_1 = N_{11} - N_{12} = 2k_{ADC}[U_X + (I_{i1} - I_{i2})R_{CL}], \qquad (4.18)$$

where $I_{i1}; I_{i2}$ are the leakage currents throughout the isolation of electrodes of 1st and 2nd key pairs respectively, $\Delta_{XL1} = \Delta_1 + \Delta_{CL} + I_{i1}R_{CL}, \Delta_{XL2} = \Delta_2 + \Delta_{CL} + I_{i2}R_{CL}$, $\Delta_{CL}, I_{i1}R_{CL}, I_{i2}R_{CL}$ are the equivalent additive CoEs, arising in CLs due to leakage current I_{i1}, I_{i2} throughout resistance R_{lz} of the *CL* wire.

Correlation analysis of (4.18) suggests that uncorrected error is determined by subtraction of the control electrode leakage currents of reverse polarity switch. In order to reduce the additive CoE, it becomes necessary to provide the high resistance insulation of reverse polarity switch control circuit. If devices have to operate at aggressive conditions and high humidity, we try to ensure the high insulation resistance. It was noticed that residual leakage currents affect the additive CoEs. The leakage current of control electrodes of polarity switch chips does not exceed ±20 % of the reverse currents of their signal electrodes [17-18]. For example, when the minimal value of insulation resistance in the digital devices is upper than 40 MΩ, the controlled voltage is 15 V, and maximal resistance of *CL* does not exceed 200 Ω, the uncorrected additive CoE is not more than a few µV that preferably meets the preconditions. We emphasize that foregoing decisions have to be based on improvement of the integrated ADC chip.

Analysis of conversion techniques has admitted the next points. To ensure the standardized metrological characteristics of MIs, the periodic calibration is needed. To correct the multiplication error alters, is expediently changing the structure of digital voltmeter by connection of precise voltage CCM with input switch. As a result, we have obtained the structure of differential voltmeter, metrological opportunities of which can be controlled permanently in situ by periodic change of CCM settings. It permits to reach flexibility regarding the conventional tested one.

Evidently, we should manufacture the voltage CCM as a separate integrated chip with the embedded facilities for the certain error component adjusting in operating condition by an additional measurement method as well as by getting rid of impacts due to the studied methods. Protection of such measure of voltage from of electromagnetic interferences is achieved by shielding.

Similarly, hermetising prevents the action of moisture and aggressive chemicals. Principal opportunities of MIs verification and achieving the CPS reliability rise whilst the voltage CCM is manufactured as a portable block. The latter is reached by periodic verifications of it. Considering this, the remote calibration of information and measuring subsystems of CPSs can be accomplished.

4.3. Remote Error Correction of Measuring Channel

CPSs are deemed to be an integral part of manufacturing systems, factories, machinery, test facilities, moving objects, vehicles etc. These facilities typically utilize a lot of physical phenomena, whose parameters are constantly changing. Each CPS is comprised of dispersed hardware components and computer software, intended to obtain information about the progress of physical processes in controlled facilities, as well as its storage, transmission, processing and production by control signals. The measurement data, received from controlled objects, would be characterized by the set of metrological parameters. The measuring channels distribution in space, permissible changes in a wide range of operating parameters and inevitable degradation of measuring circuits parameters result in a significant deterioration of the CPS measuring channel performance. Thus, an operative metrological maintenance of these channels becomes important [12].

CPS measurement data accumulation and processing is performed by means of multichannel MIs that consists of measuring sensors, CLs, CCs (Fig. 4.4). The current trends of measuring systems design seems to be the implementation the MTs that transform the received signals into electrical form aiming the direct computing [1-2]. Whereby, digital measuring information could be obtained with help of the certain methods of processing, transmission, storage, reverse transformation of control function for CPS units.

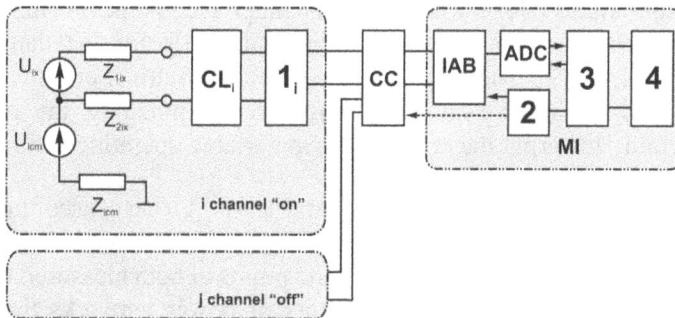

Fig. 4.4. Functional scheme of multichannel MI: 1 is the intrinsic safety barrier; 2 is the analogue control circuit; 3 is the isolation device; 4 is the instrument controller.

The low level of output sensors signals requires occurrence of amplifier that scales the previously mentioned signals to normal level for ADC operation and simultaneously converts them into the digital code necessary for MI controllers.

Measurements in sparkproof operating conditions and in dangerous environments evidently demonstrates the necessity of implementation the particular techniques. First, the inner safety barriers have to be applied at the output sensors of each measuring channel, and the analogue circuit of MI has to be electrically insulated from the digital one [30, 45-47]. The interference values often exceed the signal parameters of CCs. So, are encouraged to apply SST, IA or ID. The systematic errors that consist of additive and multiplicative CoEs emerge in measuring circuit of such DASs. Errors increase in DASs with isolated channels; therefore, it is difficult to ensure their operation by considerable time at significant temperature drift [30-34]. To correct the errors of CPSs, they are provided with calibrators of electrical quantities directly connected to measuring channel input instead of sensors [48]. To ensure the remote automatic adjustment, currently the mentioned devices are embedded into CPS measuring channels. It raises a problem of auto correction of errors of operating calibrators that have to be inexpensive due to their wide application.

Multichannel MI scheme for measured object in absence of spark and explosive environment and at the common mode voltage lower than the CCs chip breakdown voltage (~10 V), is studied. So while gauging the spark and explosive objects, it should be used the insulation blocks on the sensor outputs of every measuring channel. It could be recommended an extra insulated sensors and multichannel MIs for the dangerous objects [30, 45-47]. Therefore, the magnetic, capacitive, or optical MIs are offered to involve in measuring circuits pursuing the aim of considerably lowering the error values at variable operating conditions.

The ground loop current can be quite large ($>10^3$ A) that causes emerging the common mode voltage up to hundreds of volts. The last increase with CL length especially between the ground points of both measured facility and multichannel MI. Leakage currents of power networks that pass through measuring equipment insulation to ground loops of measured object elaborate the other source of common mode voltage. Then the application point of common mode voltage to the sensor is generally unspecified. To exclude above-mentioned drawback, the relays as CC with switching function "before turning off" for the considerable common mode voltage, can be provided. Such scheme application is inherent in a significant (>1 mV) additive CoE caused by contact EMF at temperature drift (up to ten μV/K). Thus, multichannel MI structure seems to be similar to the design shown in Fig. 4.5. To reduce significantly errors caused by CLs and CCs, the SST converters or IAs are recommended.

Three-wire sensors connection and shielding the CLs as well as multichannel MI analogue part substantially decrease the common mode voltage [1, 23, 30]. Then the block diagrams of MI significantly differ from the similar ones of multichannel MIs.

Analysis of metrological characteristics of multichannel MIs is conducted below. Here, the output sensors signals are submitted to the CC inputs through IB (if necessary) and CLs (Fig. 4.5).

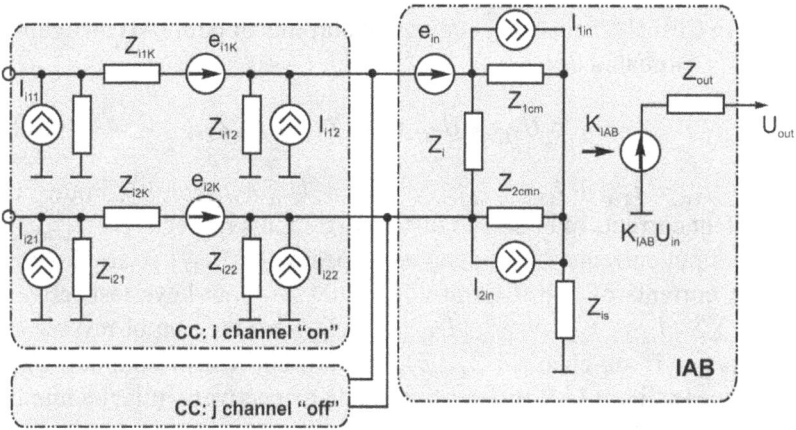

Fig. 4.5. Equivalent scheme of channel commutator and input amplifying block.

The particular common mode voltage of every measuring channel alters its determining voltage U_x. A differential circuit of IAB is provided for reducing the common mode voltage impact. Relegated to multichannel MI input the measured voltage U_{ixn} of ith on-channel is considered at a certain common mode voltage U_{icm} applying point S_o:

$$U_{ixn} = U_{ix}(1 + \delta_i) + e_{ie} + U_{il} + U_{ixc} + U_{ijx} + U_{ijxc}, \quad (4.19)$$

where U_{ix} is the output sensor voltage in i[th] on-measuring channel; $e_{ie} = e_{iCL} + e_{iIB} + e_{iCC}$ is the equivalent input offset voltage, $e_{iCL} = e_{1iCL} + e_{2iCL}$; $e_{iIB} = e_{1iIB} + e_{2iIB}$; $e_{iCC} = +e_{1ik} + e_{2ik}$; e_{1ik}; e_{2ik} is the residual voltage of the 1st and the 2nd on-keys CC respectively ($e_{iCC} = 0$ for MOS FET chip keys); $\delta_i = {Z_{ixe}}/{Z_{in}}$; $Z_{ixe} = Z_{1ixe} + Z_{2ixe}$; $Z_{1ixe} = +Z_{1ix} + Z_{1iCL} + Z_{1iIB} + Z_{1iCC}$; $Z_{2ixe} = +Z_{2ix} + Z_{2iCL} + Z_{2iIB} + Z_{2iCC}$ is the total resistance between common-mode voltage applying point and the 1st and the 2nd IAB differential inputs respectively; U_{il} is the equivalent error value caused by equivalent currents of both IAB differential inputs, U_{iec} is the equivalent error caused by common mode voltage of ith on-channel; U_{ijx} is the equivalent error caused by penetrating the measured voltages U_{jx} from other measuring off-channel; U_{ijxc} is the equivalent error value caused by penetrating the common mode voltages U_{jxc} from other measuring off-channel.

Additive CoE U_{il}, caused by equivalent currents of both IAB differential inputs, is estimated as:

$$U_{il} = U_{1il} + U_{2il} = I_{1e}Z_{1e} - I_{2e}Z_{2e}, \qquad (4.20)$$

where $I_{1e} = I_{1in} + I_{i11} + I_{12e}$, $I_{2e} = I_{2in} + I_{i21} + I_{22e}$ are the equivalent currents of both IAB differential inputs respectively; I_{1in}, I_{2in} are the input currents of same inputs respectively; I_{i11}, I_{i21} are the input reverse currents of both CCs at i^{th} channel on-input keys respectively; $I_{12e} = \sum_{i=1}^{n} I_{i12}$, $I_{22e} = \sum_{i=1}^{n} I_{i22}$, I_{i12}, I_{i22} are the output reverse CC currents for i^{th} on-channel, Z_{1e}, Z_{2e} are the equivalent common mode resistances of both IAB differential inputs respectively, n is the number of measuring channels.

We can accept that the input and output common resistances of CCs are roughly equal to each other: $Z_{i11} = Z_{i1}(1 + \delta_{i11})$; $Z_{i12} = Z_{i1}(1 + \delta_{i12})$; $Z_{i21} = Z_{i1}(1 + \delta_{i21})$; $Z_{i22} = Z_{i1}(1 + \delta_{i22})$, where $\delta_{i11}, \delta_{i12}, \delta_{i21}, \delta_{i22} \ll 1$, $\delta_{i11}, \delta_{i12}, \delta_{i21}, \delta_{i22}$ are the relative errors dispersions of the common-mode resistances estimated for CCs of ith channel. This error is assessed in a few tens of per cent. Considering the following $Z_{1ixe}, Z_{2ixe} \ll Z_{in}, Z_{1en}, Z_{2en}$ the expression for the equivalent input resistances is defined:

$$Z_{1e} \cong \frac{Z_{en}Z_{eis}}{Z_{en} + 2Z_{eis}}\left[1 + \frac{b}{Z_{en}}\left(Z_{2ixe} + \frac{Z_{1ixe}}{a^2}\right)\right] \qquad (4.21)$$

$$Z_{2e} \cong \frac{Z_{en}Z_{eis}}{Z_{en} + 2Z_{eis}}\left[1 + \frac{b}{Z_{en}}\left(Z_{1ixe} + \frac{Z_{2ixe}}{a^2}\right)\right], \qquad (4.22)$$

where $Z_{en} = \frac{Z_{1e}+Z_{2e}}{2}$, $Z_{en} = \frac{0.5Z_{i1}Z_c}{(n+1)Z_c+Z_{i1}}$, Z_{1e}, Z_{2e} are the equivalent common-mode input resistances respectively; $Z_{1e} = 1/G_{1e}$; $Z_{2ec} = 1/G_{2ec}$; $Z_{eis} = Z_{is} + Z_{icm}$; Z_{is}, Z_{icm} are the common-mode resistances of ith measuring on-channel and isolation resistances of shared measuring bus regarding the earthing point of multichannel MI, respectively; $G_{1e} = \frac{1}{Z_{i11}} + \frac{1}{Z_{1c}} + \sum_{i=1}^{n}\frac{1}{Z_{i21}}$; $G_{2e} = \frac{1}{Z_{i21}} + \frac{1}{Z_{2c}} + \sum_{i=1}^{n}\frac{1}{Z_{i22}}$; Z_{i11}; Z_{i21} are the common-mode input resistances of ith on-channel; Z_{i21}; Z_{i22} are the common-mode output resistances of ith measuring on-channel; $a = \frac{Z_{eis}}{Z_{en}+Z_{eis}}$; $b = \frac{a}{1+a}$; $Z_{1ixe} = Z_{1ix} + Z_{1iCL} + Z_{1iIB} + Z_{1iCC}$; $Z_{2ixe} = Z_{2ix} + Z_{2iCL} + Z_{2iIB} + Z_{2iCC}$, Z_{in} are the differential input resistances, Z_{1c}, Z_{2c} are the common-mode input resistances of both IAB differential inputs, respectively.

Considering (4.21) and (4.22), the additive CoE U_{il} dependent on the equivalent input currents is obtained:

$$U_{il} \cong \Delta I_e Z_{en} b + 2 I_{ein} \Delta Z_{ixe} (1 + a^2) \left(\frac{b}{a}\right)^2, \qquad (4.23)$$

where $\Delta I_e = I_{1e} - I_{2e}$; $I_{ein} = \frac{I_{1e} + I_{2e}}{2}$; $\Delta Z_{ixe} = Z_{1ixe} - Z_{2ixe}$.

Caused by common-mode voltage U_{icm} at i^{th} on-channel after few alterations, error U_{ixc}, is determined:

$$U_{ixc} \cong U_{icm} \frac{Z_{ixe}}{2 Z_{isx}} (\delta_{ixe} + \delta_{ie}), \qquad (4.24)$$

where $Z_{ixe} = Z_{1ixe} + Z_{2ixe}$; $\Delta Z_{ixe} = Z_{1ixe} - Z_{2ixe}$; $\delta_{ixe} = \frac{\Delta Z_{ixe}}{Z_{1ixe} + Z_{2ixe}}$ is the relative dispersion of both IAB total input resistance Z_{1ixe} and Z_{2ixe}; $\delta_{ie} = \frac{Z_{1e} - Z_{2e}}{2 Z_{en}}$ is the relative dispersion of both IAB equivalent input differential resistances; $Z_{isx} = Z_{icm} + Z_{is} + \frac{Z_{en}}{2}$.

Analysis envisages that error value U_{ixc} caused by common mode voltage U_{icm} inherent in additive and asymmetrical features depends on 2 differential inputs resistances of IAB (Fig. 4.5). For its reduction, one should increase the insulation resistance of IAB common bus concerning the applying point of common-mode voltage U_{icm}. Equivalent error U_{ijx} caused by penetration off-channels to i^{th} on-channel measuring voltage U_{jx} from the rest of off-channels is equal to:

$$U_{ijx} = \sum_{j=1}^{n-1} \left(U_{jx} \frac{Z_{in}}{Z_{in} + Z_{ip} + Z_{jxe}} \right), \qquad (4.25)$$

where $Z_{jxe} = Z_{jx} + Z_{jCL} + Z_{jIB}$; Z_{jx}; Z_{jCL}; Z_{jIB} are the inner resistances of sensor, CLs and IB at j^{th} CC off-channel respectively.

The origin of this, relative to the measured voltage in i^{th} on-channel error is additive one. For its adjustment the known automatic methods can be applied. Analysis of (4.25) results in increasing the error of voltage U_{ijx} in proportion to the number n of measuring channels. With their decreasing, within the multichannel MI structure, the CC of the highest off-resistances should be selected. However, this way of MI error decrease is limited substantially by the impact parameters of chip components. For example, if typical values are equal to $Z_{in} \simeq 10^9 \, \Omega$, $Z_{jp} \simeq 10^{12} \, \Omega$, $Z_{in} \ll Z_{jxe}$, and the measured voltages values approximately equal to each other $U_{jx} \cong U_{ix}$, the weighting factor of significance k_{ijx} drops 10-fold: $k_{ijx1} \cong 0.001$ at n=2 measuring channels, and $k_{ijx2} \cong 0.01$ at n=12.

Threshold value of the additive CoE U_{ijxc} impact caused by penetration of a common-mode voltage U_{jcm} of the rest off-channels to the ith on-channel, is given by:

$$U_{ijxc} = \sum_{j=1}^{n-1} \left\{ \frac{U_{jcm} Z_{ixe} Z_{jp}}{2Z_{jcm}(Z_{ip} + Z_{ieci})} (k_1 \delta_{ixe} + k_2 \delta_{ie}) \right\}, \quad (4.26)$$

where $Z_{jp} = Z_{1jp} - Z_{2jp}$ is the keys off-resistance of the jth CC disconnected channel; $Z_{ieci} = \frac{2Z_{jcm}(Z_{en} + 2Z_{is})}{2Z_{jcm} + Z_{en} + 2Z_{is}}$; Z_{jcm} is the common-mode resistance in the jth CC disconnected channel; $k_1 = \frac{2Z_{jcm} + Z_{en} + 2Z_{is}}{Z_{jcm} + Z_{en} + 2Z_{is}}$; $k_2 = \frac{2Z_{jcm} Z_{en}}{(Z_{en} + 2Z_{is})^2}$; $\delta_{ixe} = \frac{Z_{1ixe} - Z_{2ixe}}{Z_{ixe}}$ is the relative variations in equivalent resistance between the applying point of common-mode voltage and two IAB differential inputs.

Its analysis envisages that the additive CoE U_{ijxc} determined by asymmetry of input circuits of multichannel MI in i^{th} on-channel and input equivalent common mode resistances, depends on the number of measuring channels. Indeed, equilibration of input measuring circuits is time-dependent. Such schemes are symmetric for particular object and measuring current circuit parameters of MI in certain operating conditions. However, this symmetry is broken during the measurement when the circuit is reconfigured, or operating conditions are altered. In practice, exists tendention to reduce the additive CoE U_{ijxc} by ensuring the high insulation resistance Z_{is} of IAB common bus at applying point of common-mode voltage U_{jcm}. Furthermore, its value is reduced by adjustment in automatic mode.

Analysis of equations (4.19), (4.23)-(4.26) reveals that this kind of error depends also on the number of measuring channels. This is especially true for equivalent values of input offset voltage, input currents, input impedances of IAB and resistances of off-keys of CCs. Switches installation can significantly diminish the equivalent input currents and resistances impacts. However, the relay residual voltage significantly increases the additive CoE at the lowered switching channel speed of multichannel MI. Therefore we suggest to invert measurement signal at locating the switch in every channel for their residual voltages diminishing.

To reduce these CoEs, is proposed to set the smart transducer with IAB inputs located as close as possible to sensor output [1, 30]. It eliminates virtually the errors caused to CL and CC parameters as the output signals of such transducers are conditioned and signals can be supplied directly to ADC inputs.

Problem of multichannel MI design significantly complicates when the common-mode voltage U_{jcm} exceeds the breakdown voltage of electronic keys. Three-wire sensors connecting and respectively reciprocal insulation of measuring channels are recommended for these errors minimization.

Mutual insulation of measurement channels is suggested due to several reasons. The first one consists in necessity of protecting the considered MI electrical circuits against spark and/or explosive damage (Fig. 4.6). It needs additionally to connect an inner safety barrier to each measuring line. The barrier resistance can reach hundreds Ω that causes the additive CoE magnification due to voltage drop of leakage current on the mentioned resistance. Under regulations, the insulation resistance of power networks should exceed 40 MΩ, while the leakage current I_p of shielded measuring object must be ≤ 5 μA (or 220 V/40 MΩ). This current can produce the voltage drop $U_{ixiB} = I_p(Z_{ix} + Z_{iB}) = 10 \, mV$ on resistances $(Z_{ix} \approx Z_{iB} \leq 1 \, k\Omega)$ of sensor and inner safety barrier that entails the multichannel MI additive CoE increment.

Quality insulation of measuring channel diminishes the potential difference impact that emerges between the grounding points of measuring object and multichannel MI. Such difference is generated by powerful sources due to leakage current through resistances between the grounding point of device and the ground. Its value may reach the hundreds volts (electric transport, melting furnaces, etc.) that lead to challenges in switches' operation.

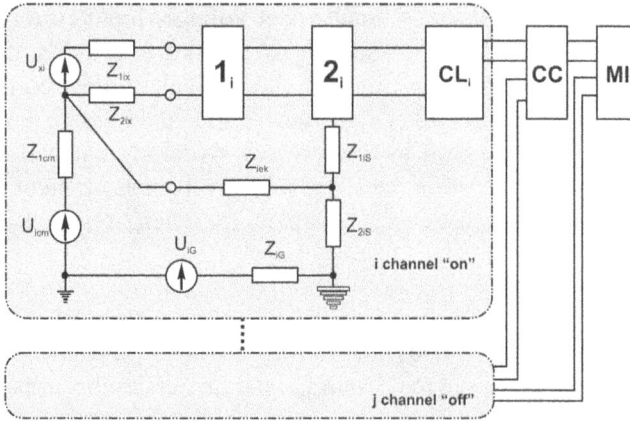

Fig. 4.6. Scheme of the multichannel MI with isolated channel:
1 is the inner safety barrier; 2 is the isolation amplifier.

To minimize the common-mode interference impact, three-wired sensors CLs apply. Here a third wire serves as the shield, which protect two information CLs in the section between sensor output and multichannel MI input. On the sensor side this shield is connected to the point of applying the common-mode voltage. Moreover, multichannel MI has to be connected to the screen at the end of CLs. Shield should be characterized by high insulation properties regarding the measuring circuit. Error caused by CL length significantly rises. Therefore, SST decreases the afore-mentioned error due to cheap electronic units implementation [1-2]. Its output voltage U_{ixis} in multichannel MI is high enough for direct interface connection with ADC. Then the relays can operate within CCs. Output voltage U_{ixs} of ith IA is given by:

$$U_{ixs} = (U_{ix} + e_{iAe} + U_{icA})k_{iA}(1 + \delta_{iA}) + e_{ioA} + U_{ijxA} \quad (4.27)$$
$$+ U_{ijcA} + U_{ine}$$

Here $e_{iAe} = e_{iA} + e_{iIB} + I_{iA}Z_{ixA}$, $Z_{ixA} = Z_{1ixA} + Z_{2ixA}$, $Z_{1ixA} = Z_{1ix} + Z_{1iIB}$, $Z_{2ixA} = Z_{2ix} + Z_{2iIB}$; e_{iA}, e_{iIB} are the offset voltage and residual voltage of ith IA respectively, $\delta_{inA} \cong Z_{ixA}/Z_{inA}$; I_{iA}, Z_{inA} are the input current and resistance of ith IA respectively; $Z_{1iis}, Z_{2iis}, Z_{3iis}$ are the insulation resistances between i[th] IA input and output, common bus and shield respectively; $\delta_{inA} \cong Z_{ixA}/Z_{inA}$; $U_{ine} = I_{i11}Z_{iCL} +$

$\left(I_{i12} + I_{in} + \sum_{j=1}^{n-1} I_{j12} \right) \left(Z_{1iCC} + Z_{iCL} \right)$ is the equivalent output voltage of IAB; e_{i0} is the offset output voltage of ith IA; U_{icA} is the equivalent input voltage of ith measuring on-channel caused by equivalent common-mode voltage; $U_{ic} = U_{icm} + U_{iG}$, U_{ijxA} is the equivalent input voltage of ith measuring on-channel caused by impact of measuring voltage U_{jx} of rest of measuring off-channels; U_{ijcA} is the equivalent input voltage of ith measuring on-channel caused by impact of equivalent common mode voltage $U_{jc} = U_{jcm} + U_{jG}$ of rest of measuring off-channels; $e_{ioA} = e_{io} + e_{iCL}$, Z_{iG}, U_{iG} are the resistance and voltage between grounding points of ith IA and measuring object at ith measuring on-channel; k_{iA} is the ith IA transform coefficient.

Input equivalent voltage U_{icA} at i^{th} measuring on-channel, due to its equivalent common-mode voltage U_{jc}, can be presented as:

$$U_{ixcA} = U_{ic} \frac{Z_{2ixe}}{Z_{icm} + Z_{iG} + Z_{3iis}} \cdot \frac{Z_{iek}}{Z_{2iis}}. \qquad (4.28)$$

Analysis of last clarifies that minimization of error voltage U_{ixcA} should be provided due to small resistance Z_{iek} of shield and high resistance Z_{2iis} regarding measuring scheme. Comparing the latter equation with (4.24) we conclude that the error U_{icxA} caused by common-mode voltage at i^{th} measuring on-channel may be reduced in Z_{iek}/Z_{2iis} times. For example, it takes place if IA AD210 type Analog Device Co can be applied, and is provided the shielding resistance $Z_{iek} \leq 10 \, \Omega$ at ordinary value of insulation resistance $Z_{2iis} \cong Z_{3iis} \cong 240V/2\mu A = 1.2 \cdot 10^8 \, \Omega$ [15]. Also, if select the common-mode voltage $U_{ic} \leq 2500 \, V$ equal to the maximum isolation voltage of the same IA type at the equivalent common-mode resistance $Z_{icm} + Z_{iG} \cong 40 \, M\Omega$, the equivalent input voltage is equal to $U_{ixcA} \leq 2500 \frac{10}{1.6 \cdot 10^8} \cdot \frac{10^3}{1.2 \cdot 10^8} \cong 41 \, nV$, which is negligibly small.

Reduced to IAB equivalent input voltage U_{ijxA} at i^{th} measuring on-channel caused by impact of measured voltage U_{jx} of the rest (n-1) off-channels of commutator is incapable to change it by comparing with value of obtained from (4.25). Threshold value of additive CoE U_{ijcA} caused by impact of equivalent common-mode voltage $U_{jc} = U_{jcm} + U_{jG}$ of all measuring off-channels at i^{th} on-channel compared to the expression (4.25) is lowered in Z_{iek}/Z_{2jiS} times. Received additive CoE becomes negligible for above-given conditions.

Analysis of (4.27) envisages that both additive and multiplicative CoEs caused by voltage bias and input current of IA, significantly affect the accuracy. In order to rise it, the manual zeroing and adjusting the transform factor of IA are involved. Unfortunately while operating, these factors impact and therefore the multichannel MI accuracy worsen substantially.

Error correction of multichannel MIs is realized in the following way. Analysis ratio of (4.19) to (4.28) helps to identify the additive CoE significant effect over the MI performance in operating conditions. It is minimized by the multichannel MI manual zeroing [49].

Usually CPS multichannel MIs are considered as distributed systems with the controlled objects located at appreciable distance from each other. So, let's suppose that it is impossible to carry out instrumental zeroing of every measuring channel at manual mode. To automate the error adjustment, it seems to be better to invert the gauging signals; input polarity switch has to be located as close as possible to sensor output [23, 50]. At IA applying this switch should be installed near-by the input amplifier (Fig. 4.7).

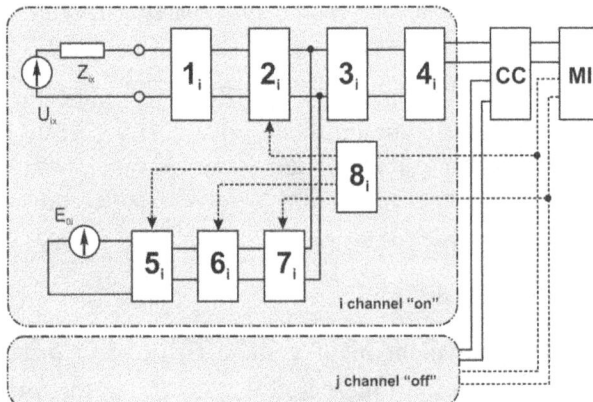

Fig. 4.7. Multichannel MI with remote errors correction: 1 is the inner safety barrier; 2 is the polarity inverse switch of measuring signal; 3 is the isolation amplifier; 4 is the isolation device; 5 is the polarity inverse switch of calibrating signal respectively; 6 is the code-control voltage divider; 7 is the switch; 8 is the control unit.

In operating conditions, the multiplicative CoE of multichannel MI is characterized by significant dispersion (up to ± 2 %) that makes the problems. We propose to perform the MI remote calibration basing on voltage CCM located in every measuring channel. Then during calibration the output voltage $U_{ik} = kE_{0i}$ of CCM feed the measuring channel input, while the sensor measuring output signal U_{ix} is disconnected. Available set of calibration codes is transmitted to voltage CCM from the multichannel MI controller. Output voltage of CCM is converted into calibration code N_{ik}, where i is the number of channel (k=1, 2,..., K, where K is the maximum number of calibration codes).

At calibrating the ith on-channel, the latter sends the N_{ik} code:

$$N_{ik} = 0.5(N_{1ik} - N_{2ik}) = 0.5k_{iA}k_{ADC}(U_{ik} + \Delta_{iAc}), \quad (4.29)$$

where N_{1ik}, N_{2ik} is the data codes for direct and reverse polarity of calibration voltage $U_{ik} = kE_{0i}$; E_{0i} is the reference voltage; k_{ADC} is the ADC transform factor; $\Delta_{IAc} = 0.5[(I_{iA} + I_{iPC})\Delta Z_{iPC} + \Delta I_{iPC}Z_{iPC}]$ is the uncorrected value of additive CoE; $I_{iPC}, \Delta I_{iPC}$ are the average value and variance of reverse currents of keys respectively; $Z_{iPC}, \Delta Z_{iPC}$ are the average resistance and resistance match of set of on-channels respectively.

While gauging the ith on-channel, the sensor signal U_{ix} is received with the measurement result code N_{ix}:

$$N_{ix} = 0.5(N_{1ix} - N_{2ix}) = 0.5k_{iA}k_{ADC}(U_{ix} + \Delta_{iAx}), \quad (4.30)$$

where N_{1ix}, N_{2ix} is the measurement result codes of sensor output signal U_{ix} for direct and reverse polarity of 2 switches connection; $\Delta_{IAx} = 0.5[(I_{iA} + I_{iPX})\Delta Z_{iPX} + \Delta I_{iPX}Z_{iXX}]$ is the uncorrected additive CoE; $I_{iPX}, \Delta I_{iPX}$ are the average value and absolute dispersion of reverse currents of keys of 2 switches respectively; $\Delta Z_{iPX} = Z_{iX} + Z_{1iB} + Z_{iPX}, Z_{iPX}, \Delta Z_{iPX}$ are the averaged resistance and resistance match of two on-channels respectively.

The determined value is transformed into:

$$N_{ix} = N_{1ik}\frac{U_{ix} + \Delta_{IAx}}{U_{ik} + \Delta_{IAc}}$$
$$= N_{1ik}\frac{U_{ix}}{U_{ik}}\left(1 + \frac{\Delta_{IAx} - \Delta_{IAc}}{U_{ik}}\right) \quad (4.31)$$

Multiplicative CoE of multichannel MI depends on reference voltage E_{oi} accuracy and on conversion coefficient k of code-controllable voltage divider. For estimating the limit of uncorrected errors we take the ordinary values inherent in ADG787 switch [51] ($I_{iA} = 30 \, nA$ max, $I_{iPX} \simeq I_{iPK} \simeq 20 \, nA$ max, $Z_{iPX} \simeq Z_{iPK} \simeq 3.35 \, \Omega$ max, $\Delta Z_{iPX,C} \simeq 0.1 \, \Omega$), $Z_{iX} + Z_{1IB} \leq 1 \, k\Omega$ max, $\Delta I_{iPX} \simeq \Delta I_{iPC} \simeq 0.05 I_{iPC} \simeq 5 \cdot 10^{-2} \cdot 2 \cdot 10^{-9} \, nA$, then $\Delta_{IAc} \simeq 4 \, nV$, $\Delta_{IAx} \simeq 0.1 \, \mu V$. Due to calibration, the adjustment of uncorrected additive CoEs is performed providing their minimization. Unadjusted constituent of additive CoE remains permanent; its value is determined while measuring by multiplying the total resistances of sensor and inner damage barrier on the difference of reverse currents of on-2nd and 5th keys. Studies have envisaged that this difference does not exceed several per cent for MOS chips. The accurate multichannel MIs can be realized in such a way.

To insure high accuracy of operating MIs, the method of remote calibration was implemented. Within it, the actual output voltage U_{ik} for k otherwise dividing factors in every measuring channel is measured at training stage of multichannel MI (on the step of installation). All K values of output voltages U_{ik} are gauged by high-precision voltmeter for every of k division factor receiving the codes array N_{uik}. Then the same voltage U_{ik} is measured by multichannel MI, and the other set of codes N_{ik} is received. High-mentioned array N_{uik} is introduced in MI memory, the appropriate calibration coefficients $K_{ik} = {N_{ik}}/{N_{uik}}$ are computed and fixed in memory, aiming their subsequent application in determining the measurement result code N_{ix}:

$$N_{ix} = K_{ik} U_{ix} \qquad (4.32)$$

Reference voltage $U_{ik} = kE_{oi}$ of code-controllable voltage divider alters throughout the period of operation. To reduce the impact-factor, the stable over time electronic components, f.i. with parameters of reference voltage $\partial E_{io}/\partial\theta \leq \pm 2 \cdot 10^{-6} \, \mu V/K$, and code-controllable voltage divider $\partial k_i/\partial\theta \leq \pm 2 \cdot 10^{-6} \, 1/K$ in the temperature range 25...85 °C, have to be selected [17-18]. Then variable values of U_{ik} and therefore K_{iK} would not exceed ±0.026 % which is satisfactory for measurements. Temperature stability of suggested multiplying voltage CCM and its particular calibration while debugging make it possible to obtain practically unchanged reference voltage values within wide range, from a few millivolts to nearly maximum value of reference voltage.

Similar is inherent in the metrological in-place verification of multichannel MI with help of portable voltage CCM. Error adjustment involves gauging the k means of output measures $U_{i\kappa}$ for every measuring channel. As regulations require, these calibration points have to be arranged throughout the measuring range. Maximum value nearby the boundary of measuring range can be used as operating standard at the ith measuring on-channel. Excluding measuring sensors the multichannel MI in-place verification, provided by means of voltage CCM, assures the particular possibility of metrological checking of all channels. Portable voltage CCM is protected against unstable operating conditions by implementation the protective and preventive techniques, and software development.

Investigations of voltage CCMs include the following peculiarities due to existing problems of their metrological maintenance. Producers offer a number of portable calibrators for effective calibrating the industrial multichannel MIs. Their main drawback is the necessity of calibration results adjustment concerning the volatile operating conditions. Therewith, the drift of metrological characteristics of calibrators exists as well as the contact EMFs emerge in multichannel MI mount points. To avoid these problems, the voltage calibrators with self-correction of errors are developed (Fig. 4.8).

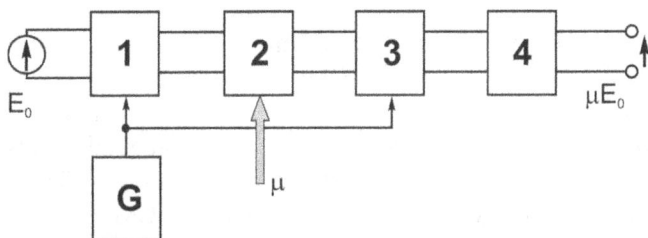

Fig. 4.8. Scheme of portable code-controllable measure with self-correction of errors: 1, 3 are the first and the second polarity inverse switches; 2 is the code-controllable voltage divider; 4 is the low-pass filter; G is the correction frequency generator, μ is the control code of voltage CCD.

In operating conditions calibrator requires the periodic manual additive CoE adjustment, which prolongs duration and complexity of metrological maintenance. In such a way the additive CoE self-correction is provided.

Additive CoE self-adjustment is based on synchronous operation of 2 polarity inverse switch. The 1st one is located on the reference voltage output and the 2nd one - on the code-controllable voltage divider output. Calibrator output voltage μE_0 averaging can be carried out both in digital and analogue forms at low-pass filter application. Then the digital processing is the summation of even number conversion results of the calibrator output voltage μE_0. For the same polarity of both inverse switches, the calibrator output voltage U_{k1i} is determined:

$$U_{k1i} = \mu_{iH}(E_{0H} + e_1)(1 + \delta_{\mu i} + \delta_E) + e_2 \qquad (4.33)$$

whilst for the different polarity, U_{k2i} is defined as:

$$U_{k2i} = \mu_{iH}(E_{0H} - e_1)(1 + \delta_{\mu i} + \delta_E) - e_2, \qquad (4.34)$$

where μ_{iH} is the nominal code of code-controllable voltage divider; E_{0H} is the nominal value of reference voltage, $\delta_{\mu i}, \delta_E$ are the relative errors of CCD voltage and reference voltage respectively; e_1, e_2 are the additive CoEs on the input and output of voltage CCD respectively. The averaged output voltage U_{ki} of calibrator for current code μ_i, is determined as:

$$U_{ki} = 0.5(U_{k1i} + U_{k2i})$$
$$= 0.5\mu_{iH}E_{0H}(1 + \delta_{\mu i} + \delta_E) \qquad (4.35)$$

Results of modeling of considered scheme coincide with experiment, in which calibrator output voltage $E_0 = 100\ mV$ and code-controllable voltage divider codes have been chosen within the range $0 - 1$ with increments 0.25. To test the additive CoE impact on obtained results, we apply $e_1, e_2 = 15\ mV$ from the stable power source. Two sets of experimental results were received: the output CCM voltage U_{k1}, in which the additive CoE value was excluded, and the same voltage U_{k2}, mV where the mentioned CoE has been included (Table 4.1).

Maximal signal of imitator of additive CoEs has been selected a priori higher than the possible maximum values of equivalent offset voltage amplifiers, installed in the calibrator. Simulator of equivalent additive CoE voltage was located in various characteristic points of calibrator, namely at inputs, outputs and all feedback loops of OAs. CCM output voltage was measured by multimeter Picotest M3511A: DCV 100 mV, accuracy 0.012 % within 1 year operation, least significant digit 1 μV.

Table 4.1. Results of CCM voltage study by means of experimental unit.

No.	μ_H	U_{k2}, (mV)	U_{k2}, (mV)
1.	0	-0.003	-0.003
2.	0.25	25.007	25.007
3.	0.5	50.015	50.014
4.	0.75	75.023	75.022
5.	1	100.031	100.031

Performed studies revealed that numeric data at the 3rd and the 4th rows of the table differ mutually not more than by absolute unit of voltmeter resolution (± 1 μV). This confirms the preliminary assumptions about the feasibility of remote auto calibration of multichannel MIs of CPSs.

4.4. Conclusions

Built-in Measuring Instruments are the essential integral subsystems of Cyber-Physical Systems. Since such Instruments contain ADCs and DACs, they have to work with predetermined metrological performance within operation period. Tasks of their regular or/and irregular metrological checking, validation, verification, calibration, and errors adjustment are permanently emerging, complicated by heterogeneous nature of the measurands.

Quality of Measuring Instruments is reasonable to evaluate by the product of two values: quantity of information, received as a result of signals physical transformation throughout the measuring circuit, and quantity of information, due to decrease in quantity of entropy as a result of measurement. The proposed functional dependence connects to each other the energy and the information properties of elements of measuring circuit. In this case, the latter is regarded as the transmission channel of analog information from sources of active/passive electrical signals to the Measuring Instruments through connection line with its own equivalent source of additive and multiplicative noise signals.

Amount of measurement data obtained from the measurement object via the analog transmission channel with the certain level of noise is proposed to estimate on the basis of Shannon–Hartley theorem. During the measurement of active electrical signals the equivalent noise, reduced to input of Measuring Instrument, additively affects the gauging result. While measuring the passive electrical signals, the impact of equivalent noise is inherent in additive-multiplicative origin. Moreover, the bandwidth of the transmission data channel depends on difference between maximum frequency of its bandwidth and frequency of adjusting transactions of bias voltage.

Taking into account both thermal noise and $1/f$ noise, the expressions for the variance of equivalent noise during the measurement of active and passive electrical quantities are specified. Here the multiplicative component, caused by the influence of noise sources of whole measuring circuit, is taken into account. To determine in these terms the quality of Measuring Instruments, is suggested to apply an information factor as logarithmic measure of uncertainty, which value is practically determined by measurement error. Auto adjustment of additive component of error significantly enhances an information efficiency of Measuring Instruments.

Analysis of general and specific methods for measurement accuracy improving suggests that the theoretical minimum value of corrected additive component of error can be achieved, if the input measuring circuit is almost unchangeable for both measurement and correction channels. Input signal inverting method is able to provide it only by means of input switching the measurement signals of both polarities.

Set of the errors inherent in multichannel Measuring Instruments with isolated channels significantly exceeds the similar one of traditional structures.

At design stage the errors remote adjustment for multichannel Measuring Instruments is suggested to carry out by means of embedding the code-controllable voltage measures. For both multi-channel Measuring Instruments and embedded code-controllable voltage measures, the additive error correction is performed by inverting input signals. Multiplicative error correction is enhanced additionally by implementation of the DAC multiplier.

For previously released multi-channel Measuring Instruments the on-line errors correction permits diminishing the additive error value to ± 1 µV, ensuring the accuracy / trueness and metrological reliability of CPS operation.

References

[1]. Smart Sensor Systems, G. Mejer (Ed.), *John Wiley & Sons Ltd,* 2008.

[2]. J. W. Gardner, V. K. Varadan, O. O. Awadelkarim, Microsensors, MEMS, and Smart Devices, *John Wiley & Sons Ltd*, Chichester, England, 2001.

[3]. M. Płóciennik et al.,. EU funded DORII project (RI-213110), European Union under the Seventh Framework Programme (FP7), *Deployment of Remote Instrumentation Infrastructure,* 2008, (http://www.nm.ifi.lmu.de/pub/Publikationen/abcd09/PDF-Version/abcd09.pdf).

[4]. F. Davoli, N. Meyer, R. Pugliese, S. Zappatore, Grid-Enabled Remote Instrumentation, *Springer*, New York, 2008.

[5]. Remote Instrumentation Infrastructure for e-Science. Approach of the DORII project, *in IEEE International Workshop on Intelligent Data Acquisition and Advanced Computing Systems: Technology and Applications*, 21-23 Sept. 2009, Rende (Cosenza), Italy, pp. 231 – 236.

[6]. M. Jurčević, H. Hegeduš, M. Golub, Generic System for Remote Testing and Calibration of Measuring Instruments: Security Architecture, *Measurement Science Review,* Vol. 10, No. 2, 2010, pp. 50-55.

[7]. R. Müller, Calibration and Verification of Remote Sensing Instruments and Observations, *Remote Sens.,* Issue 6, 2014, pp. 5692-5695.

[8]. M. M. Albu, A. Ferrero, F. Mihai, and S. Salicone, Remote Calibration Using Mobile, Multiagent Technology, *IEEE Transactions on Instrumentation and Measurement,* Vol. 54, No. 1, February 2005, pp. 24-30.

[9]. What is metrology? BIPM, 2004, Retrieved 2011-12-01 (http://www.bipm.org/en/worldwide-metrology/).

[10]. International vocabulary of metrology – Basic and general concepts and associated terms (VIM 3rd edition), JCGM 200:2012 (JCGM 200: 2012 (JCGM 200:2008 with minor corrections), http://www.bipm.org/en/publications/guides/vim.html

[11]. H. Czichos, T. Saito, L. Smith, Springer Handbook of Metrology and Testing, *Springer*, 2011.

[12]. ISO 10012:2003, Measurement Management Systems – Requirements for Measurement Process and Measuring Equipment.

[13]. ISO 9001:2000, Quality Management Systems – Requirement.

[14]. ISO/IEC 17025:2005, General Requirements for the Competence of Testing and Calibration Laboratories.

[15]. C. Renner, S. Ernst, C. Weyer, V. Turau, Prediction Accuracy of Link-Quality Estimators, Wireless Sensor Networks, in *Proceedings of 8th European Conference on EWSN,* 2011, Bonn, Germany, February 23-25, 2011, Lecture Notes in Computer Science, Vol. 6567, 2011, pp.1-16.

[16]. O. Mizuno et al., Analyzing effects of cost estimation accuracy on quality and productivity, in *Proceedings of the International Conference on Software Engineering,* April 19-25, 1998, p.410-419.

[17]. Katalog ELFA DISTRELEC (http://www.online-electronics.com.ua/catalog/).

[18]. Datasheet Catalog.com (http://www.datasheetcatalog.com/).

[19]. JCGM 100:2008. GUM 1995 with minor corrections. Guide to the expression of uncertainty in measurement. Published jointly by JCGM, 2008, 1st edition, 134 p.

[20]. M. Grabe, Measurement Uncertainties in Science and Technology, *Springer Berlin Heidelberg*, 2005.

[21]. S. Rabinovich, Measurement Errors and Uncertainties. Theory and Practice, *Springer Science & Business Media*, 2005.

[22]. E. Polishchuk, et al, Metrology and Measuring Equipment, Tutorial, Lviv, Ukraine, *Lviv Polytechnics Publishing House*, 2012 (in Ukrainian).

[23]. V. Yatsuk, P. Malachivsky, Methods of Increase of Measurement Accuracy, Lviv, *Beskyd-bit*, 2008 (in Ukrainian).

[24]. P. Novicki, A. Zograf, Estimation of Errors of Measuring Results, Leningrad, *Energoizdat*, 1991. – 524 p. (in Russian).

[25]. C. Shannon, A Mathematical Theory of Communication, *Bell System Technical Journal,* Vol. 27, Issue 3, July–October 1948, pp. 379–423.

[26]. R. V. L. Hartley, Transmission of Information, *Bell System Technical Journal,* July 1928, p. 535.

[27]. F. N. H. Robinson, Noise and fluctuations in electronic devices and circuits, *Oxford University Press,* 1974.

[28]. A. van der Ziel, Noise: sources, characterization, measurement, *Prentice Hall*, 1971, 184 p., http://www.amazon.com/Noise-Sources-Characterization-Measurement-Aldert/dp/0136231659.

[29]. Mary McCarthy, Chopping on the AD7190, AD7192, AD7193, AD7194, and AD7195, AN-1131 Application Note, *Analog Devices, Inc.,* 2011.

[30]. Data-Acquisition-Handbook, A Reference For DAQ and Analog & Digital Signal Conditioning, Third Edition, *Measurement Computing Corporation,* USA, 2012, http://www.mccdaq.com/pdfs/anpdf/Data-Acquisition-Handbook.pdf

[31]. Data Acquisition Systems, Omega Company Products, One Omega Drive, Stamford, CT 06907, 1-888-TC-OMEGA USA, (http://www.omega.com/techref/pdf/dasintro.pdf).

[32]. Data Acquisition (DAQ) Fundamentals, Application Note 007, *National Instruments Corp.*, August 1999. (http://physweb.bgu.ac.il/COURSES/SignalNoise/data_aquisition_fundamental.pdf).

[33]. Instrumentation and Measurement, *Analog Devices Inc.,* 2015, http://www.analog.com/en/applications/markets/instrumentation-and-measurement.html

[34]. Scott Wayne, Finding the Needle in a Haystack: Measuring small differential voltages in the presence of large common-mode voltages by Scott Wayne, *Analog Dialogue*, 34, 1, 2000. (http://www.analog.com/library/analogDialogue/archives/34-01/haystack/index.html).

[35]. F. M. van der Goes, G. C. M. Meijer, A universal transducer interface for capacitive and resistive sensor elements, *Analog Integrated Circuits and Signal Processing,* Vol. 14, 1997, pp. 249–260.

[36]. A. Bakker, K. Thiele, J. H. Huijsing, A CMOS nested-chopper instrumentation amplifier with 100-nV offset, *IEEE Journal of Solid-State Circuits,* Vol. 35, 2000, pp. 1877–1883.

[37]. G. C. M. Meijer, A. W. van Herwaarden, Thermal Sensors, Bristol, UK; *Institute of Physics Pub.,* Philadelphia, 1994.

[38]. P. C. de Jong, Instrumentatieversterker, *Dutch Patent Application,* 1002732, 1996.

[39]. P. C. de Jong, G. C. M. Meijer, A. H. M. van Roermund, A 300 °C dynamic-feedback instrumentation amplifier, *IEEE Journal of Solid-State Circuits,* Vol. 33, 1999, pp. 1999-2009.

[40]. G. Wang, An Accurate DEM SC instrumentation Amplifier, *Dutch Patent Application,* 1014551, 2000.

[41]. G. Wang, G. C. M. Meijer, Accurate DEM SC amplification of small differential-voltage signal with CM level from ground to VDD, in *Proceedings of the Conference on Smart Structures and Materials 2000: Smart Electronics and MEMS, SPIE 3990,* Vol. 36, June 21, 2000.

[42]. Classification of Methods of Measurements (Metrology) (http://what-when-how.com/metrology/classification-of-methods-of-measurements-metrology/).

[43]. Je. Shmorgun, V. Zorij, V. Yatsuk, V. Putsylo, V. Zdeb, Digital Voltmeter, Patent No. 966613 (USSR) / *Patent Bulletin,* 38, 1982.

[44]. ATEX directive 2014/34/EU , Decision No 768/2008/EC of the EPC of 9 July 2008 on a common framework for the marketing of products, in Brussels, 21.11.2011 under reference COM(2011) 763 final.

[45]. Explosion protection. Theory and practice, *Phoenix Contact GmbH & Co. KG,* online, https://www.phoenixcontact.com/_technical_info/ 5149416_EN_HQ_LR.pdf).

[46]. C. Lehrmann, D. Seehase, M. Sattler, M. Gruner, Latest news on explosion protection, *Kommunikation Schnell GmbH,* 2013, (http://www.vem-group.com/fileadmin/content/pdf).

[47]. Fluke Multifunction Calibration Tools, *Fluke Inc.,* 2015. (http://en-us.fluke.com/products/multifunction-calibrators/).

[48]. Precision, Wide Bandwidth 3-Port Isolation Amplifier AD 210, *Analog Devices Inc.,* (http://www.analog.com/media/en/technical-documentation/ data-sheets/AD210.pdf).

[49]. Yatsuk V, Stolyarchuk P., Mikhaleva M., Barylo G., Intelligent Data Acquisition System Error Correction in Working External Conditions, in *Proceedings of the 3rd IEEE Workshop on Intelligent Data Acquisition and Advanced Computing Systems: Technology and Applications (IDAACS`05),* Sofia, Bulgaria, Sept. 5-7, 2005, pp. 51-54.

[50]. 2.5 Ω CMOS Low Power Dual 2:1 Mux/Demux USB 1.1 Switch ADG787, *Analog Devices Inc.,* (http://www.analog.com/media/en/technical-documentation/data-sheets/ADG787.pdf).

5.

Techniques for Accuracy/Trueness Improvement

B. Stadnyk, S. Yatsyshyn, I. Mykytyn and Ya. Lutsyk

CPSs have to be equipped not only with the Measuring Instruments, tools and facilities, but also to be ensured with the reliable information. For this aim the periodic checks of their metrological parameters are assumed.

5.1. Major Metrological Characteristics of CPSs Units within Different Approaches

Each of the following factors entails that the results of measurements differ from the true values of the measurands. The quality of measurements deteriorates, and thus the quality of CPS gets worse. These factors are the next [1]: problem of object model and the measurand (it is due to simplification of measurement procedures as well as experimental and theoretical generalizations that results in idealization of object properties); mutual influence of object and MIs (for example caused by mounting the sensor at the facility); imperfection of MIs (among all other possible factors deteriorating quality of measurement result, the instrumental factor is always available); calibration of MIs (is considered below); conditions of measurement (almost impossible to determine accurately the impact functions or their values as they may by unstable over time); dynamics of variables (significant influence on the dynamic characteristics of measurands is observed in nanotechnology); mathematical simplification of sensors transfer function; volume of measurement data and conjugated computing problems (too small array of experimental data can lead to misconceptions about the course of the considered process and, conversely, too big amount of data may result not only in low-quality performance and in loss of reliability of controlled parameter, so it can be resolved involving cloud technologies); specifics of data storage and data mining.

Errors approach produces established way to the classification of errors based on their certain properties. This separation of errors defines methods of reducing their impacts and of assessing results. Errors can depend or not depend on the value of measurand. In this regard the additive, multiplicative and nonlinear errors are distinguished. Additive one is independent on the value of the measurand, and the amendment is algebraically added to the measured value. Multiplicative error increases or decreases linearly with measurand increasing; it is proportional to the product of certain factor (positive or negative) and the measured value. Nonlinear errors nonlinearly depend on the measured value. The ultimate goal of the measurement errors analysis is just assessment of boundary errors in which they are located with a certain probability. Then measurement result with intervals determined by these error boundaries with given probability, covers the true value of measurand.

In uncertainty approach of measurement result [2] on the one hand the concept of true quantity value does not use because the latter is unknown; and on the other hand a unified approach to quantitative assessment of results quality is implemented regardless of origin and method of various factors impact on the measurement result. Another quantitative characterization of measurements quality, namely uncertainty of measurement result, is introduced. Although, most of the errors approach principles are successfully utilized in hidden form. Thus both methods rely on the use of source distribution density that causes the outcome. Standard uncertainty is the uncertainty of result expressed by standard deviation. It may also be given in the form of dispersion as the square of the standard uncertainty. The standard uncertainty of type A is calculated by statistical processing of the results of series of successive observations. The standard uncertainty of type B is calculated other than in statistical way, for example basing on a priori specified source of uncertainty density distribution. The combined standard uncertainty is uncertainty that is determined in case if during measurement the effect of several uncertainty sources is simultaneously revealed or if obtained result is a certain function of other measurement results. The combined uncertainty is defined as the square root of sum of the squares of the particular standard uncertainties for the appropriate weight factors and eventual statistical relationship (correlation) between uncertainty components.

Hybrid approach of measurement result evaluation [3] that combines the error approach and uncertainty approach turned out to be the next step in the development of an integrated assessing the measurements accuracy. According to it, error is considered as the measurand with uncertainty determined by the assessment (evaluation or calculation). It reveals the possibility of simultaneous application of error and uncertainty approaches, which corresponds to hybrid approach of measurement result assessment. To wit, an error is being calculated and evaluated as a physical value whose particular coefficients are defined with some uncertainties (Fig. 5.1).

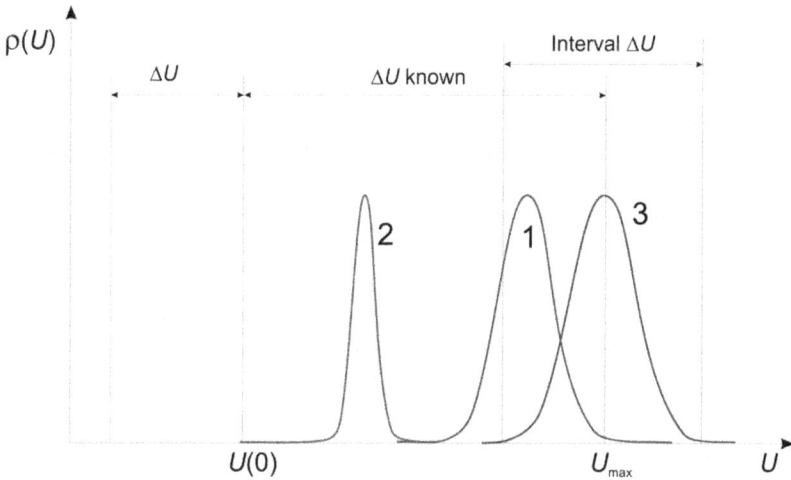

Fig. 5.1. Threshold weight of summary error and its uncertainty of result (3): systematic component due to impacts of MI and fluctuation of its properties (2); similar factor due to influence of thermometry-processed object (1).

Hybrid-thermodynamic approach of measurement result evaluation [4] is an extension of the hybrid approach towards consideration of the origins of fluctuation deviations in metrological characteristics on statistical-thermodynamic basis. Hybrid-thermodynamic approach of measurement result evaluation implies researching the total error of a temperature transducer with involving Non-equilibrium thermodynamics. In this case the threshold value of cognizable component of instrumental error systematic constituent is determined as the additive totality of multiplicative pairs of influence functions and their coefficients. Hereby, the pairs are formed so that one of the multipliers is determined by fluctuations of thermometric substance properties, and another – by that of applied outer field parameters. It corresponds to the content of Fluctuation-dissipation theorem of Irreversible thermodynamics.

The considered approach of measurement result evaluation has been developed consequently aiming to decrease the issues of MI intrusion and to improve the measurement accuracy of micro-, nano- object temperature. It roots in the threshold value determination of an instrumental error systematic component as an additional totality of influence-functions' multiplicative pairs (see below). However assessments of the origin of errors and uncertainties, based on thermodynamics, form the basis of hybrid-thermodynamic approach of measurement result evaluation. Its main reason roots in the next: the measurement result evaluation is quite good elaborated for macroobjects, having not been even established in the case of nanosamples. Nowadays the hybrid-thermodynamic approach of measurement result evaluation concerns with the study of origin sources of particular errors and influence functions and effectively applies in complicated cases of metrological reliability evaluation of MIs. In particular, the research of energy-transmission processes, based on statistical thermodynamics, enables us to determine a methodical error component as well as cognizable part of systematic component of an instrumental error component, and thus to decrease substantially the guaranteed by the producer of thermometric means a total error of measuring the temperature in working conditions.

Here the accepted IMC approach has been modified by the way of cognizing the certain components of an instrumental error through the extraction, study and evaluation of the factors influencing a MI, on the basis of statistical thermodynamic nature of their formation. The results of thermometric substance fluctuation concerning the summary influence function $\delta T_{Met\_max} = K_{\Sigma}$ of thermoelectric transducers at presence of external thermodynamic fields are determined as $K_{\Sigma} = (K_X + K_M)K_T$, where $K_X; K_M; K_T$ are the chemical, mechanical and thermal influence functions respectively caused by specific transport processes created by the external effect in thermometric substance. At the availability of fluctuations, additional impact functions (temperature, density, strain and etc. gradients) strengthen the influence actions related by the fluctuation effect of external environment up to: $K_{\Sigma}[F(T, p, V, ..., t)] = (K_X K_P + K_M K_{\Pi})K_T K_E$, where $K_P; K_{\Pi}; K_E$ are the recrystallization, porous and entropy influence functions respectively.

Joining in the pairs, where one of the multipliers is defined by the fluctuations of thermodynamic substance properties, and another – by those of the parameters of the applied outer fields caused by the thermometry-processed object, meets the content of the Fluctuation-dissipation theorem of thermodynamics. This approach is quite precious and allows determining the recognizable component of systematic error component of MI reducing significantly the guaranteed value of instrumental error.

The combined impact function of temperature measurement is defined within coverage interval received by summation from $\pm \left(\frac{K_P + K_\Pi}{K_X + K_M} + \frac{K_E}{K_T} \right)$ in the presence of two independent systematic constituents; by $\pm \left(\frac{\sqrt{K_P^2 + K_\Pi^2}}{K_X + K_M} + \frac{K_E}{K_T} \right)$ for the correlated constituents; to $\pm \left(\frac{\sqrt{K_P^2 + K_\Pi^2}}{K_X + K_M} - \frac{K_E}{K_T} \right)$ for uncorrelated values $(C_{cor} = -1)$.

As result, thermotransducers with the foreseen and manageable value of an instrumental error are developed on this basis. Thus firstly, the decrement of unrecognizable error component of nanoobject temperature measurement (absolute values, covering intervals and so on) has been reached, and secondly, the fluctuation restrictions of statistical physics for the improvement of metrological characteristics have been employed.

5.2. Reliability and Accuracy/Trueness of Measurements

One of the main metrological characteristics of MIs at periodic verification is reliability of measurement parameters that indicates the probability of not-exceeding by measurement error the permissible values with a certain probability $P(\Delta_\Sigma \leq \Delta_{al}) \geq P_{conf}$, where P is an actual value of probability; Δ_Σ is the results total error obtained by means of selected measurement units; Δ_{al} is the maximum allowable error of measurement result; P_{conf} is the given value of confidence. Impossibility of establishing the measurand's true value and accurate determination of measurement error as well as difficulty of taking into account all the possible destabilizing factors have been contributed in the normative documents published by IMEKO. To evaluate the measurement quality, the last applies the term "uncertainty" of received results, and also the recommendations to ensure the quality of both MI and of actual CPS performance or its final product.

Tolerance is the permissible limit or limits of variation in dimensions, properties, or conditions without significantly affecting functioning of equipment or a process [5]. Tolerances are specified to allow reasonable leeway for imperfections and inherent variability without compromising performance. For instance, tolerance classes of particular thermocouples type are shown in [6]. In nanotechnology the systems have dimensions in the range of nanometers. Defining the tolerances at nanoscale with suitable calibration standards for Traceability is quite difficult task for different nanomanufacturing methods. There are various integration techniques developed in the semiconductor industry that are used in nanomanufacturing.

Universal technique of errors identifying does not exist, because there is a wide variety of measurement methods, MIs, and conditions. Therefore, it should be carefully study the impact factors during the preparation of measuring experiment.

5.2.1. Methods to Improve the Accuracy, Errors and Examples of Reduction

There are developed a lot of different methods for improving accuracy. They are divided into three groups: methods of prevention of errors arising; methods of reducing the current errors; techniques of methodic errors reduction [5].

The first group includes structural and technological, protective and preventive methods. They prevent the occurrence of the error or do not allow it exceeded the permissible value. These methods are based on application of elements and components of highest quality with the most stable parameters. F.i., to reduce the temperature error, temperature-independent resistor is applied. Protective and preventive methods reduce the impact of external factors and consist in diminishing their impact on MI. Examples of such methods are: temperature control; magnetic or electrostatic shielding; stabilizing the power supply.

Methodical Error

- Methodical error of electric noise research caused by the improper technique or measurement means is one of determined components of methodical error that is due to the impossibility of increasing the integration time or bandwidth Δf for selective filter. It results in the dependence, f.i. in the close to cubic dependence of PSD $S_\omega(f)$ on frequency. The main reason is the Hrenander uncertainty principle: $t\Delta f = Const$. According to it, narrowing the filter bandwidth requires longer measurements, thus there remains the same referred component of an error. Narrowing the bandwidth at fixed duration or reducing the duration at fixed bandwidth of filter results in the significant uncertainty of noise measured PSD.

- Methodical error of electric noise research caused by the performance linearization while processing is a component of measurement error due to imperfect method or object discrepancy of model adopted for the measurement. More precisely it is caused by insufficiently correct interpretation of experimental results while further processing, or by their imperfection.

Stochastic systems are characterized by PSD $S_\omega(f)$, proportional to $1/f^\gamma$. This is the flicker-noise. Experimental data have revealed that PSD could be defined as: $S_\omega(f) = \alpha/f^\gamma$, where α is the constant; γ=0...3. For instance, our research has concluded γ=2.8 at the frequency band 3-12 Hz and γ=0.5 at 12-17 Hz for Pt; and γ=2.28 γ=0.9 for oxide resistor respectively. Considering the problem of thermal and low-frequency noises, we discuss the peculiarities of electron-phonon interaction by applying different approximations, regarding the possible types of adequate descriptions.

The measuring and processing of experimental results suggest the invariance of noise PSD $S'_\omega(f)$, cut by a filter within the certain bandwidth Δf. Thus, it does not take into account that PSD $S_\omega(f)$ is represented by the expression: $S_\omega(f) = \lim_{\Delta f \to 0} \frac{P_{el}(f,\Delta f)}{\Delta f}$, where $P_{el}(f,\Delta f)$ is the PSD at the frequency band from $f - \Delta f/2$ to $f + \Delta f/2$ reduced to approximation equation: $S'_\omega(f) \approx \frac{P_{el}(f,\Delta f)}{\Delta f}$. As a result, the additional error appears due to the linearization of previous expression

183

by the last one. It strengthens significantly the character of PSD dependence at frequency approaching to 0.

Conducted analysis for PSD spectral distribution by Debye model approximation has shown that error $\delta S_\omega(f) = {}^C/_{f^2}$ is methodical one. That is, the measured dependence of PSD noise ${}^1/_f$ is quadratically related to the frequency. Einstein model approximation within which the temperature dependence of PSD is absent (the case of thermal electric noise) allows to get rid of the methodical error $\delta S_\omega(f) = 0$.

Methodical error of temperature measuring in micro- and nano- world is an error caused by raising the significance of energy-transmission processes in the system "thermometer – controlled object" with decreasing sizes of object as well as thermometer.

Less the object we deal with, the more considerable methodical error of temperature measuring in micro- and nano- world is. Due to the intervention of sensor in energy exchange with controlled object it affects the gauge exactness, causing the emerging systematic component of methodical error. During prolong mutual contact of sensor and controlled object, while measuring, there was facilitated the determination of relative methodical error δT_{met} of temperature measurement, caused by heat transfer: $\delta T_{met} = \frac{(abh)_{sens}}{(ABH)_{ob}}$, where a, b, h are the linear dimensions of sensor, and A, B, H are the same of object. Hence, the relatively smaller sensor of MI, the smaller relative methodical error of temperature measuring. As result of prolonged thermal contact of warm sensor and cold controlled object, the latter is heated and the sensor is cooled, fixing the situation of heat exchange: $c_{ob}m_{ob}(T_x - T_0) = c_{sen}m_{sen}(T_{sen} - T_x)$. Here T_0 is the temperature of controlled object before measurement; T_x is the temperature of controlled object, which has established thermal contact with the sensor; T_{sen} is the initial temperature of sensor; $c_{ob}; m_{ob}; c_{sen}; m_{sen}$ are the specific heat and mass of the object and the sensor respectively. In this case, the sensor measures the averaged temperature of "controlled object – sensor" over the initial temperature of the first one.

Error depends on the ratio of volume or linear dimensions of sensor and controlled object. Let us consider that at comparable thermal characteristics of the object and the sensor, ratio of the volumes would be 1:1 (Fig. 5.2, a), 10:1 (Fig. 5.2, b) and 1:10 (Fig. 5.2, c) [7]. Thus, the sensor changes smoothly over time its own temperature from T_{sen} to T_x measuring the object temperature with certain error.

Expressing mass via specific density of matter w and its volume V and taking the object and the sensor uniform discoid shape (diameter D; d and height H; h, respectively), we obtain the equation of energy balance during prolonged contact of sensor and controlled object: $c_{ob}w_{ob}D^2H\Delta T_{met} = c_{sen}w_{sen}d^2h(T_{sen} - T_x)$. Dividing in $c_{ob}w_{ob}D^2H$ the left and right sides, we receive a relative methodical error of measurement:

$$\delta T_{met} = \frac{c_{sen}w_{sen}V_{sen}}{c_{ob}w_{ob}V_{ob}}\left(\frac{T_{sen}}{T_x} - 1\right)$$
$$= \frac{c_{sen}w_{sen}d^2h}{c_{ob}w_{ob}D^2H}\left(\frac{T_{sen}}{T_x} - 1\right) \tag{5.1}$$

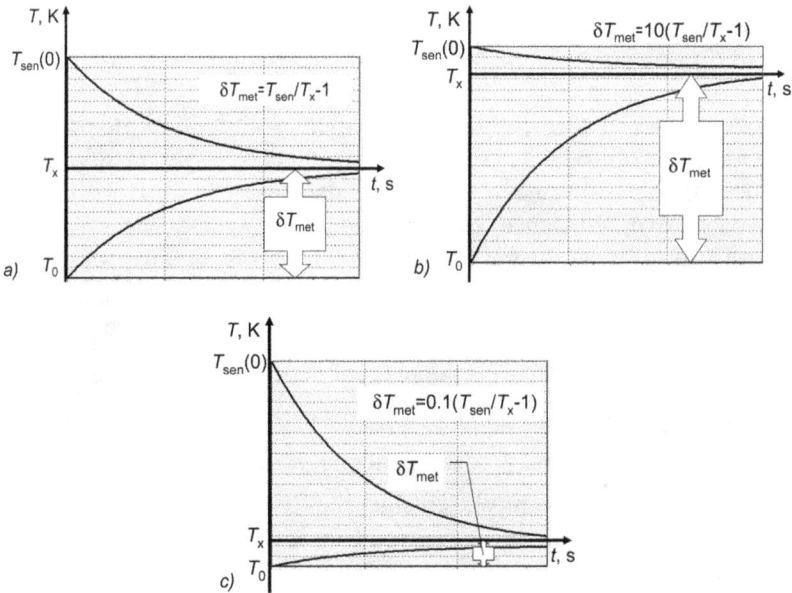

Fig. 5.2. Temperature vs. time changes of sensor cooling and object heating during prolonged thermal contact: a) $c_{ob}w_{ob}V_{ob} = c_{sen}w_{sen}V_{sen}$; b) $10c_{ob}w_{ob}V_{ob} = c_{sen}w_{sen}V_{sen}$; c) $c_{ob}w_{ob}V_{ob} = 10c_{sen}w_{sen}V_{sen}$.

Measurements of bulks assume by default that a sensor linear size does not exceed 0.1 linear size of controlled object, and the ratio of their volumes – 0.001. That defines a relative methodical error of measurement no higher than 0.1 %. This value loses in combined measurement error, including the instrumental constituent. Therefore is possible not to consider methodical error of temperature measurement. For nanosized sensors and controlled objects with comparable thermophysical properties (Fig. 5.2, b) relative methodical error is specified as $\delta T_{met} = \frac{T_{sen}}{T_x} - 1$. F.i., while controlling microobject temperature 270 K by commensurate-sized sensor of temperature 300 K, $\delta T_{met} = 0.11 = 11$ % is received. This concerns the methodical errors in nanoobjects temperature gauging with help of nanosized sensors.

Random Error

The notion "random" indicates that the measurements are inherently unpredictable, their results vary nearby the true value, and their average deviations are equal to zero for repeated measurements, performed several times with the same MI.

- Random error of temperature measurement with help of gas sensor is the random error which value is determined by volume of the gas thermosensitive substance. This error decreases to 0 if volume of thermosensitive substance of sensor increases, and vice versa: it increases if volume decreases. As the latter is mainly known (at Avogadro number $6.02 \cdot 10^{23}$ mol.$^{-1}$) it enables to express the equation with indication of numbers of gas moles n in sensitive element: $D[Q] = C_w m \frac{T_1^2}{nN_A}$, where m is the mass of thermosensitive substance. In the case of conversion to standard units of volume the next formula can be used $n = \frac{V}{22.4}$, where V is the concrete value of gas volume that is determined in m^3. Then we change the form of last equation to: $D[Q] = C_w m \frac{T_1^2}{N_A} \cdot \frac{22.4}{V}$.

Calculation of error that is specified by decreasing sensitive element chamber dimension can be done by the impact of temperature fluctuations or of heat quantity fluctuations. RMS deviation of heat quantity as function of chamber volume of sensitive element is:

$\sigma[Q] = \pm T_1 \sqrt{\frac{22.4 C_w m}{V N_A}}$. Relative RMS deviation is equal to: $\delta\sigma[Q] =$

$\pm \frac{\sigma[Q]}{Q} = \pm \frac{T_1}{T_2 - T_1} \sqrt{\frac{22.4}{C_w m V N_A}}$. Having substituted value of constants in it

we simplify the equation to: $\delta\sigma[Q] = \pm \frac{6.1 \cdot 10^{-12} T_1}{T_2 - T_1} \sqrt{\frac{1}{C_w m V}}$. Received

results of dependence of relative RMS deviations of heat quantity on the volume of sensor element under different mass indexes of its copper walls are shown in Fig. 5.3. Here is demonstrated that under significant decrease of fire sensor sensitive element dimensions (to 4 ml), the relative RMS deviation increases to ±0.007 %. Such value of random error is admissible for fire technology, where due to sensor sizes thermal inertia index is ≤ 1 s.

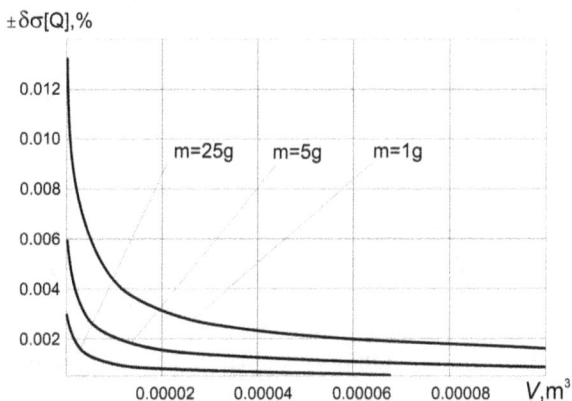

Fig. 5.3. Dependence of relative RMS deviations of heat quantity $\delta\sigma[Q]$ on volume V of sensitive element.

- Random error in noise measurement is an error that emerges in obtained results, and its value varies while performing the repeated measurements of the same quantity. The source of random error can root in the influence of proper noise of measuring systems, interferences and etc. The estimation of random error could be made due to the variance D or standard deviation $\sigma = \sqrt{D}$ of the received results. One of the methods of reducing a random error is the averaging of measurement results; particularly with the N-fold increase in the quantity of gauges, the standard deviation σ_{av} at $\sigma_x = Const$ decreases: $\sigma_{av} = \frac{\sigma_x}{\sqrt{N}}$. A random error could be reduced by enlarging the number of gauges just to the some extent. The matter is that random error is also considerably influenced by the state of object of measurement, namely thermodynamic state when the values of object parameters are the functions of time.

All these processes are specially complicated at the reducing of object size to a nano-area. In the case of the single measurements of unique properties, especially in nanotechnology, the theory of uncertainty could become expedient. Here the evaluation of a result is supposed to be made with some uncertainty determined by the effect of the same impact factors. Within the framework of an uncertainty approach the expounded above results could be reduced to extended standard uncertainty of the type A by introducing the factor $1/\sqrt{3}$.

- Random error in temperature measurement. Let us consider the possible realization of concrete gauge of certain duration concerning the object that is characterized by the given relaxation time. The most trivial case seems to be the study of relaxing thermometric properties, e.g. the research on fluctuation-dissipation changes in thermoelectric thermometers depending on annealing period at high temperature. Those changes are exponential, and we can reduce the coverage interval of transformation function drift by lowering the values of random error and increasing the reliability of measurement.

So, we may consider the response of substance as linear. Then the PSD $S_\omega(f)$ of the fixed fluctuations is proportional to the spectral absorption coefficient (Debye model): $S_\omega(f) = k_p \Pi_\omega(f) = \frac{k_p}{2\pi f}$, where k_p is the coefficient of measuring system power transfer. To wit, in consequence of stipulated application of Debye model, the frequency dependence of PSD appropriate for $1/f$ noises is gained. Stationary random processes with $1/f$-spectrum are characterized by the critical dynamics and scale-invariant fluctuation distribution. In those systems the energy of fluctuations could be accumulated at the low frequency bandwidth, increasing the probability of emergency emissions.

In the case of Einstein model, at the concentration of phonon energy on physically elementary volumes – tensile quasi-defects – an absorption coefficient is found: $\Pi = \frac{n\prime\hbar\Delta\omega}{2\pi k_B T} = \frac{n\prime\hbar\Delta f}{k_B T}$, and spectral absorption coefficient is proportional to PSD: $\Pi_\omega(f) = \frac{\Pi}{\Delta f} = \frac{n\prime\hbar}{k_B T}$, where $\Pi_\omega(f)$ is the frequency-independent which corresponds to the case of thermal noise. Experimentally fixed square character of $1/_f$–noise PSD could be caused by an instrumental measurement error; then the higher level of degree dependence up to cubic one would probably be related to the restriction of frequency-time analysis range and integration of gained signal at the measurement of substance remaining in a non-stationary disequilibrium thermodynamic state.

- Random error, dependent on quantity of noise measurements; it is a value that decreases with increasing the number of measurements in different ways depending on the type of noise. That is notified (Fig. 5.4) for "white" noise (WN), flicker-noise (FN) and "white" noise with a flicker-component (WN+FN).

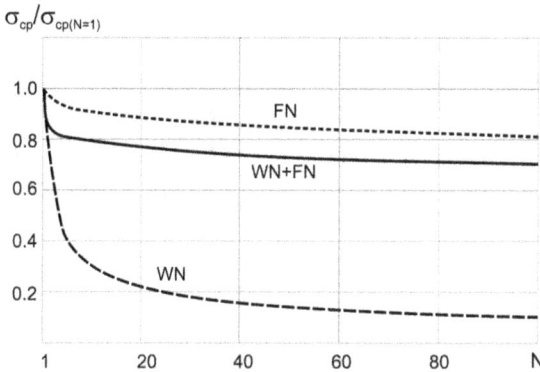

Fig. 5.4. Random error dependence on the number of gauges at $f_h = 10^{-6}$ Hz and $\Delta t = 1$ s for different types of noise.

We can see there that the averaging of the results of 100 gauges produces the 10-fold reduction of the error in the case of "white" noise, ≈ 1.2-fold reduction in the case of flicker-noise and ≈ 1.4-fold reduction for "white" noise with flicker-component. Thus, the random error could be reduced to the negligible value only if the spectrum of the measured value is the same within the frequency bandwidth from 0 to ultra-high frequencies.

The results of measurements are proved the following. Hereby, the interval of time between results of measurement is chosen from condition: $\Delta t = \dfrac{1}{f} = \dfrac{1}{2f_h}$, where f_h is the upper frequency in spectrum of measured value. Most gauges are made in the static mode of the measured value, hence $f_h \rightarrow 0$ and the flicker-component becomes of importance in the spectrum. Hereby, the error of measurement could not be reduced to infinitesimal value by the method of averaging the results of measurement.

5.2.2. Duration of Noise Signal Gauging

Duration of noise signal gauging is stipulated by random nature of the measured voltage or current noise signal. Each one has a nature of homogeneous continuous random fluctuations concerning the average that is up to zero and constitutes a random ergodic stationary process. While studying it, any moment of time can serve as starting point. Measuring the parameters of stationary process within any period of time, we should receive the same values of characteristics.

Such integral characteristics of random process as the mean of a square (variance in statistical investigations), RMS (standard deviation) and spectral density of noise signal tend to be measured firstly. Since the noise signal is a random process, its true value could be gained during the infinite time of averaging. Any restriction on the averaging time leads to the appearance of a methodic error. In the ideal case, if there are no other noise signals excepting the measured one in the measuring circuit, the standard deviation of noise signal variance σ could be calculated as: $\sigma \approx \frac{1}{\sqrt{t\Delta f}}$, where t is the duration of measurement, Δf is the bandwidth of noise signal. Results of modeling the dependence of standard deviation of noise signal variance on the time of averaging at the different values of bandwidth Δf are notified in Fig. 5.5.

To reach the relative RMS value of the variance of noise signal 0.01 % for the bandwidth $\Delta f = 100$ kHz, we should conduct measurements throughout 1000 s, and for $\Delta f = 1$ MHz – 100 s.

Taking into account that other sources of noise signals (resistance of a connecting line, amplifiers, and feedback resistors) are present in the input circuit, the dependence of RMS of noise signal variance on time of averaging becomes more complicated. Time of measurement for reaching the equal error rises as compared to an ideal case. Correspondingly, measurement of integral characteristics of noise signals could take a lot of time for averaging – up to tenth – hundredth of seconds.

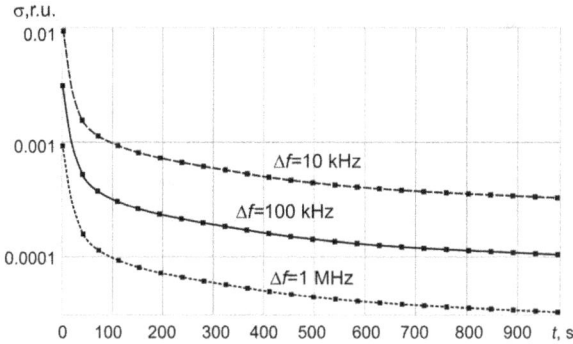

Fig. 5.5. Dependence of a standard deviation of a noise signal variance on the time of averaging.

If there is a necessity for measuring the integral characteristics within narrow bandwidth, the time of measurement rises considerably. So, to reach the relative RMS of noise signal variance 0.1 % at the bandwidth $\Delta f=10$ Hz, measurement should last approximately 30 hours.

5.3. Metrological Problems of Raman Thermometry

Raman thermometry is elaborated insufficiently due to the novelty and uniqueness of method. This problem is considered below basing on error, uncertainty and other approaches of metrology.

Within error approach, Raman thermometry considers few components of the combined error of temperature measurement. The first one is instrumental and could be determined by the accuracy of measuring device. The second one is methodical and is caused by heating the object during the process of measurement. There exists the third component that is also of instrumental type and is related to the changes in feeding parameters during the measurement. There is also the fourth component caused by the instability of a surface and adjacent layers as result of light beam effect (close to drift error).

Instrumental error with systematic and random components is caused by the fluctuation of number and frequency of scattered, especially anti-Stocks, quanta. To reduce this error, is necessary to perform the signal time-averaging. At light exciting within Raman method with lasers of different wavelengths these effects are expressing themselves in various ways. Therefore the instrumental errors are distinguished at different wavelengths. While using two lasers different wavelengths we gain two diverse instrumental errors with different results dispersion.

Errors of photocurrent measurement depend on metrological characteristics both of laser and spectrometer. Moreover, their stability is quite important: there exist instabilities caused by the errors of setting and determining the certain value of irradiation. Fortunately at serial measurements of anti-Stocks and Stocks signals the given error components compensate each other. Under the condition that photo detector sensitivities at Stocks and anti-Stocks frequencies slightly differ we could adopt that $\delta_{is} \approx \delta_{ias}$. By neglecting the slight deviation of SL 03/1 laser frequency with the wavelength 632.9910±0.0002 nM, we could advance that the instrumental error is defined as:

$$\delta T = \frac{2}{Z} = \frac{1}{ln\,{}^{i_{as}}/_{i_s} - 3\,ln\,\frac{v_i - v_0}{v_i + v_0}}, \qquad (5.2)$$

where i_{st} and i_{as} are respectively the intensity of Stocks and anti-Stocks components of scattered radiation, v_0 and v_i are the wavenumbers of reflected phonons at the given number of dispersed phonons and used laser bunch respectively.

Methodical error related to heating the researched object by laser irradiation depends on its surface energetic luminosity. Even the effect of under-powered laser leads to surface heating and arising methodical error (~27 K), which could reach much larger values for small objects. We should indicate that methodical error as well as instrumental one encloses the cognizable and incognizable components (Fig. 5.6).

The drift is caused by the changes both in the chemical composition and surface shape of measured object as consequence of intensive irradiation. It is related to complicated transform processes in surface-adjacent object layers and is attributed to the systematic constituent of instrumental error. Apart from it, the latter is treated as partly cognizable due to its different influence factors. Their estimation is carried out involving the thermodynamics of irreversible processes: the more intensive irradiation and the larger methodical error are, the stronger entropy changes and the larger drift occurs. To reduce this error, the known in metrology method of nearing to measuring point from both sides could be successfully applied. Adopting that under the condition of linear alteration in time the calculated frequency of reflected anti-Stocks component behaves similar. So in order to get rid of drift, becomes necessary to measure the anti-Stocks frequencies before and after measuring the anti-Stocks frequency, and then to average the anti-Stocks frequencies results.

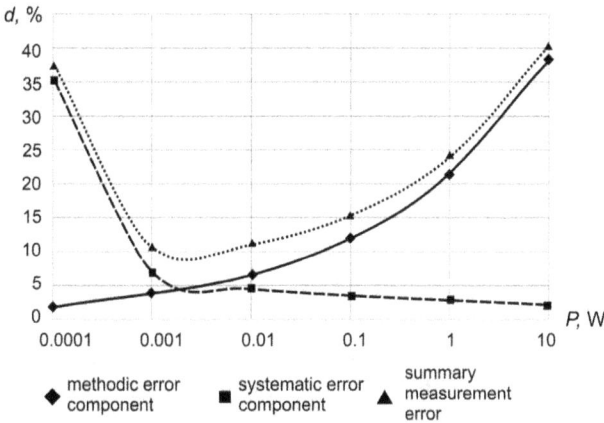

Fig. 5.6. Optimization of methodical and instrumental components in decreasing the combined error of temperature measurement at 2 mW laser power.

Within uncertainty approach in Raman thermometry, the measurement of chips temperature in the process of their manufacturing has performed only once, since the next measurements are usually realized in other conditions. For one-time measuring an error approach is not quite adequate.

The estimating concept for measurement results could be based on the uncertainty approach. Here the methodic error is being calculated and evaluated as a physical value, which certain components and coefficients could be estimated with a certain errors. To wit, this constituent of the combined error is considered with particular uncertainty. We take into consideration the peculiarities of both MIs and standard patterns. For instance, with help of Raman method the measurement of CNTs temperature within the range 30...250 °C is made. These tubes are treated as standard nanopatterns for testing and calibrating the nanotechnological means. Hereby, to study the action of seven and more possible influence factors (angle of light bunch incidence, distance to a photo-receiver, exposure time, duration of spectrum passing, power and mode of laser operation, drift characteristics and so on), 28000 gauges have been performed, enabling us to ascertain the following indices of the measurement accuracy.

Approach of errors is applied to results processing, consequently of which one of the gained results (with introduced correction to systematic error) looks as $T_{real} = 287.27\ K \pm 1.72\ K\ (\pm 0.6\ \%)$. At the same time, due to uncertainty approach, the gained result makes $T_{real} = 287.27\ K$ with ecombined standard uncertainty 0.3 % at the credence level $P = 0.95$, expanded coefficient 1.96 and efficient value of freedom degrees 130.6.

5.4. Temperature Dependent Precision Threshold

Accuracy threshold is due to the fluctuation deviations in the processes which determine the metrological characteristics of the extra-sensitive MIs. Due to such great sensitivity, the accuracy threshold becomes temperature dependent since fluctuations intensity depends on temperature.

Precision Threshold of Sensitive Balance

Main constructive unit of torsion balance is a thin thread on which a light mirror hangs. It should be noticed the same part is the basis for a ballistic galvanometer construction. Molecules thermal motion of the environment leads to irregular in time molecules bombarding of the mirror that limits instrument sensitivity and does not let to better the measurement accuracy. Thread torsion module is $a = \frac{\pi^2 d^2 G}{8l}$, where G is the shear modulus; d and l are the thread diameter and length. Then moment of force that effects on the thread is linked with rotation angle φ by the next ratio: $M = a\varphi$, and the potential energy of curled thread is $U = {a\varphi^2}/{2}$. In accordance with Boltzmann formula, the dispersion of value of angle close to which the mirror vibrates is equal to:

$$D[\overline{\varphi^2}] = D\left[\frac{\int_{-\infty}^{\infty} \varphi^2 e^{-a\varphi^2/2T} d\varphi}{\int_{-\infty}^{\infty} e^{-a\varphi^2/2T} d\varphi}\right] = \frac{T}{a}.$$ Obviously the RMS deviation of this

angle is equal to: $\sigma[\varphi] = \left(\frac{T}{a}\right)^{1/2}$. At room temperature when a $\sim 10^{-13}$ J, the mirror rotation angle RMS deviation is determined as $\sim 10^{-4}$ radian.

194

This is a real limit of single measurement sensitivity for practically all MIs in nanometrology.

By the same way the fluctuations impact on metrological parameters of a spring balance with coefficient of elasticity k and equilibrium stretching X_0 is considered. Mass center oscillations occur in it as result of the temperature fluctuations presence. That's why counting of equilibrium position of the pointer X_0 cannot be made more accurate than with the RMS deviation of absolute value of the instrumental error random component: $\sigma[X_0] = \sqrt{(\Delta X_0)^2} = \pm\sqrt{T/c}$, where c is the constant that conjugates mechanical qualities and sizes. On this basis let determine the RMS deviations of absolute and relative value of instrumental error random component of mass determination:

$$\sigma[m] = \pm\frac{c}{g}\Delta X_0 = \pm\frac{k}{g}\sqrt{\frac{T}{c}}$$

$$\delta\sigma[m] = \pm\frac{\sigma[m]}{m} = \pm\frac{1}{mg}\sqrt{\frac{T}{c}}$$

(5.3)

Hence, the instrumental error random component is smaller as the spring is weaker. However, in this case the equilibrium stretching increases: $X_0 = \frac{mg}{k}$. It specifies practical inconvenience of balance construction. Hereby temperature dependent fluctuations limit the metrological characteristics of balance.

Precision Threshold of Sensitive Ballistic Galvanometer

In electrical measurements, fluctuations specify the absolute error independent of MI perfection state. So far as the ballistic galvanometer is used as supersensitive mean for small values of impulse current measurement, it is considered in details. The galvanometer current I is measured by the mirror deviation angle φ. In equilibrium state when spring forces moment cφ is equal to electromagnetic forces effect moment γI, the mirror rotation angle is $\varphi_0 = \gamma I/c$ (here c, γ are the constants).

RMS deviation estimated by the mirror rotation angle is in line by its content with the similar estimation of the instrumental error random component of spring balance. In such a way it was derived the RMS deviations of absolute and relative values of the instrumental error random component of the current determination:

$$\sigma[I] = \pm \frac{c}{\gamma} \sqrt{\frac{T}{c}}$$

(5.4)

$$\delta\sigma[I] = \pm \frac{\sigma[I]}{I} = \pm \frac{\sqrt{cT}}{\gamma I}$$

Hence, for the galvanometer accuracy improvement it needs to take smaller value of constant c and higher value of constant γ (otherwise to increase the number of winds in the galvanometer current coil). This leads to the equilibrium angles φ_0 deviation that contradicts with springiness demands of the hanging thread deformation $\varphi_0 \ll \pi/2$. Therefore, temperature fluctuations limit the galvanometer accuracy: $A = 1/|\delta|$. Then the accuracy limit which can be gained in measurements is determined by assigned in advance sensitivity.

5.5. Dynamic Error and Instrumental Error

It is considered (on example of thermotransducer) as an error caused by heat inertia of transducer and inertia of measuring device, and is equal to the difference between transducer's error in variable temperature mode and its static error.

- Dynamic error emerges when the transducer has not enough time to follow the rapid temperature changes of controlled object. Exactly this delay characterizes thermal inertia index. If to consider that temperature in the cross area of transducer is uniform and heat removal and radiation exchange are absent, it becomes possible to study the nature of change in temperature on the basis of elementary theory of thermal inertia for uniform transducer and express it by expression:

$$\Delta T_{dyn}(\tau) = T_T(\tau) - T_0(\tau) = -\frac{1}{m} \cdot \frac{dT_T(\tau)}{d\tau},$$

(5.5)

where, $T_T(\tau)$ is the temperature of sensitive element of transducer, $T_0(\tau)$ is the object temperature, τ is the time, m is the parameter which characterizes the rate of heat exchange due to convection (Fig. 5.7, a-d).

- Instrumental error of NT is a component of measurement error due to its intrinsic properties. It may contain few components, including error of measurement and the error caused by the interaction of transducer with the object of measurement. For example, 100 Ω (at 27.15 K) sensitive elements of NTs were made from pure Ni, Pt, Cu; alloys (Ni-Cr and composites of various oxides). Research has been performed at reference points of ITS (4.2 K; 77 K...273.15 K) according to IMEKO method, and at higher temperatures. Revealed deviations from linearity of calibration characteristics as the relative error δT increase from Cu (0.05 %) to Ni (0.26 %) sensitive elements. Mentioned deviations have not been fixed for elements made from transition metal alloys and composites.

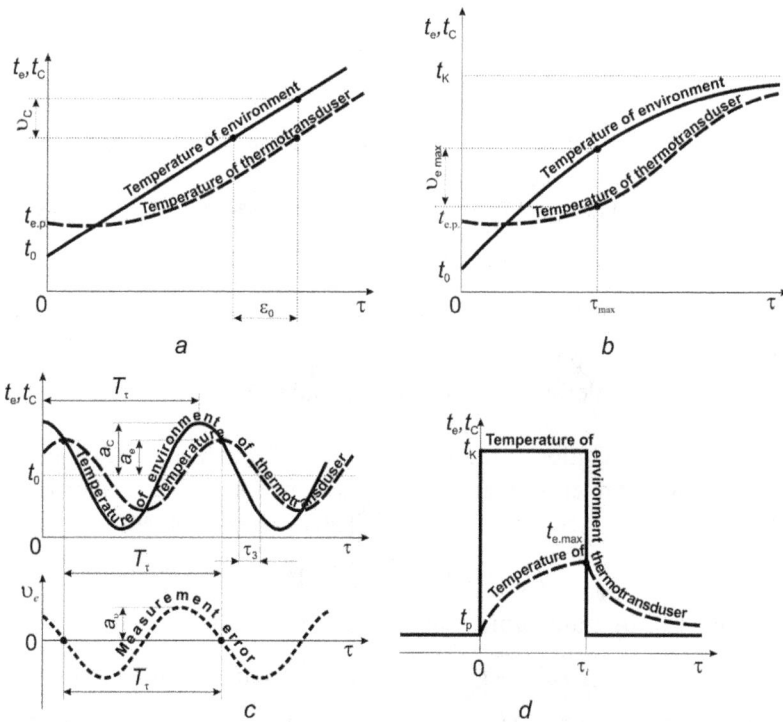

Fig. 5.7. Variations in thermotransducer's temperature under ambient temperature changes [7].

197

Analysis of measurement error of NT has shown the additional component existence that goes beyond a basic acceptable error. This error is caused by structural processes in sensitive element due to its manufacturing (bending, tension). The constancy of research temperature – 77 K – does not mean the thermodynamic equilibrium state of sensitive element and environment. The relaxation of nonequilibrium thermodynamic state depends on several factors (temperature, time, type and concentration of defects). Otherwise, condition of thermodynamic equilibrium, at which Nyquist formula is derived, has been broken in the case of a real NT wound at 300 K and used at low temperatures.

Temperature dependence of electrical noise power is derived directly from the basic equation of thermodynamics. In the stationary nonequilibrium state, the thermometer calibration characteristic nonlinearity appears due to violation of energetic processes of "environment – thermometric substance" exchange. It is expressed by the instrumental (absolute ΔT and relative δT) error as:

$$\Delta T = T_c - T_r = (b_c - b_r)P_{el}; \ \delta T = \frac{b_c - b_r}{b_r} = \frac{\Delta b}{b_r}, \quad (5.6)$$

where T_c and T_r are the estimated and real temperature respectively; b_c and b_r are the constants of estimated and actual calibration characteristics. The prolonged use of NT leads to maximizing the constant $b \sim \left(\frac{dS}{dt}\right)^{-1}$ at minimizing the entropy dissipative flow: $\frac{dS_\vartheta}{dt} = \frac{\Delta S}{\tau_p} = min$. Nonstationary nonequilibrium thermodynamic state corresponds to the power change of nonequilibrium electrical noise in the elastic-plastic deformed thermometric substance.

Therefore its relaxation effects lead to error emergence. Substance of density ρ rapidly releases the previously accumulated elastic energy with appearing microcracks of length 2l. Thus, relaxation constant τ_1 is estimated as $\tau_1 \sim l\sqrt{\frac{2\rho E}{\sigma^2}}$, where $\frac{\sigma^2}{2E}$ is the density of elastic energy. The latter can be transformed into surface microcracks energy with its relaxation constant τ_2: $\tau_2 \sim l\sqrt{\frac{\rho l}{\gamma}}$, where γ is the specific energy of surface tension; or removed from the relaxation place with constant τ_3, linked to thermal diffusivity a: $\tau_3 \sim \frac{l^2}{a}$. At temperatures lower than 20-

30 K, thermal diffusivity in 100 and more times is higher than at 300 K. So τ_3 is much smaller in comparison with the relaxation constant of motion τ_4 $\left(\tau_4 \sim \lambda/\omega\right)$ or reproduction τ_5 of dislocations ($\tau_5 \sim \left(\frac{\rho l^3}{n L_d E_d}\right)^{1/2}$), where λ is the effective length of dislocation run; ω is the dislocation velocity; L_d is the typical size; E_d is the dislocation energy, referred to one interatomic distance.

Effect of the mentioned above constants $\tau_1 \dots \tau_5$ is combined, and depends on the temperature and background of substance, forming the total relaxation constant $\tau_{n.st} = \frac{1}{\sum_{i=1}^{n} 1/\tau_i}$. Consideration of competitive effects of constant τ_2 due to microcrack formation, and constant τ_3 due to heat removal from the place of energy relaxation produces the modified constant $\tau_{st} = \frac{\tau_2 \tau_3}{\tau_2 + \tau_3}$. The joint effect of these two mechanisms creates the reasons for changes in the electrical noise power and thus changes in the readings of NT. Hence, the error of thermometer, whose sensitive element is in stationary nonequilibrium state, is determined by the competitive action of two major in those conditions dissipation processes that form deviation from the calibration characteristics: $\delta T = \tau_2/\tau_1 = A(ad)^{1/2} c/\chi$, where C is the sound velocity; a is the grain size; d is the atomic size, χ is the thermal diffusivity. Hence, the lower speed of sound and higher thermal diffusivity, the more efficient work mechanism for a heat removal and the less noticeable influence of dislocations on the electrical noise power and consequently on NT error.

Described before concerns pure metals and is not related to alloys and composites due to significance of the process of dislocation multiplication (constant τ_5) that occurs in their blades and is accompanied with the microcrack formation. Finally, high temperatures up to a melting point are matched with diffusion removal at the relaxation constant τ_4. That is, in the high-temperature case, one should consider the competitive action of two relaxation mechanisms: diffusion mechanism and formation of microcracks in the deformed local substance microvolumes. Introduced before criterion is varied at the high temperatures to: $\tau_4/\tau_1 \sim (ad)^{1/2} c/D$. Here coefficient of diffusion D which increases exponentially with temperature means deviations. It proves the absence of calibration characteristics at high temperatures.

- Instrumental error of thermoelectric thermotransducer. Its study has been completed by elaboration of algorithmic principles of thermotransducer error minimization realized on basis of thermodynamic forces and fluxes consideration in sensitive substance (Fig. 5.8).

Consequent evaluation of certain influence functions due to complicated transfer processes described by corresponding freedom degrees in basic equation of thermodynamics is realized. Preliminary algorithm settings comprise the values of: 1) transformation function and its dispersion; 2) Temperature range, environment, exploitation time and mode; 3) Peculiarities of sensor substances and thermotransducers manufacturing. Moving along the chart, from magnetic freedom degree, alternately estimate the effects of influence functions of all possible degrees on the transformation function [4].

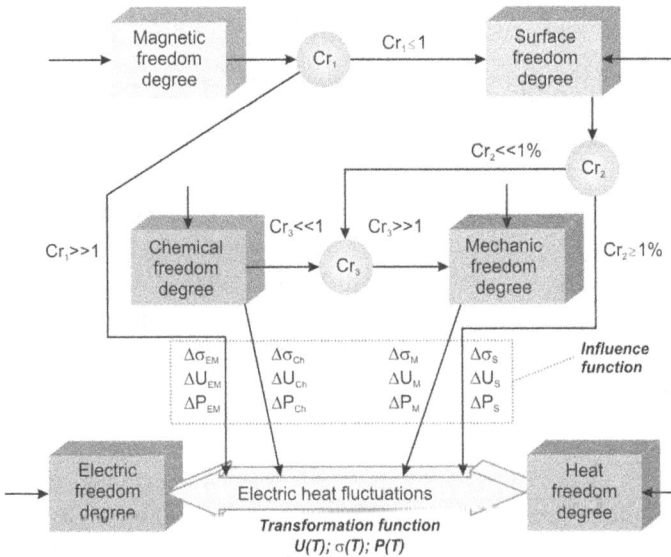

Fig. 5.8. Transformation and influence functions of thermotransducers: thermoelectric ΔU; resistive $\Delta \sigma$; noise ΔP.

5.6. Methods of Correction and Statistical Minimization of Errors

Mentioned methods are directed at reducing of already existing errors. Adjustment (or functional minimization) is considered to be the method that reduces errors, mainly systematic ones, by means of analytical or experimental study. Under statistical minimization we understand reducing the expected but not identified measurement errors; it is carried out both during and after the measurement (error reduction by spatial or temporal averaging). Examples include: reducing the random errors of the multiple measurement results by time or spatial averaging; reduction of quantization error [1].

For MIs calibration, the direct measurement by verified MI of outgoing signal or by multivalued measure with determination of the error as a difference can usually apply. Correction methods of systematic error constituent are realized by operator or automatically in off-line mode when, f.i., self-calibration is carried out.

Errors adjustment with the operator participation can be fulfilled in 2 ways. The first one is the calibration of MI (Fig. 5.9).

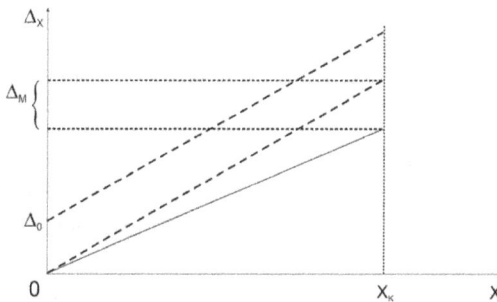

Fig. 5.9. Measuring instrument error adjustment by calibration [1].

To correct the dominant error additive and multiplicative constituents the instrument calibration is usually performed in two points of scale: at shorted input and at supplying the measure's output signal to input of verified MI. At shorted input, the operator sets the zero readouts of the mentioned device, then at connected measure with help of calibration unit sets the readout that corresponds to submitted value of measure.

Calibration without action on the MI is performed by introducing amendments, or during the measurement results processing. For example to correct error additive constituent, two measurements are performed; two indications are recorded: $y_1 = kx + \Delta_0$ and $y_2 = \Delta_0$, and the measurement result is computed as $y = y_1 - y_2 = kx$. Automatic correction aims the amendments introduction into device structure or into measurement algorithm.

There exist some specific measurements which include the next methods of error adjustments.

Auxiliary measuring method is the version of invariance principle, according to which are needed as many additional measuring channels as the impact values exist. Iterative method consists in the multiple specifying of adjustment results performed by successive approximations. Therefore it requires the precise feedback transducer. Method of standards establishes the real conversion performance by connecting a set of standards (or one multi-level standard) to input of MI.

When to turn off the physical quantity from device input or realize its set of measures is impossible, the test methods are applicable. The latest generate the test values involving both measured and model quantities.

Reducing the Methodical Errors

Due to complexity of setting the correct experiment, inevitably arise methodic errors caused by inadequate of considered method to real conditions of measurement. They can include an incorrect transmission function and mismatching the characteristics of different MIs. To correct methodical errors, detailed study of conditions and nature of these instruments should be performed. To reduce some methodical errors, the special measurement methods have been developed: method of substitution, method of error compensation by sign, method of contradistinction, method of symmetrical observation, etc. [5].

Method of substitution consists in submitting the initially measured value to input of MI. Subsequently this value is replaced by the appropriate measure of known value at which the readout of instrument remains unchangeable.

Method of error compensation by sign consists in double measuring of the same value at variable measuring conditions in a way that unchangeable systematic error would be included in the measurement result with the opposite sign.

Method of contradistinction consists in the double gauging the measured value; firstly it is compared with the value that is reproduced by measure; and before the second comparison these two values are mutually changed in measuring circuit. So, result of measurement becomes independent of the transfer factor of measuring circuit.

Method of symmetrical observation is as follows. First value X is measured, then after a time Δt full or partial substitution with measure of known value X_M is performed; over the time interval Δt measurements is repeated again. This excludes the permanent and linearly-dependent systematic error constituents.

Eliminating the Systematic Errors

During analysis of adjusting methods, the absolute value of the combined error of MI is conveniently to divide in three components:

– Additive component Δa, independent of X; it is also named "zero error" (it occurs if MI registers the certain readout when the latter should be zero) and causes concurrent shift of the MI characteristics; this kind of error can be easily detected at $X = 0$;

– Multiplicative component $\Delta_M = \delta_s X$, proportional to X. It is known as "sensitivity error" that causes the MI specifications rotation concerning the zero of coordinates; this kind of error be easily detected while applying the measure or scale transducer;

– Error $\Delta_{non-lin}$ of non-linearity of MI characteristics that non-linearly depends on X and may be efficiently detected while applying the multi-level measure or scale transducer in measuring circuit.

Common Methods

- Amendments. Action of systematic and other regular (e.g., linear-increasing over time error or drift) influences on the received result is reduced by using appropriate types of result adjustment or by introducing the amendments. For this, a variety of methods is developed; most of them are based on performance of additional measurements with applying the so-called standard values, i.e. quantities with certified magnitude. Adjusted result (x_{am}) is obtained by adding to the measured value (x) the amendment p which is equal to the corresponding systematic error component with opposite sign: $x_{am} = x + p$.

Introduction of amendments is compulsory for each stage of results processing. However, it is virtually impossible to fully explore the effects of systematic or regular impacts, at least due to the absence of ideal MIs, presence of random influences or time restrictions. Therefore complete correction of systematic effects is impossible. Error can be reduced by adjusting the results only if the relationship between the impact factor and output value is known. Cold-junction compensator is a brief example of such device (Fig. 5.10).

It carries out the compensation of cold-junction temperature of thermocouple, or adjusts its shifted readouts. Electronic means can also compensate the similar errors for thermocouples of various types, and so reach the improvement of accuracy. Also, bridge scheme is designed so that, when changing ambient temperature and therefore cold-junction temperature, it could provide adding the voltage proportional to mentioned temperature to thermo-EMF.

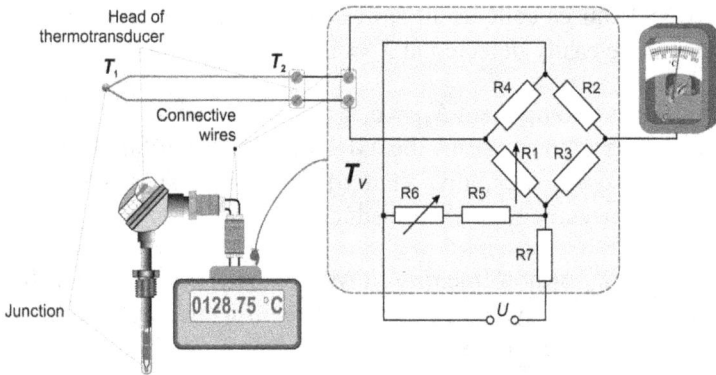

Fig. 5.10. Universal automatic gauge with cold-junction compensation T_V with help of the set of resistance thermometers.

- Processing the Measurement Results. When $Y = kX + \Delta Y_a$, the additive error ΔY_a can be excluded by performing one additional observation at $X = 0$, and the following subtraction. If the additive error exists at $X = X_1$, the output value of device is equal to $Y_D(X_1) = kX_1 + \Delta Y_a$. Then at $X = 0$: $Y_0 = \Delta Y_a$. After subtraction we get adjusted value of measurand:

$$Y_{adj}(X_1) = Y_\Pi(X_1) - Y(0) = kX_1 + \Delta Y_a - Y(0)$$
$$= X_1, \tag{5.7}$$

Multiplicative error δ_m is excluded via single-channel fixed measure by calibrating the MI at the given value X_0 and subsequent dividing and multiplying: $Y_1 = kX$; $Y_2 = kX_0$, whereof $X = X_0 \, {Y_1}/{Y_2}$. If additive and multiplicative constituents of error in MI readings exist, they are also excluded via similar measure of fixed value by means of two additional measurements at X equal to 0; X_0. So, correct result Y_{cor} is defined by subsequent computing:

$$Y_{adj} = X_0 \frac{Y_2 - Y_0}{Y_1 - Y_0} \tag{5.8}$$

In the case of nonlinear transfer function $Y = k_n(1 + \delta_s + \varepsilon X)X + \Delta Y_a$ of MI, the problem arises of selecting the optimal calibration value X_{cal}. Firstly, you must reach the readout of MI at zero $(Y = 0)$ for $X = 0$. During instrument calibration its transfer function is approximated by the linear dependence:

$$Y_k = k_n(1 + \delta_k)X_k = k_n(1 + \delta_s + \varepsilon X_k)X_k, \tag{5.9}$$

where δ_a is the adjusted relative error at $X = X_k$. Let us divide at $k_n X_k$ both sides of the equality and bring relative error of MI to input $\delta_k = \delta_s + \varepsilon X_k$. Its absolute error would be ΔX (Fig. 5.11):

$$\Delta X = \frac{\Delta Y}{k_n} = \frac{k_n X(1 + \delta_s + \varepsilon X) - k_n X(1 + \delta_k)}{k_n} \tag{5.10}$$
$$= \varepsilon X(X - X_k).$$

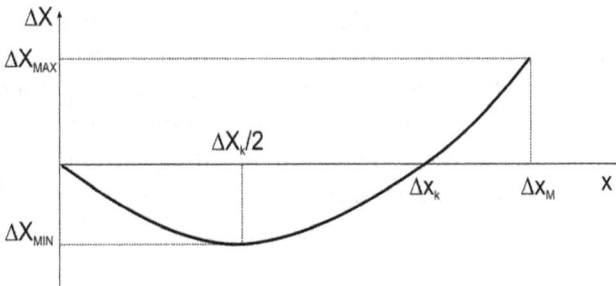

Fig. 5.11. Absolute error of MI reduced to input [1].

Minimum and maximum errors are respectively defined at the top of parabola $\left(X = {X_k}/{2}\right)$ and at the end of measuring range $X = X_m$:

$$\Delta X_{min} = \frac{\varepsilon X_k}{2}\left(\frac{X_k}{2} - X_k\right);$$

(5.11)

$$\Delta X_{max} = \varepsilon X_m(X_m - X_k)$$

The condition of minimizing error over the entire range of MI is $|\Delta X_{min}| = |\Delta X_{max}|$, and at $X_k = \alpha X_m$. We obtain quadratic equation $\alpha^2 + 4\alpha - 4 = 0$, the physical meaning of which has just positive root: $\alpha = 2(\sqrt{2} - 1) = 0.82$. So, for a quadratic approximation of transfer function of MI its calibration should be performed at the point $X_k = 0.82X_m$.

Calibration of MIs is performed by changing its sensitivity, or by altering the tilt of its characteristic. It is especially effective at predominance of error multiplicative component and can be implemented by operator for circuits with measures, with working standard MI, or with model reverse converter and calculating unit.

If there is multiplicative error component δ_m, the equation for MI readout is presented as $\alpha = k_1 X = k_0(1 + \delta_m)X = k_0 X + k_0 \delta_m X$, where k_0 is nominal transfer factor. While calibrating (at applied to input of such device the measured value X_0), the operator changes factor k of device until at $X = X_0$ readout becomes equal to α_0. The last is usually marked on the scale with red tag, or corresponds to the end value of scale. As a result of the calibration, coefficient becomes equal to $k_0 = {}^{\alpha_0}/_{X_0}$.

Calibrating the MI with help of the working standard is mostly performed at X-value close to X_0. During calibration, the coefficient is changed as long till readout of calibrated MI do not match the readout of standard which (with model reverse converter and computing chip) is especially suitable for calibration of measuring transducers. While their calibrating, we change coefficient k till the error Δp in output of reader would be established at the zero. In this case, $X = X_k = {}^Y/_{k_0} = {}^{kX}/_{k_0}$.

Special Methods

Special methods for improving the accuracy are: by-sign-errors compensation; method of contradistinction; method of symmetrical observations; method of transposition.

- Method of Errors Compensation by Sign. Two measurements of the same value are fulfilled with such changeable measurement conditions, but in the second time, the unchangeable systematic error of measurement has to be included in result with opposite sign. So, in the measurement of voltage the result is received by two cycles (for identical and opposite polarities of additionally installed switches). Then if for measuring time the parasitic EMFs are immutable, the result becomes independent to their values.

- Method of Contradistinction. Measured value X is compared twice with adjustable measure X_k, and these two values (X and X_k) are swapped before the second measurement. For instance, classic metrological task is the definition of mass in inaccurate balance. Here the result of the measurement $M_x = \sqrt{M_{N1}M_{N2}}$ is derived, considering the system of two equations (Fig. 5.12):

$$M_x L_1 = M_{N1} L_2; \qquad\qquad (5.12)$$

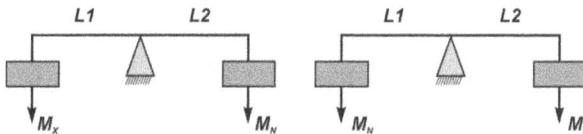

Fig. 5.12. Improving the accuracy of masses definition by means of inaccurate balance.

- Recent example of method application in stepless Z-shift regulation of nanomashines. During the creation of hydraulic positioning unit along the axis Z of nanomachine with providing stepless shift and position control possibility along this axis, it is suggested to use hydraulic potential. To this purpose the U-shaped hydraulic construction is proposed with ends of large (D=20 mm) and small (d=0.3 mm) diameter. Diameters ratio is equal to: $D/d = 66.66$. Spontaneous or enforced liquid level shift, for instance, in the small diameter tube, is detected by means of micrometer head (Fig. 5.13).

In the motionless fixed tube its pivot sinks by the shift of micrometer head pressing hydraulic liquid to the wide cylinder. Consequently the liquid level is increased in this end – on Δh_1. It results to weak but appreciable liquid level increase in the wide end. In accordance with joined vessels law the level increases on ΔH_1 in the wide end. Floating plate mounted in this wide end lifts the same as the studied nanoobject mounted on it. In this case the condition of liquid quantity invariability under pouring from one end to another one can be described by: $\frac{\pi}{4}D^2\Delta H = \frac{\pi}{4}d^2\Delta h$, we use $A = {D^2}/{d^2} = 4444.4$ as constant for this unit construction. So, we have got the formula specifying the level drops changes in wide (ΔH_1) and narrow (Δh_1) device ends, as $\Delta H_1 = {\Delta h_1}/{A}$. With respect to the error approach [6] the relative errors of liquid level change in both device ends are linked between each other by: $\delta(\Delta H_1) = \delta(\Delta h_1) + \delta A$, where $\delta(\Delta h_1)$ is the relative error specified by the inaccuracy of the drop level measurement by the micrometer head; δA is the relative error specified by the inaccuracy of value A.

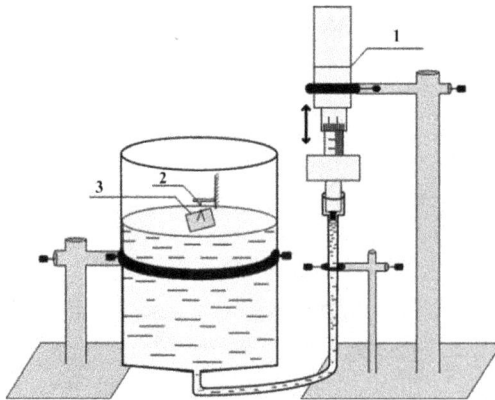

Fig. 5.13. The unit of nanosized objects hydraulic positioning: 1- shift micrometer head; 2 – cantilever; 3 – float with mounted research nanoobject.

The first error component is determined by the following way. So far as liquid level measurement drop in the narrow end is 5 µm, then the device measurement step is determined in the wide end is ${5.0\ \mu m}/{4444.4} = 1.1\ nm$. Micrometer head absolute error in accordance with passport is ±2.5 µm. Its value included to the result of the level shift in wide end is ±0.55 nm. The step measurement refinement result is equal to 1.10±0.55 nm. In the case of liquid level shift gauges in the narrow end with error ±1.0 µm, it seems possible to reach the relative measurement error of hydraulic shift ±20 %.

Hereinafter the unknown second component of the MI error – the relative error of constant A value determination – is considered below.

- Metrological experience of error systematic component minimization. Method of contradistinction can be used for accuracy improving by multiplicative error component minimizing. Its peculiarity consists in that the measured quantity H_X is compared twice with regulated measure - H_{N1}; H_{N2} (before the 2nd measurement it is rearranged with the measure). Consequently this quantity value with the eliminated measurement error multiplicative component can be gained as: $H_X = \sqrt{H_{N1}H_{N2}}$.

To realize the multiplicative error component minimization we perform the hydraulic device calibration (Fig. 5.14).

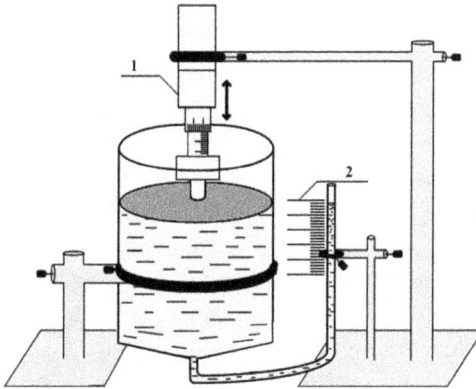

Fig. 5.14. Calibration unit for hydraulic positioning: 1- micrometer head; 2 – linear scale of the liquid level shift.

This operation is fundamentally opposite to direct measuring operation. Thereby, device enables to set the liquid level, and nanoobject can be mounted on the floating platform with absolute error which is slightly more than atoms size.

- Method of symmetrical observations. It is applied for correcting the additive and progressive (linear-variable over time) components of errors. Three measurements are usually carried out at regular time intervals Δt:

$$Y_1 = X + \Delta Y_a;$$

$$Y_2 = X_k + \Delta Y_a + \Delta_1 \Delta t; \qquad (5.13)$$

$$Y_3 = X + \Delta Y_a + \Delta_1 2\Delta t.$$

Errors of measurement results are determined by:

$$\Delta_1 = \frac{Y_3 - Y_1}{2\Delta t};$$

$$(5.14)$$

$$\Delta Y_a = \frac{Y_2 - X_k - (Y_3 - Y_1)}{2};$$

where ΔY_a is the additive error component of MI; Δ_1 is its rate of error change; X_k is the value of measure. The adjusted value of obtained result is determined from the 1st expression of mentioned system:

$$X = Y_1 - \Delta Y_a = Y_1 - Y_2 + X_k + \frac{Y_3 - Y_1}{2}. \qquad (5.15)$$

- Method of Transposition. It is applied when the experimenter does not have a complete set of MIs to eliminate the errors that arise in them. Let the transformation equation of MI is $Y = f(X)$ and let's consider two basic varieties: its replacement with variable measure and with adjustable scale converter. The first one is used at the exact measurements. Method is implemented in two stages. At the first stage, signal X is fed to input, and output signal Y_1 is fed to a memory element. At the second stage, the signal of a variable value X_N is submitted from regulated measure's output; it changes as long as signal Y_2 does not become equal to Y_1. When using method of substitution, the additive and multiplicative errors of MI do not bring contribution in result. Request is imposed to factor k that consists only in temporary stability, since permanence of k must be provided within a small interval of time which is equal to expectancy of 2-stage measurement. Method with adjustable scale converter is realized on the basis of set of elementary means. It can be recommended if the unambiguous non-adjustable measure X_0, adjustable scale converter for value X are available, and comparison unit fits only for value $k_H X$.

The method of transposition is widespread in the bridge circuit measurements, where firstly the resistance R_x is measured by bridge circuit; then it is substituted by multi-level measure R_N. Under the theory of bridge circuits, the error of resistance measuring by method of substitution δ_x equals to δ_N (error of measure) at its full replacement by measure: $R_x = R_N$.

This method is considered quite valuable in metrology, especially at the measures calibration. As example, the method of Ω size transfer from State standard to 1 Ω, 10 Ω and 100 Ω secondary measures is realized in 8 stages with help of ratios measure containing ten 10-Ω resistors; the latter can be connected in parallel (1 Ω), in series (100 Ω) and in series-parallel (10 Ω).

Moreover, importance of above method we can underline with next linked option, namely with implementation of exact measure of electrical resistance on the basis of conductance quantum in CPS self-checking operation. Such a standard is able to replace older one in the modern standardization practice.

Consumers of metrological services of the State Institutes of Metrology and Standardization, who are in great interest in transfer of proposed Resistance unit to CPS working standard, aim the subsequent accuracy improvement of CPS's products. We have considered earlier that the appropriate prototype of resistance measure (12906 Ω) could be applied for calibrating the MIs of high accuracy.

Elsewise, we have obtained the reference point of Ohmmeter scale important for its calibrating in the high accuracy class. In this way, it can be realized the self-check, self-calibration of MIs and therefore self-validation of gauging data. Advantage of the similar methods of metrological self-check is evident; it was demonstrated in [8] on examples of checking the temperature, pressure and other kinds of smart sensors. By continuous controlling the reliability of metrological data and basing on the self-checking results for previous time duration, the forecasting of device's metrological state is developed as well as CPS's state prediction.

5.7. Measurement Inexactness in (Nano)thermometry

Measurement inexactness is the deviation from the true value of a result gained from measurements and assigned to the measurand, or the measurement result minus true value of a measurand.

Peculiarities of thermometric materials manufacturing lead to the emergence of a transformation function dispersion even within the same party of thermometers (Fig. 5.15, curve P_0).

In turn, the manufacturer takes into account that irrespective of operating conditions in thermometric materials occurring the internal processes. They lead to a significant enlargement of transform function coverage interval (Fig. 5.15, curve P_2). The mentioned function changes under impacts during operation. The correct technology of thermometers stipulates that these changes should not extend beyond the guaranteed by producer instrumental error for the scheduled operating conditions. This extended error modifies as a result of studies that allowed selecting a systematic component $\Delta U_{syst} = \Delta U_{known}$ with its own sign and narrowing the coverage interval of random component (Fig. 5.15, curve P_1). The specified consideration of error is effective if it has been identified a trend of transform function drift during the operation. Most often it occurs in the initial 10 hours of operation when changes are directed and intensive.

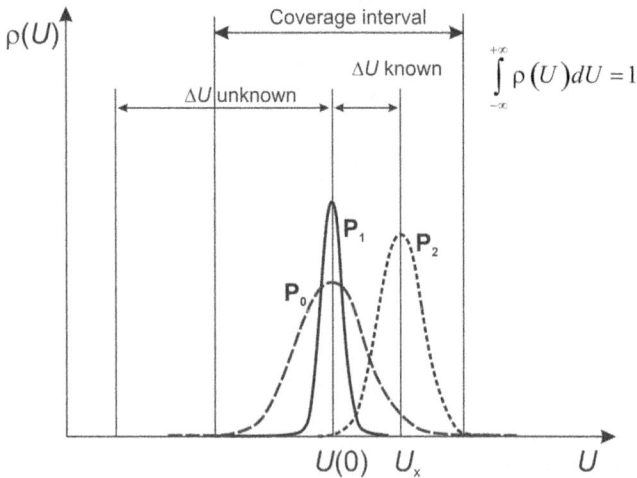

Fig. 5.15. Instrumental error and dispersion of transformation function.

Identification of the standard characteristics at the nanolevel is related to the predetermined temperature. To measure it, the high-accuracy thermometers have to be improved in terms of sensitivity. In the world of nanotechnology the combination of measuring technologies and theoretical research is getting more significant, since:

212

a) It concerns a single, non-repeated measurement where a classical error approach could not be applicable;

b) MI is getting more important; its intrusion during the energetic exchange disturbs the dimension of the studied value.

Moreover, measurement inexactness of the gained results and that of experimental investigations are two different notions. The experimental accuracy evaluation itself implies not just the study of experimental results accuracy, but an estimation of nanosamples as well as the selection of possible and correlated research methods.

On contrary to macroobjects where the improvement of research measurement inexactness could be reached by the increment in the experiment extent, the improvement of measurement conditions and the minimization of outer factors influence become important in the case of nanoobject. The problem of correlation between expediency and significance of MI energetic intrusion with the purpose to determine the quantitative object characteristics begins to dominate. As well are frequent the problem of reproducibility of the gained research results with the help of various sets of MIs of the same direction at different scientific centers.

5.8. Coriolis Mass Flowmeter and its Uncertainty Management

In the case of engineering on Coriolis Effect basis, quite spread are the CMF [9]. Their TF is the dependence of liquid mass or volume on flowing velocity through the specified cut. It is determined by the comparison of time characteristics of two identical sensors in the CMF input and output. More phase difference of high mentioned characteristics corresponds to faster controlled environment flowing. So, there is a dependence on the hydrodynamic regime of the current environment flowing through the CMF, its viscosity and temperature, etc.

Oval Corp. presents the CMF Ultra Mass MK 2 [10] with its 1.0 % of the readout error under the nonessential zero-shit error as an outstanding achievement in the branch of liquid and gas flow metrology. The process of liquid pouring from one vessel to another one or gas pumping is the transitional process accordingly to aero-, hydrodynamic and heat engineering conceptions. Therefore Maersk Oil Trading Company within flowmeters mounting on tank barges is proud of the next achievement: the CMF error doesn't exceed ±0.5 % [11].

- Error estimation and CMF accuracy problems. Within metrological conception [12], the CMF consists of transducers with appropriate sensors and peripheral devices and the microprocessor unit of received signals processing. The CMF sensors determine the flow velocity, temperature and provide information in form of output signals to the microprocessor that carries out the function of the brain of the measurement device and system in total providing access to the display, main menu and output device of processed information for the interaction with other systems, for instance, with the filling system. Peripheral devices provide monitoring, warning signalization and other functions, for instance, periodic processes management and the function of liquid density more accurate determination, etc.

It is important to understand the correlation or non-correlation of viscous and temperature components of the CMF instrumental error. Let's consider for this moment referring to [13] that the CMF instrumental error mentioned components are non-correlated. The same is exposed in [14] with the set of instrumental error correcting means (for correcting an instrumental error according to a sum of respective compensative instrumental-error values defined by the temperature), the temperature difference correcting means, and the resonant frequency correcting means. The elasticity of metal tubes changes with temperature; they become more elastic as get warmer. To eliminate the corresponding measurement error, the tube temperature is continuously measured by additional temperature sensor and is used to continuously compensate for variations in tube elasticity.

The TT error temperature component is the error specified by the temperature regime of liquid/gas flowing. It depends on the temperature of the controlled liquid/gas; on the CMF body outlet temperature (that is provided by [14] with the help of temperature detecting means of the inner tube and of the outer tube); temperature correcting means for compensating an instrumental error according to a change of temperature of the inner tube; temperature difference correcting means for compensating an instrumental error according to a difference between a temperature of the inner tube and a temperature of the outer tube.

The main problems of the CMF accuracy improvement lie in that hydrodynamic, thermal, electrical and other processes in the CMF which are not studied completely that specifies an unsatisfied TF reproducibility. Four platinum film resistance thermometers mounted in the CMF for the instrumental error maximum decrease are more accurate than thermocouples in the measurement stationary mode thanks to good accuracy (better than thermocouples), good interchangeability and long-term stability.

The accurate measurements of thermometers electrical resistance predict the usage of 3- or 4- wires measurement systems (last one is unaffected by extension wires resistance). However, resistance thermometers demand the preliminary warming up from power sources in the measurement process that leads to emerging some additional errors. Under the presence of 4 resistance thermometers with 4-wires measurement scheme of every resistance and the necessity of thermometer preliminary warming results in increasing the number of wires to 6. So, up to 24 wires are led just from the resistance thermometers to the device microprocessor switchboard. Consequently the construction is being unwarrantedly complicated. The resistances measurement duration increases due to preliminary heating necessity. The additional component appears not only of instrumental error but also of dynamic one specified by the possible controlled substance parameters change throughout the measurement period. As a result, some questions appear regarding such CMF usage in short-term measurements, for instance while beer bottling.

5.8.1. Accuracy Problems on Example of CMF with Straight Tube

- Flowmeter construction. Let's consider in details the straight tube CMF construction assembled of the measuring and reference tubes. The special reference frequency generator creates one tube vibrations relative to another one. These tubes reciprocal position will change on the condition of liquid movement through the CMF. Then the effect of tubes "beating" appears proportionally with the velocity of liquid movement (quantity) that passes through the CMF.

CMF has two separate contours for temperature measurement. The 1st contour is in the form of the resistance thermometer which has a thermal contact with the measuring tube and it gauges the liquid temperature passing through the CMF measuring tube. The 2nd contour (for the temperature measurement) consists of three analogical sensors: two are located on the reference tube and the third is on the body. These sensors are used for temperatures difference correction between the body, the reference tube and the measuring tube. Such CMF provides highly accurate value measurement of the flow and the liquid density under conditions of the changeable surrounding temperature that is mainly different from the liquid storage temperature in a reservoir.

Impact factors analysis shows that there is viscosity values dispersion of the control flow under the nominal measurement temperature, and the flow viscosity changes during the measurement on account of its intaking mainly from some storage layers where the velocity value is extreme because of the durable terrestrial gravity effect. As result the viscosity component of the CMF instrumental error measurement emerges which is specified by the studied liquid. It is determined by transfer function changes appearing, for instance, in case of the volume reservoir emptying and getting into the CMF of the liquid with viscosity changeable over the time. So far as this instrumental error component is partly specified by the intaking conditions of the studied substance, it isn't often mentioned by the CMF producer.

- Instrumental error temperature component of the measurement by the means of Pt100 film sensors. The CMF temperature CoE value in the gas flow gauging is underlined by the [15], where, for instance, it is tempting to investigate the temperature impact (only 13.7 %) at its tripling $\left(\frac{60}{20}\right)$. But the ratio of the absolute temperatures, important for flow measurement, has changed at $\frac{(60 + 273)}{(20 + 273)} = 1.137$, or the same growth.

As a result of producing peculiarities the CMF transfer function of one lot is characterized by some dispersion. In turn, the producer takes into account that during 2 sensors mounting in the fuel quantification scheme and 4 temperature sensors in the adjusting scheme, the CMF TF low uncertainty cannot be attained. Individual calibration of every CMF becomes necessary. It enables to minimize [16] the instrumental error to ±0.1 % in the most accurate CMF types. It is obviously that calibration is effective only in some before prescribed instances, e.g. in the limited temperature range of the CMF operation that includes beforehand indicated temperatures of the studied substance and the CMF, etc. Though most producers don't mention, the calibration is done merely in one temperature point of TF under the fixed impacts, the number of which is quite significant. Besides, the studied substance can change its own viscosity during the measurement cycle.

- Dynamic error. This methodical error appears in practice [17] due to the resistance sensor heat inertia. It is caused by the sensor heating from gauged substance and is inherent in the additional component due to internal self-heating while gauging. As a current must be passed through the sensor to obtain a voltage signal for the electronics, there needs a small amount of generated power which causes the sensor to warm up and thereby changing its resistance. A large current gives a nice big signal for the electronics but also a larger self-heating error. Small current reduces this error but lower electronics drift is required to minimize errors of the circuit. The best trade off depends on the application however of currents of the 1 mA-order or less. Self-heating errors are larger for gas temperatures measurements due to poor heat withdrawal from the sensor [7]. Simultaneously the heating alternation of each of 4 resistance sensors increases the CMF dynamic error more than 4 times.

As the resistance sensors in industrial applications are commonly used the film sensors which the market was in white-ware applications. Film sensors have a platinum pattern fixed onto its surface that gives 100 or 1000 Ω resistance at 0 °C. They are known as Pt100 and Pt1000 respectively. Industrial Pt100 detectors are divided into four tolerance classes under [18]: AA, A, B and C. The new standard makes a difference between wire wound sensors and film sensors. Experience shows that film sensors can't be handled for as wide temperature ranges as wire wound sensors under the different tolerance classes. Film sensors are used often in industrial applications. Closer tolerances at other temperatures can only be achieved through calibration that determines the specific properties of the individual sensor. This is caused by low quality of platinum specified in IEC 60751 and reduced for the thermometric material (alloy of platinum with added palladium - performed for compliance with the traditional DIN standard). The alloy gives rise to departures from the ideal curve, necessitating a safety margin corresponding to the slope of curve A or, at worst, of curve B. Tolerances for assembled Pt100 sensors including film sensors are given in IEC 60751:2008. For example, film sensor type B has performance tolerance ±0.5 °C at 50 °C or ±1 %, though [19] informs about typical inaccuracies 0.2 %, 0.1 % of such sensors. It should notice that accuracy and performance tolerance are absolutely different in effect. The last one

217

concerns the sensors interchangeability and is especially important in case of identical sensors usage in one CMF.

- Temperature component of CMF instrumental error of measurement by means of thermocouples. To replace Pt resistance sensors, well studied Pt-Rh thermocouples are proposed. For temperature range 243...1000 K thermocouples type S (Pt–10 % Rd/Pt) are used. At first sight they are worse than high-mentioned resistance sensors. Their calibration characteristics reproducibility is worse the similar one for second class thermocouples (±1.5 °C at 0...600 °C [6]) against ±0.1 % in wire wound platinum resistance sensors. Nevertheless, the film sensors are characterized by reproducibility of ±1 % that is slightly worse compared to characteristics of mentioned thermocouples.

Table 5.1 [19] describes briefly the main differences between thermocouples and Pt100 resistance sensors in general terms. Let's pay attention that in 90 % cases the CMF operates at the room temperatures, where thermocouple instability is quite low and ageing affect - comparatively high. It provides significant advantage concerning the thermocouples. Latest have other essential advantages: the methodical error absence caused by current heat at measurement cycle; this mentioned procedure time is economized and the dynamic error is considerably lesser; the CMF electrical structure is significantly simplified due to reduction of leading-in wires number from temperature MIs. The sensors signals measurement is more simple and accurate so far since the compensation method of thermocouple voltage measurement inherent in much better accuracy than the potentiometric method of resistance sensors signal measurement. Then the additional resistance sensors power supply is unnecessary.

Table 5.1. Comparing thermocouple vs resistance sensor.

Property	Thermocouple	Pt100 Sensor
Measuring range	Large: –200<T< 1000 °C	Limited: –200< T<600 °C
Stability	Not good, especially in high temperature	Excellent. annual drift <0.01 °C
Measuring location	From hot junction to reference location	Across entire Pt-wire length
Ageing	Significant in high temperature	Insignificant, see Stability.

Response time	Very short possible <1 s	Not as short as corresponding TC
Excitation power	None	Significant <1 mW
Physical strength	Very good	Limited
Pricing	Slightly lower than corresponding Pt100 s	Slightly higher than corresponding TC

- Thermal response time. The response time of platinum resistance sensors is 60 s. It means the following. During the first minute the thermometer is warming up to the temperature 63.2 % of the applied temperature drop/jump value. After 3 minutes it reaches 95 % of the temperature drop/jump and after 5 minutes reaches 99 %. Very small and quick Pt100 film sensors type TF101N thermal response times referring to manufacturer data [20] are: $t_{0.9}$ in air - 10 s, in water <1 s.

Because it takes longer for Pt100 sensor to warm up, latter generally have a higher response time than thermocouple. The last one with the unprotected junction (made by welding two 0.125 mm diameter thermoelectrodes) manufactured like the sphere of 0.2 mm diameter is characterized by the thermal response time ~1.0 s in the air and 0.04 s in the water (Table 5.2) [21]. Otherwise, it is characterized by the significantly lesser inertia in comparison to the film resistance sensor.

Table 5.2. Response time* of thermocouples.

Wire size, mm	Still air 427 °C / 38 °C	60 ft./sec air 427 °C / 38 °C	Still H₂O 93 °C / 38 °C
0.025	0.05 s	0.004 s	0.002 s
0.125	1.0 s	0.08 s	0.04 s
0.381	10.0 s	0.80 s	0.40 s
0.75	40.0 s	3.2 s	1.6 s

*The time constant is defined as the time required reaching 63.2 % of an instantaneous temperature drop/jump.

The measurement resistance procedure of every of 4 sensors includes the following steps: the current is consecutively led to every of them, after the sensor warming the voltage jump is measured. It takes usually 7...10 minutes for Pt100 sensors produced of the bobbin wound wires. Only in this case the taken platinum resistance sensors readings can be considered by microprocessor. At the same time methodical error appears caused by measurement current warming and by the necessity of the extra time for this warming that is equivalent to thermal response time additional increase.

5.8.2. CMF Performance Improvement Under the S-type Thermocouples Usage

CMF operation theory is based on the temperatures difference measurement of two points. It can be successfully done by one thermocouple which hot and cold junctions are located at the mentioned points. Uniform thermoelectric inhomogeneity of S-type thermocouple in accordance with [6] is ±1.5 °C. The inhomogeneity is estimated [22] as difference between maximum and minimum values of the thermo-EMF concerning comparison the certified reference material on every 5 m length of wire in the bobbin at the temperature 900±20 °C... 1300±20 °C. It is specified by the real imperfect technology of the wire producing. So tolerance classes of calibration characteristics have been proposed. They are intended to ensure interchangeability of thermocouples without special calibration testing. A unique advantage of thermocouples is that 'bobbin calibration tests' may be performed with samples from a given bobbin of wire, and those test results will apply to all thermoelectrodes made from that same bobbin. In this way significantly closer control of tolerance variations may be gained without incurring the high cost of testing a large number of particular thermoelectrodes and therefore the thermocouples [23].

The necessity of electric signal output from the thermocouple demands the double compensation wires usage that leads to a number of other problems. That's why the better decision is considered to measure the temperatures difference of 2 points by the means of the differential thermocouple that consists of 2 towards switched thermocouples. Then leading-in wires are done from the same thermoelectrodes. Naturally such wire is the best one of two by the thermoelectric inhomogeneity, in our case – Pt wire.

However, the thermoelectric inhomogeneity of 2 already connected thermocouples increases in $\sqrt{2}$ times, so to ±2.12 °C (the thermoelectric inhomogeneity of 4 thermocouples used in the construction will reach $\pm 1.5 \times 2 = \pm 3$ °C) that cannot meet technical demands.

220

Scientific and technological solutions are based on the technology knowledge of thermocouple wires production of platinum and platinum group alloys. Taking into account that in the bobbin of thermocouple wire there are equal by their impact local and extensive inhomogeneities, the action of the last inhomogeneities can be practically eliminated by the significant decreasing of thermoelectrodes length [23]. Especially it concerns the next cases: a) the usage of small parts of thermoelectrodes, for instance, the length about 0.05 m that is less than 1 % of length under which the thermoelectric inhomogeneity is studied; b) the usage in two and more thermocouples of thermoelectrodes from neighboring piece of bobbin. Then for the wire of Pt–10 % Rd the thermoelectric inhomogeneity of such wire length mounted in small CMFs can be decreased in some times: from $\leq \pm 8$ μV to $\leq \pm 2$ μV.

Thus, the differential thermocouple is produced for the CMF in Π-shape where 2 vertical thermoelectrodes are made from Pt wire with thermoelectric inhomogeneity $\leq \pm 2$ μV and between them there is a short horizontal section of Pt - 10 % Rd wire with inhomogeneity $\leq \pm 2$ μV. As result, the S-type thermocouple inhomogeneity that is determined as RMS deviation of non-correlated values of 2 different wires inhomogeneity is $\pm 2\sqrt{2}$ μV or ±0.52 °C. This is the reproducibility of temperatures difference measurement in 2 CMF neighboring points by the means of replaceable differential S-type thermocouples.

Totally the CMF is equipped by 4 ordinary or better by 2 differential thermocouples. The first differential thermocouple consists of two towards switched thermocouples that measure correspondingly the temperatures of input and output of reference tube. In second pair the other 2 thermocouples are switched towards – for temperatures difference measurement on the CMF measuring tube and on its body, so one more differential thermocouple is realized. This fulfilment enables to introduce temperature corrections: for the first thermocouple – on gauging the liquid temperature changes within the tube as result of the changeable thermal-hydro dynamical flowing regime; for the second thermocouple – on gauging the flowing substance temperature changes as a result of passing through the CMF.

5.9. Conclusions

Basing on the current metrological experience of complex technical objects we suggest the improvement of efficiency of Cyber-Physical

Systems at equipping them with enhanced metrological subsystems. The latest have to provide exact measuring the performance, including the gauging the actuators that actually together with sensors ensure the necessary CPS operation mode.

Qualitative metrological instruments, their efficient metrological supervision and insurance enable us to enhance the accuracy of metrological subsystems. However, to improve the CPS accuracy and to raise substantially the quality of manufactured products by some orders, providing in-place the man-out-loop metrological procedures such as self-checking, self-validation, self-adjustment and so on becomes crucial.

Distinctive feature of such procedures could be implementation of special methods of minimizing the different kinds of random and systematic errors, for instance, through the introduction the methods of contradistinction, of transposition or/and others. These techniques are envisaged on the examples of a number of modern measuring instruments and devices, for instance on Coriolis mass flowmeters as the bright sample of multi-parameter smart devices.

References

[1]. V. Yatsuk, P. Malachivski, Methods of improving the measurement accuracy, *Beskyd-Bit,* Lviv, 2008 (in Ukrainian).
[2]. ICGM 104: 2009, Evaluation of measurement data - An introduction to the 'Guide to the expression of uncertainty in measurement' and related documents, *JCGM,* 2009.
[3]. K. Ranev, Hybrid model of result processing, in *Proceedings of International Conference on Metrology,* Minsk, Belorus, *BELGIM,* 2009, pp. 24-31.
[4]. B. Stadnyk, S. Yatsyshyn, Accuracy and metrological reliability enhancing of thermoelectric transducers, *Sensors & Transducers,* Vol. 123, Issue 12, December 2010, pp. 69-75.
[5]. B. Stadnyk, S. Yatsyshyn, Ya. Lutsyk, Metrological Array of Cyber-Physical Systems. Part 14: Basics of Metrology and Techniques for Accuracy Improvement, *Sensors & Transducers,* Vol. 196, Issue 1, January 2016, pp. 7-23.
[6]. CSN EN 60584-1. Ed.2. Thermocouples - Part 1: EMF specifications and tolerance, Category 258331, 2013.
[7]. S. Yatsyshyn, B. Stadnyk, Ya. Lutsyk, L. Buniak, Handbook of Thermometry and Nanothermometry, *IFSA Publishing,* Spain, 2015.

[8]. The Total Calibration Solution, *CPS. Instrumentation & Calibration Experts,* 24 Sep 2014 (http://www.cps.co.nz/blog-display.aspx).

[9]. Flow and level measurements, Omega Engineering, A Technical Reference Series, *Transactions in Measurement and Control,* Vol. 4, (www.omega.com/literature/transactions/transactions_vol_iv.pdf).

[10]. www.oval.co.jp/

[11]. J. Rosenkrans, Bunkering in the 21-st century: The case for Coriolis flowmeters and experiences thus far, *Wilhelmsen Premier Marine Fuels Bunker Seminar,* Oslo, Oct. 14, 2010.

[12]. M. Kazahaya, A Mathematical Model and Error Analysis of Coriolis Mass Flowmeters, *IEEE Transactions on Instrum. and Measurement,* Vol. 60, Issue 4, 2011, pp. 1163 – 1174.

[13]. V. Voytov, Diagnostics of hydraulic machines running gear, Scientific *Bulletin of Nat. Forestry Engineering University,* Lviv, Ukraine, 20, 3, 2010, pp. 74-77 (in Ukrainian).

[14]. T. Endo et al., Error correcting Coriolis flowmeters, G01F1/84, *Patent US* 5796012 A.

[15]. D. Spitzer, The Consumer Guide to Coriolis Mass Flowmeters, Seminar. Spitzer and Boyes, LLC, 112 p., 2004, http://www.spitzerandboyes.com/wp-content/uploads/2014/03/Slides-Flow-Coriolis.pdf

[16]. MicroMotion. Flow and Density Measurement, *Emerson Process Management,* (http://www2.emersonprocess.com/en-us/brands/micro motion/pages/coriolis-flow-density-measurement.aspx).

[17]. N. Yaryshev, Theoretical basis of measurement of nonstationary temperature, *Energy-atom-Publishing house,* Leningrad, USSR, 1990 (in Russian).

[18]. Resistance thermometer theory and practice, 14 p. http://educypedia. karadimov.info/library/RTD.pdf

[19]. IEC 60751:2008, Industrial platinum resistance thermometers and platinum temperature sensors, IEC WebStore (https://webstore.iec.ch/publication/3400#).

[20]. http://www.thermo-electra.com/nl/producten/

[21]. Resistance thermometers and thermocouples. Intrinsically safe designs, Ex i. Operating instructions, *WIKA Alexander Wiegand SE & Co,* 2010, (http://www.wika.com.ar/upload/OI_TR_TC_Ex_i_en_de_6309.pdf).

[22]. I. Levin, I. Rogelberg, Thermoelectric inhomogeneity of the wires and the methods of its measurement, in Investigation of thermocouple alloys, Proceedings of State Institute of non-ferrous metals machining, *Metallurgy,* Moscow, USSR, 1971, pp. 93-106 (in Russian).

[23]. R. Park, Thermocouple Fundamentals, Course Tech. Temp. 2-1 (http://www.advindsys.com/ApNotes/tcfundamentals.pdf)

6.

Frequency, Noise and Spectrum Metrology

S. Yatsyshyn, B. Stadnyk, Z. Kolodiy and O. Seheda

Circle of application problems, related to frequency research, is steadily expanding. This is predefined by the fact that the most reliable and precise are the experimental results of measuring just the frequency characteristics in science and technology. It is well-known that standard uncertainty of frequency (time) measurement is several orders less than uncertainty of any other SI unit. Advances in sensor technology and applications, information investigation and management have emerged new opportunities for CPS long-term monitoring and data reporting. Providing high reliability, the frequency characteristics can significantly enhance the quality and quantity of information, available for decision makers. Beyond advances in "passive" spectrum monitoring technologies, there is a vast and increasing amount of "active" data being collected [1].

In the field of Optical Metrology [2] a challenge is to develop measurement techniques and standards, to meet the needs of next-generation advanced manufacturing, which will rely on nanometer scale materials and technologies. It is difficult to provide samples on which precision instruments can be calibrated on nanoscale. Calibration standards are important for repeatability to be ensured. It is difficult to select a universal calibration artefact with which we can achieve repeatability on nanoscale. At nanoscale, while calibrating, care needs to be taken for the influence of external factors (noise, vibration, motion) and internal factors such as interaction between the artefact and equipment which can cause significant deviations. The most universal nanothermometry method, Raman method, apt for the direct temperature measurement of micro- and nano-objects within the range 100 nm – 100 µm, could be distinguished within this from cryogen till midhigh temperatures. In addition, it does not demand calibration; therefore the mentioned method is particularly considered below.

At this stage of technology development the scientific community focuses in CPSs on dealing with frequency fiber-optical studies [3], believing their crucial role in improving performance, since it turned out that fibers are characterized by wide but finite bandwidth. However, there are a number of other frequency-dependent characteristics of CPSs that can be useful. For instance, we are successively considering here the content of some selected tasks of metrology associated with the use of MIs as CPSs' information-measuring subsystems, whose work is based on measurements of frequency-bound characteristics, furthermore of the wide range from infra low frequencies ($^1/_f$ noise) to light-emission frequencies. Such studies may be objective not only for particular scientific goals, and for the intense investigations and developments of testing real-time methods of large distributed systems monitoring, and for predicting the determining shifts in CPS's behavior. According to [3] "a thorough understanding of such failure mechanisms will enable designing more robust and resilient future communications" and production systems.

6.1. Frequency-Phase Techniques in Thermometry

- Nuclear quadrupole resonance thermometer is the thermometer, the action of which is based on temperature dependence of nuclear quadrupole resonance frequency of thermosensitive substance, namely on phenomenon of resonance absorption of electromagnetic waves by a nuclear system. Frequency of electromagnetic waves corresponds to the energy difference of quadrupole levels. Temperature dependence of this frequency is a physical property of the thermometric substance characterized by high steepness (units of kHz/K that permits to achieve a high sensitivity threshold of temperature changes ≤ 0.001 K); by exclusive stability that allows to construct on its base a high precision thermometers (error $10^{-5}...10^{-6}$ %) of high reproducibility [2].

- Quartz thermometer is thermometer, the work of which is based on temperature dependence of the resonance frequency of piezo crystal vibrations. This dependence is described as $f(T) = f_0(1 + \alpha T + \beta T^2 + \gamma T^3)$, where f_0 is the frequency at known temperature (mainly 0 °C); α, β, γ are the coefficients. Quartz thermometer is a high-precision temperature sensor. It measures temperature by the frequency gauges of a quartz crystal vibrator. With the proper direction of cutting out the piezo crystal plate it can be achieved the linear thermometric scale (at

$\beta = \gamma = 0$) as a result of a linear temperature coefficient of frequency. So, temperature measurement could be essentially reduced to frequency measurement which is substantially more precise.

According to its operation principle the quartz thermometer actually belongs to acoustic resonance thermometers. Quartz thermometers are mainly created with the high-frequency resonators of Y-cuts, which use shear vibrations. Operation frequency is within a band 1…30 MHz Vibration on main frequency (1.1 MHz) or on the 5th overtones (5.3 MHz) is also used. Temperature coefficients of resonator frequencies are within $(20…95) \cdot 10^{-6}$ 1/°C. Sensitiveness of thermometer can be raised by application of frequency multipliers. Resonators are performed vacuumed or sealed. The first are characterized by the lesser drift, second - by the lower thermal inertia index. Resolution is $10^{-4}…10^{-6}$ K; operating temperatures vary from 80 to 250 °C with error 40 mK.

Quartz thermometers surpass other thermometers by their metrological characteristics (accuracy, inertia, sensitiveness, settling ability), but inherent in their own shortcomings: limited temperature range of measurement and hysteresis of indices related to temperature prehistory of resonator. Physical nature of hysteresis is not fully clarified, but experimentally is established possibility to reduce energy of hysteresis tenfold by specimens exposition alternately in liquid nitrogen and warm thermostat (150 °C) for 1 hour, or by imposition of a permanent electric field at elevated temperature (170 °C).

High linearity of quartz thermometer makes it possible to achieve sufficient accuracy over an important temperature range that contains only one convenient temperature reference point for calibration that is a triple point of water [4].

- Radiometric thermometer is the semiconductor transducer of temperature-frequency type, based on the temperature dependence of semiconductor device impedance. These devices apply semiconductors with negative differential resistance, including S-type diode, unijunction transistor, etc. on which are built auto generating radiometric thermotransducers. The informative parameter of such thermotransducers is an oscillation frequency of generator signal, which substantially depends on temperature. For instance, the sensitivity of a radiometric thermotransducer made on a unijunction transistor is approximately 1 kHz/K. Examples of radiometric thermotransducers on bipolar transistors are given below (Fig. 6.1).

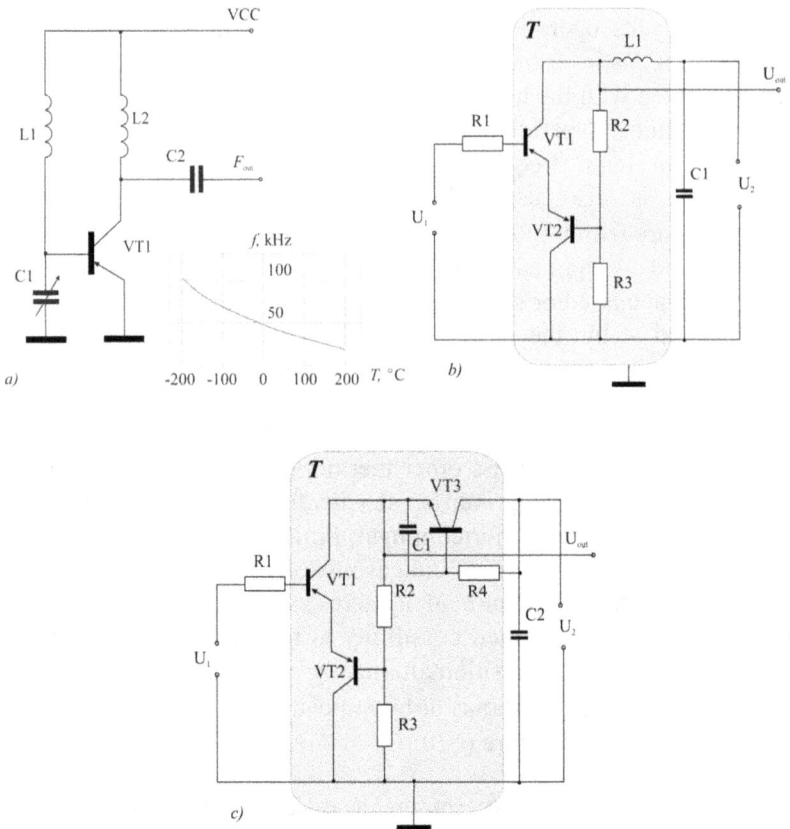

Fig. 6.1. Radiometric thermotransducers on bipolar transistors.

In the radiometric thermotransducer, the frequency rises at temperature fall due to the reduction of base-emitter junction capacitance, and the transducer sensitivity is ~3.5 kHz/K.

Radiometric thermotransducer circuits can be implemented in integrated performance. The thermotransducer TF shown in Fig. 6.1, b is represented by the expression:

$$f = \frac{1}{2\pi} \sqrt{\frac{2[C_{jbe}(T) + C_{jbc}(T)]}{C_{jbe}(T)C_{jbc}(T)L}},$$

where C_{jbc} is the base-collector junction capacitance, C_{jbe} is the barrier layer capacitance, L is the inductance. The change of frequency at a temperature rise in the range 170...370 K is ~1100...700 kHz. The radiometric thermotransducer TF in Fig. 6.1, c can be described by the equation:

$$f = \frac{1}{2\pi} \sqrt{\frac{2[C_{jbe}(T) + C_{jbc}(T)]}{C_{jbe}(T)C_{jbc}(T)L_{equ}(T)}},$$

where L_{equ} is the equivalent inductance, and the sensitiveness of such radiometric thermotransducer is ~4.5 kHz/K.

Heat flux transducers and infrared radiometers have been proven in thousands of applications for over thirty years – in ground and flight aerospace testing, fire testing, heat flux standards for flammability testing, heat transfer research, materials development, and furnace development.

- Radio pyrometer is the pyrometer, which operates in microwaves range (0.5 mm...100 cm). To concentrate radiation the radio pyrometers apply classic elements of radio engineering: antennas, waveguides and etc. It is known the cosmic radio pyrometer in which controllable noise diode was switched against the input noise signal.

- Josephson junction noise thermometer is the thermometer, operation principle of which is built on the Josephson Effect. The latter can be obtained by means of Josephson element, which contains two superconducting plates separated by a thin oxide layer. An element is switched in measuring circuit through a point contact between sharpened wire and plate of the superconducting material. Charge carriers of element may exist as the electrons that are scattered by lattice, and as Cooper pairs that create the superconductivity effect, but are not scattered by a lattice [5]. Due to tunnelling, electrons as well as Cooper pairs would flowed through oxide layer resulting in emerging a stationary (DC of superconductivity that not exceed particular value I_0 can pass through the oxide layer without voltage drop) and a non-stationary Josephson Effect.

Non-stationary Josephson Effect consists in emerging the oscillations at frequency $f_0 = \frac{2e}{h} u_0$, where e is the electron charge, h is the Planck constant, while the DC voltage u_0 is applied to Josephson element. Therefore the latter becomes a generator of sinusoidal current. Thermal noise that arises in the Josephson element creates the current pulsations resulting in monochromatic frequency blurs of signal. Since the thermal noise inherent in a normal distribution, then extend of signal frequency mode has a typical bypass of Gaussian curve. Half-width of the spectral line of thermal noise in Josephson element, measured by radio spectrometer, is given by the expression $\Delta f_0 = 4\pi k_B T r \left(\frac{2e}{h}\right)^2 \left(1 + \frac{Ir}{u_0}\right)$, where T is the thermodynamic temperature, k_B is the Boltzmann constant, r is the resistance, which shunts the element in measuring scheme, I is the current that passes through the element.

An example of the implication for absolute noise thermometry is NIST development [6] of Josephson junction noise thermometer. Here an absolute cryogenic temperature scale is being defined at the using the linewidth of radiation emitted by a resistively shunted Josephson junction.

6.2. Noise Metrology

Noise method [7] is becoming intensively applied in nanotechnology and nanobiotechnology [8], to study processes taking place inside the research-objects and determine their characteristics. Quite thorough research of electric noise [9], particularly of $1/f\gamma$ noise [10], and their theoretical analysis have been conducted in [11]. However, at the absence of insufficient metrological elaboration and a coherent approach to the phenomenon description, an application of a noise method in nanotechnology is not always stipulated. The gained results are not satisfactory interpreted, which obviously does not foster the development of the mentioned areas.

Practically, the noise investigations including electrical noise research of new substances, schemes and objects (nanotransistors) are mainly made not by metrologists but physicists, chemists and other specialists who do not have sufficient metrological experience; on the sets that frequently demand special calibration; using methods remaining in a state of metrological elaboration. The mentioned finally leads to the uncertainty accumulation. In the conditions of durable development of nanotechnology with its unrepeatable measurements it causes the substantial decrease in the reliability of information extracted from the results of a noise method for technology purposes.

6.2.1. Fluctuations and Thermodynamics, Proper Noise and Thermometry

Metrologically correct approach to noise research consists in methodologically correct methods development for studying the noise phenomena (thermal, shot and flicker noise) in particular the integral characteristics of noise signals in the input circuits. It is ascertained that to reduce the influence on the result of measurement of:

- Flicker noise, we should work within the bandwidth above 1–20 kHz where its level is negligible;

- Thermal noise, we should provide the conditions when a noise measurand is as much as possible in comparison with uninformative thermo-noise, and use the correlation method of measurement;

- Shot noise, we should select the active elements with small values of noise currents and voltages, and use the correlation method of measurement.

Research methodology is first of all the methodology of measurements that includes the following aspects:

a) Assurance of the increased interference resistance due to the low amplitude value of the measured signals;

b) Separation of the bandwidth for intended investigations;

c) Predetermination of the general duration of an experiment, particularly while undertaking low-frequency research on $1/f$ noise with regulation of a basic reasonable error of method;

d) Selection of the methods of processing the gained signals (particularly, application of the high-speed Fourier transformation that includes the differentiation of a signal into frequency spectra; separation of interference frequency, for instance, the frequency of feeding power with its multiple frequencies; and their elimination from the re-formed signal);

e) Selection of optimal noise-characteristics research methods in terms of quality and negative issues;

f) Choice of methods apt to characterize the researched materials by known informative quality, durability, reliability, uncertainty and other parameters.

The research of integral characteristics of noise signals appearing in sensitive elements enables comprehensive studying into physical nature both of noise processes occurring in substance, and non-noise substance parameters and characteristics. Exactly such an approach to noise research most completely renders its significance for modern electronics. Hereby the improvement in the means of measuring technics, development of new conceptions of conducting the gauges themselves, automatization of a measuring process, implementation of interference-resistant technologies of gaining the output signals and etc. are considered. The optimization of sensor resistance is proposed in order to satisfy the needed ratio of spectral densities of the measured noise signal and noise interference with taking into account the proper input noise current and voltage of an input circuit.

Theoretical and experimental research in the area of measuring the integral characteristics of noise signals reveals the chain of problems, namely:

- Durable measuring caused by random nature of a noise signal;

- Influence of uninformative noise signals in the input circuit on the results of measurement;

- Low level of a noise signal;

- High requirements to interference-resistance of the MIs [12].

Each of the considered problems leads to the appearance of whole chain of CoEs of measuring the integral characteristics of noise signals. These components depend on the method of measurement, metrological characteristics of MIs and their proper noise characteristics (noise voltage and current). The analysis of possible error sources, minimization of their influence on the result of measurement, consideration of the specificity of the measurand, satisfaction of appropriate ratio "noise-signal being measured – uninformative noise-signal" and etc. enable gaining the reliable results of measuring the integral characteristics of noise signals.

Defining Role of Fluctuations in Forming a Noise Thermometer Transformation Function

Attracting of the noise method to substance description at the macro-, micro- or nano- level becomes more effective and exact if a sensitive element of NT is made from the mentioned substance. Then the whole metrological experience of manufacturing and exploiting the sensitive elements of NTs as well as developing their metrological supply and assistance [13] could be used for the nanometrological support of nanotechnology efforts. Despite of type of such substance (metals or semiconductors; crystal or amorphous; solid, liquid or gaseous) as a main cell where an informative signal is generated, this system is characterized by the heterogeneity of parameters. The latter reveal themselves in the form of physical, chemical, electrical and other noises being caused by fluctuation deviations. Hereby with reducing the size of the researched substance samples, noises are behaving themselves more and more actively which is successfully described with the help of nanothermodynamics [14].

NT is the only metrological device in which noise voltage represents a main metrological characteristic. It lets determine the thermodynamic temperature T due to the power $\bar{E}^2$ of the fixed electric thermal noise, making the electric, heat and other transfer processes that take place in a thermodynamic substance available for studying. To describe the work of NT, it has been proposed the classical formula of Nyquist where the mean square of the noise voltage $\bar{E}^2$ is calculated like:

$$\bar{E}^2 = 4hfR(f)\left[exp\frac{hf}{k_BT} - 1\right]^{-1}, \qquad (6.1)$$

where h is the Planck constant; f is the frequency; k_B is the Boltzmann constant; R is the nominal value of a noisy resistor. The given formula is substantiated on the basis of statistic thermodynamics [15]. In general, the fluctuation-dissipation deviations of certain parameters form the theoretical basis and practical foundations of numerous applied scientific disciplines such as electrochemistry, materials and metals sciences, thermoelectricity, and so on. The analysis of (6.1) has shown that the quantum corrections could become substantial at the high and the low temperatures. We deal with 300 K and the frequencies over 60 GHz or at 4.2 K – with 800 MHz and higher when the contribution of quantum correction exceeds 1 %. Thus, within the range of frequencies 10^3-10^7 Hz commonly used in noise thermometry, quantum corrections could make the marked contribution only at the super low temperatures. At $hf \ll k_B T$ and $R(f) = R = Const$ equation (6.1) changes to:

$$\frac{\bar{E}^2}{R} = P = 4k_B T\Delta f, \tag{6.2}$$

From equation (6.2) we could state that the concrete and precise determination of RMS voltage $\bar{E}$ and hence electric power P implies the usage of a studied object with a priori known value of electric resistance R. At precisely defined Boltzmann constant [16] a relative instrumental error of the measured voltage δe is formed as the sum of notified values: the relative errors of determining object resistance and temperature, as also the relative error of specifying frequency bandwidth of measurement. It indicates the undoubted advantages (resistance and temperature are known with high accuracy) of employing the material of a NT sensitive element as sensitive substance in systematic noise studies. Another advantage of the proposed modification of a noise method – a passive kind of method – is non-invasiveness, i.e. the absence of any proper distortions of research results or/and methodical component of the error. The expression (6.2) successfully relates the property of a thermodynamic system – noise (electric fluctuations) power P – with disequilibrium process that is characterized by an electric parameter – resistance. It is valid for any resistance without regard to its nature. For example, the $p - n$ junction within a certain temperature range could be treated as the electric resistance $R = {dU}/{dI}$ - of the same origin.

Although there are noticed the exceptions characterizing the violation of equilibrium conditions. For example, plasma supported by a discharge at the direct current not always stays in equilibrium. Therefore we could not state a priori that the conducting plasma generates thermal noise relevant to its electron temperature. The mechanism of disequilibrium noise appearance is also immanent for the bulk substances. The defects of structures (vacancies, dislocations and grain borders) actively participate in the process of microstress relaxation. Their interaction with electrons reflects at the noise spectrum. For example, due to the conduction electron dragging, the excess electric charges appear around verge dislocations in metal. Their chaotic movement is interpreted as additional noise [17].

Studying fluctuations through the alteration in NT transformation function. Paradox of highly precise temperature measurement with the help of NTs roots in neglecting changes of thermodynamic state both in thermometric substance and controlled medium while gauging within the wide temperature range. It essentially affects a NT transformation function. The sensitive element of the latter gets into a disequilibrium thermodynamic state even if it was in equilibrium before. The reverse transition is far not always possible since real non-equilibrium transfer processes have the certain trend and are accompanied with irreversible phenomena of fluctuation-dissipation genesis. Therefore the transformation function (6.2) of NTs alters, and consequently their instrumental errors increase [18].

To study the nature of these phenomena due to the alteration in a NT error, a thermodynamic approach [19] has been employed. It supposes that NT thermodynamic substance is chosen as the researched system. The system boundaries are determined by its surface, and properties reveal themselves at the interaction with environment. The system is isolated if an environmental effect on substance is negligible. In the case of a NT, especially at the investigation of bio-objects, the researched object itself, e.g. a cell, becomes the part of such a sensitive element. The totality of possible states of the system could be divided into two: stationary and non-stationary states. Processes accompanied with dissipative effects are treated as irreversible or spontaneous. A rise in entropy of the isolated system remaining in a non-stationary disequilibrium state is quite logical. It lasts till the system reaches a non-stationary equilibrium state. Mathematically it could be expressed by the law of entropy rise: $dS \geq 0$. In non-isolated system remaining in a non-stationary disequilibrium state, consequently of dissipation effects inside the system, an entropy generation occurs. Processes in homogeneous systems whose states are determined by entropy, volume, and mass and described thermodynamically at any moment of time are studied the most thoroughly. Those systems interact with environment by heat, mechanic or/and mass exchanges. The isolated system attempts to reach a certain stationary state which consequently of its isolation is unambiguously equilibrated. Evolution of an isolated system to the equilibrated state is called relaxation, and occurring there processes are relaxation ones and irreversible by their nature.

The main underlying principle of statistic non-equilibrium thermodynamics [15] is as follows. Macroscopic quasi equilibrated state of the system is considered by means of thermodynamics as a significant fluctuation. Then gradients of temperature, density, volume and so on formed in the investigated thermodynamic system by the external factors are subjected to the same statistical regularities as gradients caused consequently by action of fluctuations. The mentioned approach is valuable by its capability of both considering the thermodynamic system of investigated substance in the context of environment, and lending an insight into the fluctuation phenomena occurring there. Otherwise, studying higher level fluctuation-noise processes we can extract information on characteristics of a lower level and vice versa. In this way we might determine or even predict the reliability of chips and other elements due to the ascertained features of noise characteristics [20].

Electric genesis of a NT signal is determined following from the basic equation of thermodynamics [21]. Hereby, the main degrees of freedom for a NT transformation function are electrochemical and thermal ones - see (6.2). The thermometric material itself due to the nature of its manufacture and usage could refer to the closed homogeneous thermodynamic system. At the complicated analysis of processes inside this substance we should also consider a mechanical degree of freedom. For such a system with independent thermodynamic degrees, the basic equation of thermodynamics takes the form of power balance:

$$T\frac{dS}{dt} = \frac{dU}{dt} + p\frac{dV}{dt} - \eta\frac{dN}{dt}, \qquad (6.3)$$

where S is the entropy; T is the temperature; U is the inner energy; p is the pressure; V is the volume; η is the electrochemical potential. Evidently, the given equation is suitable to describe processes taking place in investigated substance at its deformation during manufacturing or exploiting a NT. It would be also valuable for amorphous alloys or bio-objects possessing substantial internal energy. In other cases, e.g. for elastic continuum with dislocations, the changes in inner energy could be neglected: $dU/dt \to 0$. Therefore the component: pdV/dt could be substituted [22] with the product of tensors (or product of scalars in the case of elastic continuum) of deformations σ and stresses ε:

$$p\frac{dV}{dt} = \sigma\frac{d\varepsilon}{dt}, \qquad (6.4)$$

The introduction of stress-deformation effects into consideration of the noise processes inside electrothermometric sensitive elements is grounded [23] by the simultaneous and relevant action of electric and

236

"mechanic" noises. We treat it similarly to two allied merges of revealing the substance structuring at micro-, nanoscopic level that are integrated by the term "electromechanical noise". The expounded before could be successfully described by the equation applicable to the researched substance:

$$T\frac{dS}{dt} = -\eta\frac{dN}{dt} + \sigma\frac{d\varepsilon}{dt}. \tag{6.5}$$

Hereby, under electrochemical noise we mean the appearance and development of a charge-potential relief that causes the continuous change in a stress-deformation field and takes place in the researched substance at the nano-, micro- and macro- levels. The electrochemical potential η could be defined [21] by the chemical potential μ and electric potential φ: $\eta = \mu + e\varphi$, where e is the electron charge. Therefore equation (6.5) is modified due to separation of electrochemical noise in electric noise and "chemical" noise related to NTs and thermoelectric thermometers respectively:

$$T\frac{dS}{dt} = -\mu\frac{dN}{dt} - \varphi\frac{dQ}{dt} + \sigma\frac{d\varepsilon}{dt}. \tag{6.6}$$

Joint processes of electric and heat transfer in thermometric substances enable to determine the type of TF of the mentioned thermometers (Fig. 6.2).

So electric and heat degrees are the main degrees of freedom in determining the TF $P(T)$ of a NT. Consideration of other pairs of conjugated thermodynamic forces and flows (mechanical and chemical degrees of freedom) allows to qualify their influence as fluctuation-dissipation deviations of characteristics caused by the mentioned noise types. The last ones usually form the appropriate influence functions. For example, selective gas sensing is developing nowadays with the help of pristine graphene [24] as sensitive substance. Hereby some gases induce clear peaks in the well reproducible noise spectra, which could be used as their signatures in sensors. So, it has been possible to characterize it by simultaneous low-frequency noise of gate and drain currents of AlGaN/GaN transistors [25].

Other degrees of freedom, for instance the superficial degree that is essential for nanobiotechnology, are responsible for their own influence functions. Their effect can be evaluated through the transfer processes, due to responsible for this thermodynamic forces and flows.

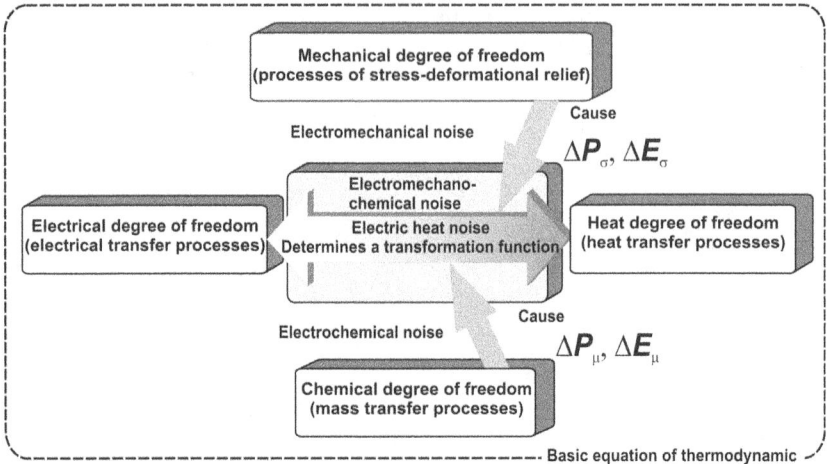

Fig. 6.2. Participation of different noise types in forming transformation functions of a noise thermometer and the relevant freedom degrees of the basic equation of thermodynamics.

- Processes in thermometric substance of a noise thermometer. We could neglect the "chemical" noise consequently of temperature gradient absence along the NT structure. The same concerns electromechanical noise (under condition of quality design and precise production). As a result, there remains only electric noise related to the entropy dissipation in (6.6). The equation of the NT transformation function, quite close to (6.2), has been deduced. It relates the power of electric noise P_{el} with the thermodynamic temperature T through the dissipation rate of entropy dS/dt:

$$T\frac{dS}{dt} = -e\varphi\frac{dN}{dt} = P_{el}, \qquad (6.7)$$

where e is the electron charge; N is the quantity of charge carriers. Chemical and mechanical degrees (Fig. 6.2) determine the influence functions by the effect on electric noise. Thus, at the deformation of thermometric substance under the action of electromechanical noise, a transformation function of a NT tends to change as regards to (6.7). For example, energy accumulated in the deformation process (consequently of wire reeling of a SE) substantially affects a TF through "mechanical" noise of the electrical power P_σ forming the influence function directly deductable from (6.6) with regard to stresses and deformations:

$$T = bP_{el} - bP_\sigma, \tag{6.8}$$

where $b = \left(\frac{dS}{dt}\right)^{-1}$ is the constant of TF of a NT inversely proportional to the dissipation rate of entropy. The given constant is usually determined experimentally at the stage of calibrating NT during simultaneous control of electric noise power / voltage and temperature of its SE with the help of additional gauging means. Evidently, a dissipation rate of entropy inside the researched substance could be estimated following a statistical thermodynamic approach [26]. Hereby, we should follow from the most probable physical processes under the given conditions:

$$\frac{dS}{dt} \approx \frac{\Delta S}{\tau}, \tag{6.9}$$

where ΔS is the entropy changes taking place consequently of a relaxation process with the constant τ. Noise spectrum peculiarities become most evident at low temperatures where it could be easily provided the separation of certain influence factors [27].

For example, at 77 K and lower when the velocity of diffusion processes is substantially decelerated, the capability of generating the excessive noise inside *Ni* and *Cu* caused by the distortions during plastic deformation (by wounding and stretching at the manufacture of a NT sensitive element) has been estimated. The energy that can be accumulated by a crystalline lattice is 3.4 and 2.1 kJ/kg respectively for these metals [28]. It causes the non-stationary electric noise generation with the power 10^{-8} W. So, the non-stationary state inside sensor's substance could last till 10^8 s, which considerably exceeds the gauging threshold duration.

The restoration of a thermodynamic equilibrium could occurs involving different mechanisms: the accumulated elastic energy could be transformed into the energy of a microcrack surface with the relaxation constant τ_1 or removed as heat energy from the relaxation zone due to the thermal diffusivity with the constant τ_2. Hereby, the constant of the given relaxation process is much smaller in comparison with the relaxation constants of dislocation movement and their duplication. At the room temperatures the dissipation process in the form of microcracks' formation becomes predominant. Lowering temperature to a nitrogen reference point strengthens the mechanism of heat dissipation from the places of previously accumulated energy which is gaining special significance.

We have conducted the experimental research of NT proper noise (Fig. 6.3), e. g. at 273 ... 4.2 K when the TF drift ~0.03 % of its nominal values was fixed. Considering the mentioned competitive processes of relaxation of the previously accumulated elastic energy through the microcracks' formation with the constant τ_m and through heat removal with the constant τ_h in (6.9), we conclude [6]:

$$\delta T = \frac{\Delta T}{T_n} = \frac{\tau_m}{\tau_h}. \qquad (6.10)$$

The ratio of two constants is determined in [29] as a factor of special importance at the materials science research of mechanical characteristics:

$$\frac{\tau_m}{\tau_h} = A \frac{c\sqrt{ad}}{\chi}, \qquad (6.11)$$

where c is the sound speed; a is the grain size; d is the atom size; χ is the thermal diffusivity; A is the constant.

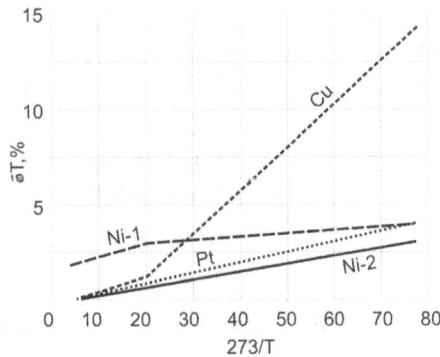

Fig. 6.3. Inverse temperature dependences of instrumental error of NTs with different thermometric substances (Ni-1 is annealed i Ni-2 is hardened).

At the error of measuring the noise powers ≤0.92 % at 4.2 K; ≤0.5 % at 77 K within the range 4.2...77 K, it is revealed the deviations from linearity for SEs made from pure Ni, Cu and Pt; the absence of which is fixed for composites and alloys with low thermal diffusivity and considerable sound speed. The results of undertaken research of NTs with *Ni, Pt, Cu* sensitive elements at 77 K have shown substantial deviations of their electric noise power from the values determined after Nyquist equation). The proportionality of the fixed deviations to the ratio of relaxation times supports the correctness of the proposed reasoning. The ascertained values of the relative changes of a TF (see an equation (6.10) and a vertical axis of Fig. 6.4), with considering the values of sound speed, grain sizes, atom sizes and thermal diffusivity (see equation (6.11) and a horizontal axis of Fig. 6.4), apt to align.

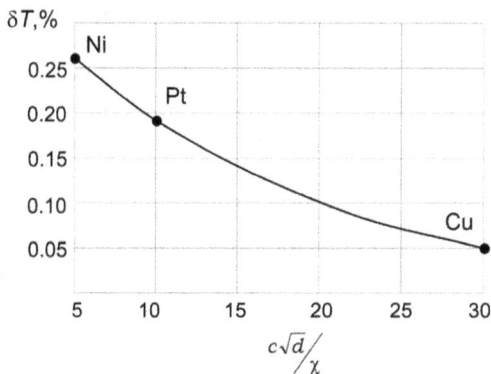

Fig. 6.4. Instrumental error $\delta T = \delta T \left(c\sqrt{d}/\chi \right)$ of NT at 77 K depending on the researched thermometric substance: theory (———) and experiment (•). The values of: thermal diffusivity – $24.6\cdot10^{-6}$, $18.0\cdot10^{-6}$, $112.5\cdot10^{-6}$ m^2/s for Pt, Ni, Cu respectively [30]; sound speed – $2.7\cdot10^3$, $4.8\cdot10^3$, $3.8\cdot10^3$ m/s [31]; atom size – 3.92; 3.52; 3.61 Å [32] with an average grain size 3 µM.

It proves the correctness of involving a thermodynamic approach to the description of fluctuation-noise effects.

The average temperatures are characterized by the relaxation mechanism that is based on the motion and multiplication of dislocations. Finally, high temperatures up to a melting point are matched with diffusion removal. Then one should consider the competitive action of two relaxation mechanisms: diffusion mechanism and formation of microcracks in the deformed local substance microvolumes. The role of diffusion is emphasized by the research of $1/f\gamma$ noise [20].

The higher self-diffusion coefficient, the longer the microcracks formed due to increased substance plasticity with the same value of mechanical stresses. Smaller cracks as compared to critical are quickly healed due to diffusion of vacancies from the area of microcrack to nearby located sinks. The introduced plasticity criterion at the high temperatures is varied to: $\tau_4/\tau_1 \sim c/D$, since in (6.11) the thermal diffusivity χ is replaced by the coefficient of diffusion D which increases exponentially with temperature. It means the practical absence of possible alters of calibrated transformation characteristics at high temperatures.

6.2.2. Problems and Methodology of Noise Measurements

Evolving the method of research on the electrical noise of substances of different phase-states (liquid, solid or gaseous) and linear sizes, starting with macro-sizes and finishing with nano-sizes, requires to interpret the noise phenomenology as the fact that is determined by interaction within the limits of the substance itself and that between the substance and a measuring device. It highlights the need for the distinct understanding of both research subjects and methods. For example, with studying the electrical noise of CNT of the electric resistance 12.6 kΩ [33], a nanotube could remain in the superconductive state [34], and the value of resistance (12.9 kΩ according to [14]) is supposed to be determined by the resistance of nanotube contacts with bringing-in wires. It means that the measured noise could be referred to nanotubes just actually characterizing their contact areas.

Beside the need for awareness of monitored object noise-characteristics, some metrological experience is necessitated. This is the experience in manufacturing and exploiting NTs with sensitive elements whose materials are used for research on information parameters by means of noise peculiarities. Apart from, the availability of metrological support for the very NTs when the main monitored parameter is electric noise characteristics (voltage, current, power) could be adopted for the nanometrological assistance of materials science and industry efforts of nanotechnology exactly while studying electrical noise of different substances [8] as well as components of manufactured wares of nanoelectronics [2].

Methodology of investigating the integral characteristics of noise signals. Research methodology is first of all the methodology of measurements that includes the following aspects:

a) Assurance of the increased interference resistance due to the low amplitude value of the measured signals;

b) Separation of the bandwidth for intended investigations;

c) Predetermination of the general length of an experiment, particularly while undertaking low-frequency research on $1/f$ noise with regulation of a basic reasonable error of method;

d) Selection of the methods of processing the gained signals (particularly, application of the high-speed Fourier transformation that includes the differentiation of a signal into frequency spectra; separation of interference frequency, for instance, the frequency of feeding power with its multiple frequencies; and their elimination from the re-formed signal);

e) Selection of optimal noise-characteristics research methods in terms of quality and negative issues;

f) Choice of methods apt to characterize the researched materials by known informative quality, durability, reliability, uncertainty and other parameters.

The research of integral characteristics of noise signals appearing in SEs enables comprehensive studying into physical nature both of noise processes occurring in substance, and non-noise substance parameters and characteristics. Exactly such an approach to noise research most completely renders its significance for modern electronics. Hereby the improvement in the means of measuring technics, development of new conceptions of conducting the gauges themselves, automatization of a measuring process, implementation of interference-resistant technologies of gaining the output signals and etc. pose essential requirements to the methodology of measuring noise characteristics. Theoretical and experimental research in the area of measuring the integral characteristics of noise signals reveals the chain of problems, namely:

- Durable measuring caused by random nature of a noise signal;

- Influence of uninformative noise signals in the input circuit on the results of measurement;

- Low level of the measured noise signal;

- High requirements to interference-resistance of the MIs.

The latest means a reasonable request of the high precision measurement of informative noise signal on the quite powerful noise-interference background.

The researchers should ensure increasing the noise immunity of the means of measuring the integral characteristics of noise signals. The low level of the measured noise signal leads to the necessity for applying the wide-band amplifiers with the coefficient of amplification more than 10000. At such an amplification coefficient the high-sensitive input of the device gauging part with a wide bandwidth of the signal being measured has been gained. It has entailed the high requirements to the noise immunity (interference-resistance). The source of outer interference could be the subject of measurement itself; or/and power network feeding the MI; or/and computer connected with the MI by installation wires and so on. Inner interference could arise due to a high coefficient of amplification – the influence of high-volt output circuits on the input high-sensitive circuits consequently of parasite resistive, inductive and capacitance connections.

Minimization of the noise interference influence on the measurement result could be made by means of screening and grounding in the schemes of analog signal conversion, optimal parting and galvanic segregation of common conductors of digital and analog schemes etc. For any construction the analysis of the most vulnerable areas in relation to interference should be conducted. Besides, the certain constructive decisions supporting the steady work of schemes should be made. Screening and grounding in the schemes of analog signal conversion far not completely obviate the noise interference effect. Then the filtration methods for marking the measured signal off the input one which could be a superposition of the measured signal itself and interference should be applied.

In most areas of science and technics, the measurements of determined signals are conducted, whilst in the case of noise values the random signals are of interest. The interference could be both determined and random. Besides, taking into consideration the fact that measurements are performed within the certain bandwidth, the determined interference could reveal itself at the various frequencies. Therefore we should use the filters of complicated configuration that combine a band filter for forming the work-bandwidth of a measuring device, and rejecter filters for some frequencies which could be numerous. Taking into account the specific conditions of measurement as well as design difficulties and working principles of such filters, the synthesis of digital intelligent filters with the usage of rapid Fourier transformation which meets the enumerated requirements is optimal.

Durable measuring period is stipulated by random nature of the measured signal. The measured signal is mainly noise voltage or current. Each of them has a nature of homogeneous continuous random fluctuations concerning the average that is up to zero and constitutes a random ergodic stationary process. Studying it, any moment of time can serve as starting point. Measuring the parameters of a stationary process within any period of time, we should receive the same values of its characteristics.

Such integral characteristics of a random process as the mean of a square (variance in statistical investigations), mean square value (standard deviation) and spectral density of a noise signal tend to be measured at the first place. Since a noise signal is a random process, the true value could be gained during the infinite time of averaging. Any restriction on the averaging time leads to the appearance of a methodical error. In the ideal case, if there are no other noise signals except the measured one in the measuring circuit, the standard deviation of a noise signal variance σ [35] could be calculated as:

$$\sigma \approx \frac{1}{\sqrt{t\Delta f}},\qquad(6.12)$$

where t is the time of measurement, Δf is the bandwidth of a noise signal.

Results of modeling the dependence of standard deviation of a noise signal variance on the time of averaging at different values of the bandwidth Δf are notified in Fig. 5.5.

To reach the relative mean square value of the variance of a noise signal 0.01 % for the bandwidth $\Delta f = 100\ kHz$, we should conduct measurements for 1000 s, and for $\Delta f = 1\ MHz - 100$ s. Taking into account that other sources of noise signals are present in the input circuit (resistance of a connecting line, amplifiers, feedback resistors), the dependence of mean square value of a noise signal variance on the time of averaging becomes more complicated. Time of measurement for reaching the equal error rises as compared to an ideal case. Correspondently, the measurement of integral characteristics of noise signals could require the certain duration of averaging: tenth – hundredth of seconds. If there is a necessity for measuring the integral characteristics within narrow bandwidth, the time of measurement rises considerably. So to reach the relative mean square of noise signal variance 0.1 % at the bandwidth of a noise signal $\Delta f = 10\ Hz$, time of measurement has to be approximately 30 hours.

- Analysis of uninformative noise signal influence on the result of measurement within the input circuit. In addition to the noise signal being measured, uninformative noise signals exist in the input circuit due to their emersion in connective lines. In a general case, noises determine the lower limit of measurement. The ratio "noise – signal" should be reduced to minimum for any measurement [36]. Influence of uninformative noise signals (in the input circuit) on the measurement results within the input circuit is considered to be maximal as compared with the connective line and the input circuit of a measuring part, and etc.

A signal is determinable in the majority of measurements. Thermal, shot, generation-recombination noises and flicker noise are regarded within the scope of scientific work. Thorough investigation of the nature of these noises and influential factors changing the noise-parameters enables not only conscious development of methods of mitigating the levels of those noises but also their employment in the analysis of different systems state. In our case, the signal being measured is random, i.e. has a noise nature which complicates the extraction of the particular measured signal at the background of a noise interference; signal and interference could be commeasurable in terms of their level.

Thermal noise is the noise determined by the current and voltage fluctuations due to the thermal motion of charge carriers in conductors. The relation of these fluctuations is defined by the Nyquist formula, according to which the mean square voltage value of conductor's resistance R being in the heat equilibrium at temperature T equals to $\overline{U}^2 = 4k_B T R \Delta v$, where Δv is the frequency band, within which voltage fluctuations are measured, k_B is the Boltzmann constant [7].

Thermal noise appears at the expense of random motion of charge carriers in any conductor. Consequently of this motion, the randomly variable electro-motive force arises at the ends of the conductor. The similar phenomenon is observed in the conducting channel of field transistors. Thermal noise is decisive in any device of electrical nature that remains in thermo-equilibrium with the environment. Generation-recombination noise appears when free carriers are generated or recombined in the semiconducting substance. Fluctuating speeds of generation and recombination could be considered as consecutions of independent randomly appeared events, and therefore the process could be regarded as a shot noise.

For the thermal noise, the spectral density $S_{Utn}(\omega)$ of the disconnected circuit noise-voltage [37] and that $S_{Itn}(\omega)$ of short-connected circuit noise-current are correspondingly equal to:

$$S_{Utn}(\omega) = \frac{4k_B T R}{1 + \omega^2 \tau_e^2}; \ S_{Itn}(\omega) = \frac{4k_B T}{R(1 + \omega^2 \tau_e^2)}, \quad (6.13)$$

where τ_e is the mean time of free path of electrons in substance between collisions, ω is the circular frequency.

For all frequencies of practical relevance, the item $\omega^2 \tau_e^2$ is negligibly small. In this case the expression (6.13) is shaped as:

$$S_{Utn}(\omega) = 4k_B T R; \ S_{Itn}(\omega) = \frac{4k_B T}{R}. \quad (6.14)$$

To decrease the influence of uninformative thermal noise, the measurement conditions should be assured providing a noise-measurand is as much as possible in comparison with uninformative thermal noise hereby satisfying the needed ratio "measured noise signal – noise interference". Besides, the influence of uninformative thermal noise on the result of measurement could be minimized following the correlation measurement method [38].

Shot noise, always present in the case a noise-phenomenon, could be regarded as succession of independent random events. Due to the random nature and mutual independence of the motion start of certain charges that arrive in the device working area, overcoming the potential barrier, the spectral shot noise density does not depend on frequency (white noise) and is described by the Schottky equation that is often called Schottky emission: $\overline{i^2}/{\Delta f} = 2qI$, where $\overline{i^2}$ is the mean square value of current fluctuations, Δf is the frequency band, within which noises are measured, q is the elementary charge, I is the current.

At frequencies for which the charge flight time through operation area is commensurate with the period of fluctuations, the spectral shot noise density in circuit connected with the mentioned area, begins to diminish with the frequency increase. This effect is defined by the current pulses spectral composition of duration τ generated in circuit by the every flying charge. While considering the mentioned effects the shot noise can be given by Schottky formula $\overline{\iota^2} = 2qIF_t^2\Delta f$, where the factor F_t^2 depends on frequency and flight time. The deviation from this formula exists also in the case when a current is limited by the spatial charges.

For instance, with releasing electrons by thermo- or photo- cathodes, electron emission constitutes the consecution of such events. To wit, shot noise is inherent in emission currents. The phenomena of crossing the *p-n* junction by charge carriers (electrons or holes) make the succession of independent events in transistors. Therefore these currents reveal the features of shot noise. It is also valid for the transitions between two energetic levels, for example, with generating and recombining the carriers in a semiconductor or with emission of laser photons. Input circuits of MIs could be constructed with the use of bipolar and field transistors. For bipolar transistors on low-frequencies, when $\alpha(\omega) = \alpha_0$, the spectral densities of collector, emitter and base noise currents are: $S_{Ic}(\omega) = 2q_eI_c$, , $S_{Ie}(\omega) = 2q_eI_e$, $S_{Ib}(\omega) = 2q_eI_b$, where q_e is the electron charge, I_c is the collector current, I_e is the emitter current, I_b is the base current. For field transistors the spectral densities of channel current and gate current are: $S_{Ic}(\omega) = \frac{2}{3}4k_BTG_m$; $S_{Ig}(\omega) = \frac{4\omega^2C_i^2}{15G_m}4k_BT + 2q_eI_g$, where G_m is the mutual (transferring) conductivity, I_g is the gate current, C_i is the input capacitance of a field transistor. The influence of shot noise could be mitigated by marking off active elements with small values of noise currents and voltages, and by applying the correlation measurement method.

Flicker noise that could appear consequently of different reasons is characterized by special spectral density nonlinearly rising within the band of low frequencies. The main property of flicker-noise is resemblance of spectral density to $1/f$. In the case of a semiconductor element, the spectral density $S_{Inp}(\omega)$ of current random component and the same $S_{Unp}(\omega)$ of random voltage component on the coal resistor make: $S_{Inp}(\omega) = \frac{2\pi\alpha I_0^2}{N_0\omega}$, $S_{Unp}(\omega) = \frac{2\pi A I_r^2}{\omega}$, where α, A are the constants, I_0, I_r are the currents passing through the semiconductor element and resistor, respectively, N_0 is the equilibrated amount of carriers in a semiconductor.

The influence of shot noise could be mitigated by marking off active elements with small values of noise currents and voltages, and by applying the correlation measurement method. Flicker noise that could appear consequently of different reasons is characterized by special spectral density nonlinearly rising within the area of low frequencies. The main property of flicker-noise is resemblance of spectral density to $1/f$. In the case of a semiconductor element, the spectral density $S_{Inp}(\omega)$ of current random component and the same $S_{Unp}(\omega)$ of random voltage component on the coal resistor make:

$$S_{Inp}(\omega) = \frac{2\pi\alpha I_0^2}{N_0\omega}; \ S_{Unp}(\omega) = \frac{2\pi A I_r^2}{\omega}, \qquad (6.15)$$

where α, A are the constants, I_0, I_r are the currents passing through the semiconductor element and resistor, respectively, N_0 is the equilibrated amount of carriers in a semiconductor.

The typical dependences of spectral density of thermal, shot and flicker noises are represented in Fig. 6.5.

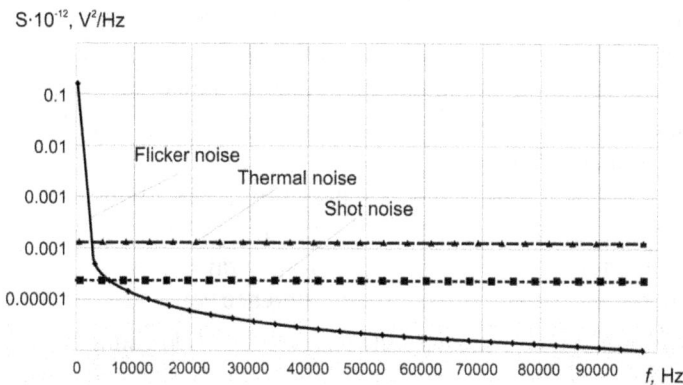

Fig. 6.5. Typical shapes of spectral density of different noise signals.

Under some conditions the level of thermal noise could be lower than that of shot noise, and flicker noise may acquire the dominated value to the frequencies 1...3 kHz. To reduce the influence of flicker noise on the result of measurement, the operating bandwidth should be above 1...20 kHz, depending on the conditions of measurement, sensor resistance value and etc. If necessary to conduct flicker-noise measurements at low frequencies, the need for the considerable increase in measurement duration should be taken into account.

Low level of a noise signal influences on the measurement results and it becomes quite important whereas the estimated below voltage may be about 1 μV. Therefore while creating the means of measuring the integral characteristics of noise signals it should be used the wide-band amplifiers with the coefficient of level 1000 – 10000, depending on the value of a noise signal. Mainly the latter is of a very low level. So for example the average value of a square of noise voltage calculated after the Nyquist formula and the mean square value of noise voltage for the bandwidth Δf=100 kHz for a sensor of resistance 100 Ω at the room temperature are:

$$\overline{e_x^2(t)} = 0.16 \cdot 10^{-12}\ V^2; \quad U_x = \sqrt{\overline{e_x^2(t)}} = 0.4 \cdot 10^{-6}\ V.$$ Optimal choice of noise parameters of amplification elements enables reaching the necessary ratio "signal being measured – noise interference", decreasing the error of measurement and improving the metrological characteristics of the MIs. For the relation of spectral densities of the measurand S_x to the noise interference S_{nz} prevailing 10, the relative standard deviation of a methodical error for different averaging times is practically stable. The further improvement of the ratio "noise signal being measured – noise interference" does not provide the considerable decrease in the methodical error of temperature measurement, and 10 could be treated as the nominal ratio. Manufacturers of OAs norm the values of noise voltage and current as the square root of spectral density of noise voltage and current: $U_{oa} = \sqrt{S_{Uoa}}$; $I_{oa} = \sqrt{S_{Ioa}}$, where U_{oa} is the spectral density of noise voltage mean-square-value of OA, S_{Uoa} is the spectral density of OA noise voltage, I_{oa} is the spectral density of mean-square-value noise current of OA, S_{Ioa} is the spectral density of OA noise current. Correspondingly, the ratio of spectral densities of the noise signal under measurement to noise interference could be estimated after: $\frac{S_x}{S_{oa}} = \frac{4k_B T_x R_x}{U_{oa}^2} = 10$. To satisfy the given ratio while using the sensor of the nominal resistance 100 Ω at the room temperature, an amplifier with the following parameters should be used: $U_{oa} = \sqrt{\frac{4k_B T_x R_x}{10}} = 0.41 \frac{nV}{\sqrt{Hz}}$. On

the other hand, we could increase the value of sensor resistance, and hence the level of a measured noise signal, hereby assuring the needed ratio of measured noise signal and noise interference spectral densities. With using OA in the input circuit (e.g. LT1028 with $U_{oa} = 0.85 \frac{nV}{\sqrt{Hz}}$), the value of sensor resistance should be equal to $R_x = \frac{10 U_{oa}}{4 k_B T_x} = 430 \, \Omega$. It is a real approach to the problem solution. However, we also should consider the proper input noise current of OA whose effect on the result of measurement rises with an increase in resistance of a sensor and eventually could become predominating. To wit, while using OA the special attention should be paid to such OA parameters as proper noise voltage and input noise currents.

The usage of a correlation amplifier enables minimizing the influence of OA noise voltage. Minimization of the influence of a noise current could be reached by reducing the nominal sensor resistance and selecting OA with small values of noise current. Consequently the question of optimizing the input circuit and hence the means of measuring the noise in general has become vital.

- Increasing the noise immunity of measuring the integral characteristics of noise signals. As there was mentioned above, the low level of the measured noise signal leads to the necessity for applying the wide-band amplifiers with the coefficient of amplification more than 10000. At such an amplification coefficient the high-sensitive input of the device gauging part with a wide bandwidth of the signal being measured has been gained. It has entailed the high requirements to the noise immunity (interference-resistance).

The source of outer interference could be the subject of measurement itself; or/and power network feeding the measurement device; or/and computer connected with the measuring device by installation wires and so on. Inner interference could arise due to a high coefficient of amplification – the influence of high-volt output circuits on the input high-sensitive circuits consequently of parasite resistive, inductive and capacitance connections. Minimization of the noise interference influence on the measurement result could be made by means of screening and grounding in the schemes of analog signal conversion, optimal parting and galvanic segregation of common conductors of digital and analog schemes etc. For any construction the analysis of the most vulnerable areas in relation to interference should be conducted. Besides, the certain constructive decisions supporting the steady work of schemes should be made. Screening and grounding in the schemes of analog signal conversion far not completely obviate the noise interference effect. Then the filtration methods for marking the measured signal off the input one which could be a superposition of the measured signal itself and interference should be applied.

In most areas of science and technics, the measurements of determined signals are conducted, whilst in the case of noise values the random signals are of interest. The interference could be both determined and random. Besides, taking into consideration the fact that measurements are performed within the wide bandwidth, the determined interference could reveal itself at the various frequencies. Therefore we should use the filters of complicated configuration that combine a band filter for forming the work-bandwidth of a measuring device, and rejecter filters for some frequencies which could be numerous. Taking into account the specific conditions of measurement as well as difficult construction and working principles of such filters, the synthesis of digital intelligent filters with the usage of rapid Fourier transformation [39] which meets the enumerated requirements is optimal.

Each of considered problems leads to appearance of the whole chain of error components of measuring the integral characteristics of noise signals. These components depend on the method of measurement, metrological characteristics of measuring means and their proper noise characteristics. Analysis of possible error sources, minimization of their influence on the result of measurement, consideration of the specificity of the measurand, satisfaction of appropriate ratio "noise-signal being measured – uninformative noise-signal" and etc. enable gaining the reliable results of measuring the integral characteristics of noise signals.

- Electrical noise at electrical, mechanical and thermal loadings. The sources of electrical noise appearance are different, so the bandwidth and spectral density exist. Essential and considered below are current frequency-dependent noise at the low frequencies and thermal noise at the high frequencies.

The research [40] of frequency dependences of noise PSD of metal contacts and thin semiconducting films allows us to specify the nature and origin of the mentioned noise by dint of attracting electron microscopy and other structure-sensitive methods. Particularly, the PSD frequency-dependence at the low frequencies reminds the similar dependence of the fatigue limit studied by the method of internal friction.

Here the $1/f$ dependence of PSD is characteristic for the low frequencies 15...200 Hz. With a rise in frequency it becomes frequency-independent which is related to the marked heat-waste during cyclic deformation.

Continuous energetic feeding of the substance by passing the electric currents of considerable density enables simultaneous observing and segregating of equilibrated and disequilibrated noise components of $1/f\gamma$ type, which are researched through the spectrum form index γ, through the apparition of 2- and 3- multiple frequency-satellites and through the ratio of components. Thus, with an increase in current, the spectrum form index γ rises from 1 to 2 and higher which is revealed at 320...410 K in the films of aluminium and its alloys with silicon (current density makes 0.3 ... 2.5 10^4 A/mm^2). The same refers to the study on CNT [33]. To reveal $1/f^2$ noise at the direct current, the measurement has been conducted within the bandwidth 0.01...1.0 Hz, since it was masked at higher frequencies by the equilibrated $1/f$ noise with the spectrum form index $\gamma \approx 1$.

Equilibrated $1/f$ noise of metal films including non-deformed and annealed ones appears due to motion fluctuation of charge carriers while their dissipating on the lattice. The processes of substance electrotransfer to different stocks in the volume (volume diffusion) or on the grain borders (surface diffusion) entail the irreversible structural changes in the films emitting disequilibrated $1/f^2$ noise. The given component called as the electromigration $1/f^2$ noise has non-stationary nature and is practically absent at the alternating current and room temperature. With an increase in temperature, the electromigration processes leading to enlargement of the spectrum form index γ to 2 tend to intensify. In general, the dependence of average noise-power density $S(f) \sim 1/f^2$ on the direct current density is described by the expression: $S(f) = j^3 BT^{-1}f^{-2}exp\left(-{E_a}/{k_B T}\right)$, where B is the constant; E_a is the energy of electromigration process activation corresponding to that of matrix atom self-diffusion activation. For instance, in the case of polycrystalline aluminium film it is equal to 0.6 eV at 327...396 K. In the metal conductors of submicron dimensions with a monocrystalline or bamboo structure, in which the average longitudinal grain size considerably prevails the film width and therefore mass transfer by the grain boundaries is absent, the values of activation energy, determined from the temperature dependences of average $1/_{f^2}$ - noise-power density, are much larger as compared to polycrystalline films. They make 0.8 eV and are assigned to the energy of activating the diffusion along dislocations.

There are known cases of appearance of disequilibrated noise in the films whose structure differs from thermodynamically equilibrated one: freshly made films, deformed or irradiated films. The similar is observed in the aluminium films on a polyimide substrate where for frequency 20 Hz while applying an external bending force (within the area of elastic deformation), the average power density of $1/_{f^\gamma}$ noise was rising approximately 30 as much at the increasing of efforts from 0 to 120 MPa [19]. Within the carbon fibers of the diameter 6 μm and of the length 10 cm with the electric resistance 41.3 kΩ at the efforts of extension up to 250 MPa, a 20-fold increase in average $1/_{f^\gamma}$ noise power density is fixed (electric resistance rises negligibly). Moreover, irreversible changes in the noise level are related to the plastic deformation of fibers.

In length of time the structures of films being exposed to temperature or electric current are being ordered: the inner energy and the amount of defects are decreasing exponentially. Consequently, non-stationary noise is smoothly converted in the stationary one during relaxation time decreasing with a rise in temperature. Moreover, the spectrum form index γ of the given disequilibrated noise was changing within the limits 2...3. For only just precipitated films of chromium a decrease in the spectrum form index γ is also observed [40] but from 2.5...3 to 0.7...1.2 consequently of vacuum annealing duration 30...45 minutes. It has proved the necessity for attracting the disequilibrated mechanisms of clarification the reasons for the noise level decrease in the process of chromium films aging. An increase in a specific material volume consequently of combining the disequilibrated vacancies into complexes or into closed micro- or submicro- pores under pressure of internal mechanical stresses may be one of those reasons [41].

- The case of equilibrated noise (current does not pass through the researched substance). Thermal noise prevails at the frequencies above 100 Hz and its nature is uncorrelated. The analysis of PSD frequency dependences has revealed the considerable influence of dissipation of the accumulated energy [42]. The frequency-dependent $1/_f$ noise of low-frequency range (lower than 80 Hz) corresponds to the reversed transformation of energy into phonons at the tensile defects. Their PSD is: $S(f) = {P_{el}}/{\Delta f} = {c}/{af}$.

The case of disequilibrated noise (the ratio of amplitudes of feeding voltage and noise voltage makes $\sim 10^{-7}$ and of their powers 10^{-14}) has been studied in [40] on the molybdenum films of the thickness 247 and 560 nm. The average density of harmonic tones amplitude fluctuation power of a response signal concerning the test effect of sine voltage of the bandwidth 10...1000 Hz, and also on the effect of direct current 0.45 mA have been considered. The latter has provided the total PSD of equilibrated and disequilibrated noise. The mechanism of multi-phonon capture [43] has been activated on the tensile defects as the phonon traps. The dependence of the index γ within the limits 2...3 on the frequency is explained by the influence of electro and mass transfer.

- The research of dynamics of the change in noise voltage at the thermal shock. At the considerable speed of temperature alteration all transfer processes are much complicated. For example, in substance kept at the certain temperature and rapidly moved into the environment with higher temperature, surface-volume mechanical tensions capable of accumulating the inner energy appear consequently of forming the dislocation ensembles. All real crystals have their defects distributed due to distortions in the atom allocation providing atom sizes exceed considerably the crystalline lattice constant. At dynamical temperature mode the presence of such structural defects could lead to the change in noise characteristics.

At the rapid heating the sensors of NTs the transient noise process caused by thermodynamic disequilibrium has been detected. Such a behavior of a noise signal could be revealed at the expense of internal changes intensified consequently of applying an uneven temperature gradient on the substance with inner defects. The research on the behavior of a noise signal in the dynamic temperature mode (Fig. 6.6) has been made following the methods of rapid transferring of a sensor from the environment of one temperature into that of higher temperature. In the moment of time ~ 170 s from the beginning of measurement, the temperature of the researched environment is changing abruptly from 288 K to 368 K. The temperature that was fixed during the transient process and determined by the noise power exceeds the temperature value of the researched medium almost 1.5 as much.

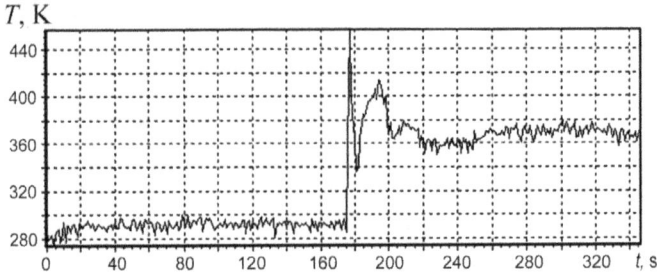

(a)

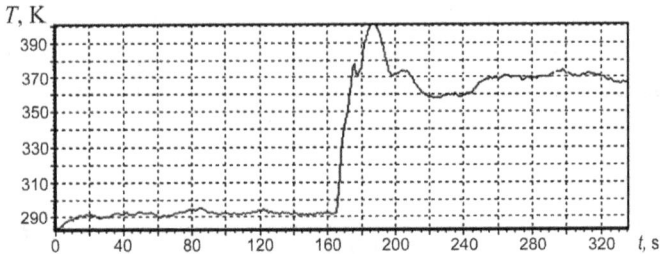

(b)

Fig. 6.6. Process of altering the indices of a noise thermometer at temperature jump from 288 K to 368 K (bandwidth of a noise signal: 10–110 kHz; averaging time: (a) 1 s; (b) 10 s.

While studying the resistance thermometers and thermoelectric thermometers, the similar deviations from the equilibrated indices have also been observed (Fig. 6.7). Temperature exceeding has been stimulated by an increase in temperature jump ΔT. Besides, this process was getting more intensive while a two-stage jump $(\Delta T = 600 + 600\ K)$. Those changes could be related to the appearance and relaxation of mechanical stresses causing the changes in thermoelectric power and taking place nearby structure defects.

Moreover, the probability of passing the researched substance from one thermodynamic state into another at negligible temperature changes is determined by the value of a temperature jump, entropy and time. At the considerable temperature changes, entropy remains to be the most decisive factor determining all the transient processes inside the concrete substance. We tend to consider further the researched sensitive substance as a subsystem (thermodynamic system with minimal outer influences), forming the part of a larger transducer's system and being the smallest part of much larger system of the monitored environment.

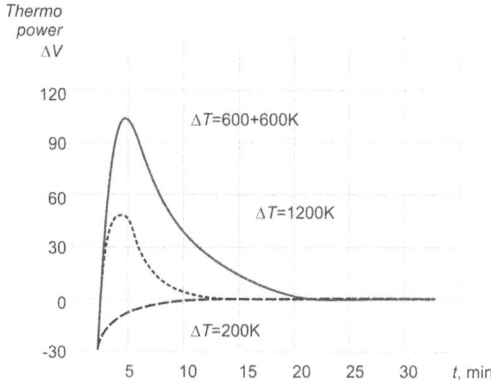

Fig. 6.7. Transient thermo-process according to the indices of differentially connected two thermocouples at the different temperature jump ΔT.

The probability of the transition of the given subsystem from the state X_0 to the state $X_0 + \Delta X$ is determined by the prehistory of substance: $dP \sim exp\left[-\frac{\Delta W(X)}{k_B T}\right]$, where ΔW is the work equal to the change in inner energy ΔU that the outer environment should apply to the given subsystem to put it out of state with the parameter X_0 to the state with the parameter $X_0 + \Delta X$. In general, $\Delta W(X)$ is considered as a measure of fluctuation probability of the parameter X. Probability of independent transition of the given subsystem from the state X_0 into the state $X_0 + \Delta X$ (or probability of fluctuations) is the larger, the smaller ΔW is. Considering this equation at $\Delta U = T\Delta S + S\Delta T$ (S is the entropy) the link of internal energy and the heat degree of freedom TS could come to: $dP \sim exp\left(-\frac{\Delta S}{k_B}\right) exp\left(-\frac{S\Delta T}{k_B T}\right)$. According to the law of minimal speed of entropy production that is treated as the more general case of the widely known statement of saving invariable entropy $(dS \to 0)$: $\frac{dS}{dt} = min$. It means that the entropy of substance subsystem primarily taken out from the state of thermodynamic equilibrium consequently of environment temperature change from T_0 to T, is increasing linearly with time: $S(t) = S_0 + \Delta S = S_0 + gt$, where S_0 is the entropy in the moment of time $t{=}0$; g is the constant determined by the speed of entropy change for the given subsystem. Then the probability of its transition from one stage into another will be: $dP = C exp\left(-\frac{gt}{k_B}\right) exp\left(-\frac{S_0(T-T_0)}{k_B T}\right) = C_1 exp(-a_1 t) exp(-\frac{a_2}{T})$, where $a_1 = \frac{g}{k_B}$; $a_2 = \frac{S_0 T_0}{k_B}$; $C_1 = C exp\left(-\frac{S_0}{k_B}\right)$, where C is the constant.

Consequently, the probability of spontaneous transition is proportional to the power of noise. Moreover, the parameters of a temperature jump are included in the coefficient a_2 whilst the technological parameters, determining the primary entropy value and its change at the dissipation of a thermal shock, are defined by the coefficients C_1 and a_1. Under the condition $T \gg T_0$, i.e. that in the initial state the given substance is remaining at the low and even room temperatures, we can gain: $dP = C_2 exp(-a_1 t)$. It means that the changes in considered noise voltage depend also on temperature and on a substance prehistory (e.g. on the mechanical processing or/and on vibrations).

6.3. Non-Invasive Diagnostics

CPS development foresees, in particular, the progress in modeling tools and data sets that enable researchers accurately model the scale and dynamics of current and future Internet control systems. The same holds for models and tools for measuring and predicting the performance of secure network-based CPS [3]. In this direction the method of non-destructive noise spectroscopy is considered as perspective one [44].

6.3.1. 1/f Noise Studying for Diagnostics

In scientific world the application of methods of $1/f$ noise studying for diagnostics of phenomena of biological [45], chemical [46], social and technical [47-50] origin is widely known. Methods of non-invasive diagnostics are vital for wide circle of producers – from companies engaged in production of relevantly simple items (for instance, transistors, diodes etc.) and even to airline companies and space agencies for which the reliability of entire modules is important. Non-invasive diagnostics of systems enables to receive information about state of studying system by $1/f$ noise, specific of its evolution, and peculiarities of its structural organization on basis of analysis of generated signals without outside effects on it, or without temperature, electrical, mechanical impacts on this system.

Stochastic signal is completely described by the distribution law by means of which moments of the first and higher order can be determined. The analysis of these moments enables to receive information about stochastic signal origin. However in many cases the distribution law of stochastic signals of real systems is unknown or for its determination it is necessary to conduct additional studies connected to time and resources consuming.

In general, signals generated by natural systems and in many cases by technical systems, are inherent in stochastic character. The examples of stochastic signals generated by the system are its parameters fluctuations which often called the systems noise. For electrical systems there are fluctuations of electrical resistance, fluctuations (noise) of resistance on resistor clips; for biological systems – parameters fluctuations which characterize vital activity of human body and other objects (blood pressure, electrical activity of brain, heart and muscles, cells reproduction velocity, their size etc.) [45]; for technical systems – fluctuations which look like acoustic signals (acoustic noise) and appear during the operation of mechanical [47] and thermal systems, and are also caused by internal frictions in bulks; for environmental system – these are fluctuating changes of atmospheric pressure, geomagnetic field, solar spectrum [51], temperature etc.

The problem of information receiving from stochastic signals of natural and technical systems can be solved in another way. The point is that distribution law is built on the basis of stochastic signals analysis *S(t)* presented in form of temporal series. Simultaneously in many cases it is known the spectral distribution of stochastic signals (or it can be easily received), which can contain a few bit information than time distribution.

Information that withheld in stochastic signals can be estimated quantitatively through informational entropy which for case of continual stochastic signal *x(t)* is determined as:

$$H_x = - \int_{-\infty}^{\infty} p(x) \log p(x) \, dx, \qquad (6.16)$$

where H_x is the differential entropy; $p(x)$ is the probability density of continual signal value presented in form of temporal series.

But such information estimation doesn't make possible to distinguish two stochastic signals which have similar statistic parameters. Two stochastic signals with normal distribution law and similar dispersion σ^2 (for instance, thermal noise of electronics element and encephalogram signal) will have from (6.16) the same quantity of information $H_{max} = \log \sqrt{2\pi e \sigma^2}$. In practice, such signals are distinguished both by frequency spectrum and information interpretation embedded in these signals. By the first one it can be concluded about the system state (electric system with uniform energy spectrum of the voltage (current) fluctuations is in equilibrium thermodynamic state [7]) and also about processes occurring in it [52]. Besides that, practical unrepeatability in frequency dependencies of real systems energy spectra, especially in the low-frequency band, enables to distinguish studied systems by their peculiarities.

Energy spectrum of real systems stochastic signals is not identical in the whole frequency range – practically in all cases of real systems stochastic signals in the low-frequency band, the signal power is higher than in high-frequency band – flicker noise appears in the energy spectrum of stochastic signals. Today experimental researches have proven that flicker noise is sensitive to inner structure of studied systems [48-49, 53-55], so flicker noise spectrum analysis can be applied for systems of non-invasive diagnostics [56-59].

It is considered in [60] the perspective of use of the chaotic series of different nature variables received under research of various processes and structures. To receive information about the studied system they are presented mainly as the temporal or spatial series and charts. At the same time the emphasis is placed on analyzing flicker noise power spectrum dependences of the signal formed by the sequence of δ-functions. Information about processes going on inside the system is proposed to find out under interpolation of resulting power spectrum; the latter is determined in [60] on the basis of experimental values of temporal series:

$$S(f) = \frac{S(0)}{1 + (2\pi f T_0)^n},\qquad(6.17)$$

where $S(0)$, T_0 and n are the phenomenological parameters ("passport parameters") which help to distinguish studied complicated structures or dynamics of studied evolution of open dissipative systems. Parameter n characterizes "memory loss" (correlations) velocity in sequence of bursts on time intervals; parameter T_0 has content of time correlation; $S(0)$ is the spectral density in environmental frequencies. So under $n=4$ inside the studied system the turbulent diffusion is going on; under $n = \frac{5}{3}$ the turbulence is fully developed, and etc. Comparison of "passport parameters" values received during analysis of temporal series with their values determined for partial cases, enables to present qualitatively the character of these complicated processes that specify the studied evolution.

Flicker noise spectroscopy method proposed in [60] inherent in apparent defects: is awkward to determinate (or better to fit) parameters T_0 and n under which the expression (6.17) approximates the real spectrum with good accuracy.

Conducted research is based on the expression for energy spectrum of isolated system parameters fluctuations (one of conditions of system parameters measurement is its isolation) which is in quasi-equilibrium state $S_{NB}(f) = \frac{exp(f \cdot \tau)}{exp(f \cdot \tau) - 1} S_B(f)$, where f is the frequency, τ is the relaxation time, $S_B(f)$ is the energy spectrum of the system fluctuations under $\tau \to \infty$. While relaxation with constant τ which distinguishes one quasi-equilibrium state from another one through system fluctuations, the energy releases. The latter could be determined by integral power:

$$W_{FN} = \int_{f_1}^{f_2} S_{NB}(f) df - \int_{f_1}^{f_2} S_B(f) df, \qquad (6.18)$$

where $f_1 \to 0$, f_2 are the frequencies which $S_{NB}(f) \cong S_B(f)$. Value $S_B(f)$ can be defined by experimental data of fluctuation study and considered as energy spectrum in high-frequency band (Fig. 6.8).

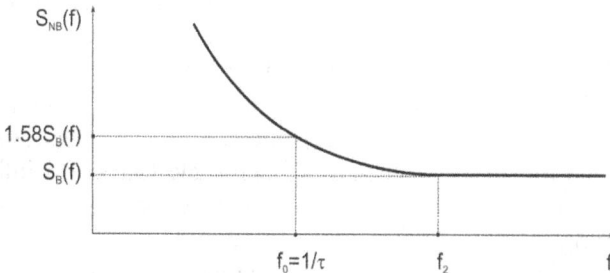

Fig. 6.8. Fluctuations (noise) energy spectrum of quasi-equilibrium systems.

Also τ can be determined by the energy spectrum $S_{NB}(f)$ (Fig. 6.9) under $-f_0 = 1/\tau$, $S_{NB}(f_0) = 1.58 S_B(f)$. Here the informative parameter is a relaxation time τ that depends on peculiarities of the studied inner structure.

To find out the information essence that is inherent in the studied system stochastic signals the experiments had been conducted by modeling the dependences of system relaxation time τ from its proper structure peculiarities. In Fig. 6.9 is shown the model of chaotic movement of particles forming the system; their structure elements are non-transparent partitions (quantity, size and position can be changed), Also are given appropriate energy spectra $S(f)$. By means of non-transparent partitions can be imitated the studied system heterogeneity, for instance structure defects.

Relaxation time τ of simulated systems was determined as shown in the Fig. 6.9. Analysis of given below models and their energy spectra shows the following. On the assumption of the absence of the system structural heterogeneities (partitions of Fig. 6.9, a) fluctuations energy spectrum is the same in the whole frequency range $\left(S_{NB}(f) = S_B(f)\right)$ and relaxation time tends to infinity. With the same quantity of partitions ($N=10$) chaotic movement energy spectra of particles forming system are different (Fig. 6.9, b; Fig. 6.9, c).

The systems in which partitions are disposed orderly distinguish by higher relaxation time (Fig. 6.9, b). Relaxation time decreases at order infringement (Fig. 6.9, c; Fig. 6.9, d). Experimental researches results confirm the impact of the studied system inner structure on altering its energy spectrum. For instance, in metal-oxide field-effect transistors till and after energy processing the noise energy spectra become different [61]. As known during substance heat treatment the internal structure is changing. Noise energy spectra of same type of transistors are distinguished too [50]. The impact of graphene transistors structure defects on the noise spectrum is studied in [62]. Given modeling results enable to concern that by relaxation time analysis τ of energy spectrum of the fluctuation flicker-component is possible to receive information about the system state (at $\tau \rightarrow \infty$ the system is stabilized $(S_{NB}(f) = S_B(f) = Const)$. Also it could be find out about peculiarities of inner structure (relaxation time τ of the systems with less quantity of defects (Fig. 6.9, c) is bigger than τ of systems with greater quantity of defects (Fig. 6.9, d). Furthermore, it clarifies inner structure dynamics (relaxation time τ of the systems with ordered defects (Fig. 6.9, b) is bigger than that with disordered defects (Fig. 6.9, c).

262

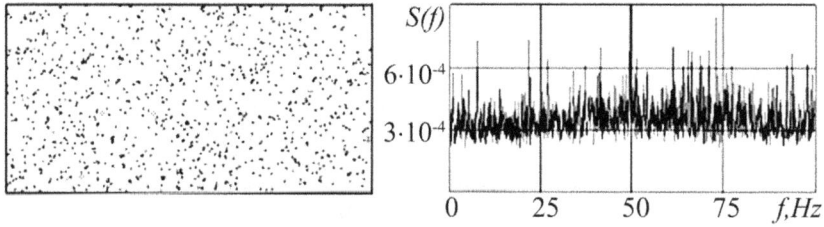

(a) $\tau \to \infty$

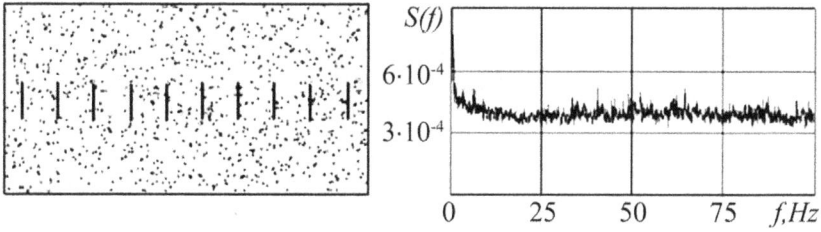

(b) $\tau \cong 0.6\ s$

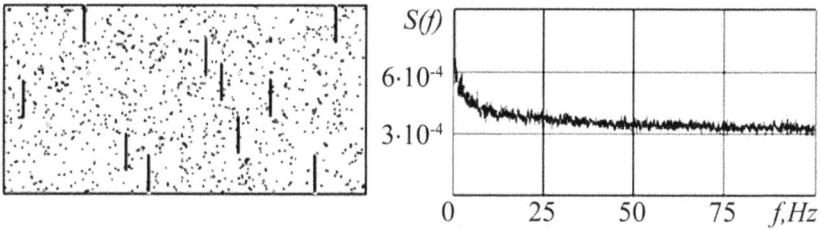

(c) $\tau \cong 0.3\ s$

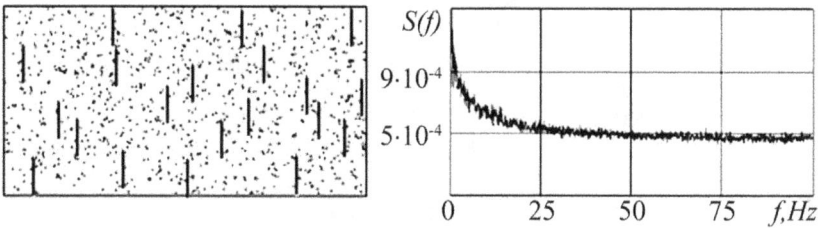

(d) $\tau \cong 0.2\ s$

Fig. 6.9. Models of the systems with inner heterogeneities and energy spectra of chaotic movement of particles forming systems

In real systems the greater quantity of structure defects increases probability of failure. So taking into account parameter τ of informational essence of the stochastic signals energy spectrum it can be proposed the method of studied systems reliability diagnostics: by τ comparing of one-type systems let to estimate their relative reliability. The systems with greater τ-value are more reliable. This value decrease attests oncoming failure state of the system. As parameter τ connected with energy spectrum $S_{NB}(f)$, in principle the methods can be simplified not calculating τ-value. While comparing the intensity of flicker-component of the energy spectrum fluctuation of one-type systems, more reliable one is the system with lower intensity of flicker-component fluctuations (τ is bigger). If there are determined one-type systems, gauging of fluctuations level by measuring τ of each enables to determine mean intensity of the flicker-component fluctuations. System, the intensity of flicker-component fluctuations of which considerably exceeds the average level (its τ-value is far less than mean), is inherent in higher possibility of failure. Systems analysis by means of generated stochastic signals on the basis of parameter τ of their energy spectrum is easier comparing with similar for parameters T_0 and n (see (6.17), since the latter have to be selected, and parameter τ can be determined from experimental energy spectrum $S_{NB}(f)$. By means of parameter τ it becomes possible to compare information quantity which retains in stochastic signals of real systems. Drawing the analogy of the information parameter determination by the distribution of the continual signal (see (6.16)) the differential entropy can be defined as [67]:

$$
\begin{aligned}
H_H(\tau) &= \int_0^\infty P_H \, log \frac{1}{P_H} \, df \\
&= \int_0^\infty \frac{exp(f \cdot \tau) - 1}{exp(f \cdot \tau)} \, log \frac{exp(f \cdot \tau)}{exp(f \cdot \tau) - 1} \, df,
\end{aligned}
\tag{6.19}
$$

where $P_H = \frac{exp(f \cdot \tau) - 1}{exp(f \cdot \tau)}$ is the spectral distribution of fluctuations probability. As opposed to (6.16), equation (6.19) considers the spectral distribution of the energy spectrum of the real systems parameters fluctuations. According to (6.19) more information is contained in systems with smaller value τ (in which the larger number of structural defects exists).

6.3.2. Passive Method of Electronic Elements Quality Characterization

Passive method of electronic elements quality characterization is one of the few non-destructive methods of electronic elements diagnostics by dint of elements proper electric noise. The proper noise studying as method of non-destructive testing could lie into the basis of passive noise spectroscopy. It could be realized the most precisely owing to the usage of NYs as calibrated measuring devices.

Reliability of solid electronic elements is immediately related to the state of their inner structure - the number of crystal lattice defects and their type (microcracks, dislocations, interstitial atoms, defects and etc.) - and hence additional inner tension caused by defects presence. Nowadays, defect ascertainment in electronic elements structure comes down to endeavors to activate the hidden defects of an element (integral schemes of a module) before its deliverance to a consumer. Those stimuli could be different variants of electro-, thermo- or combined trials, energy-scrapings and so on. An element is subjected to considerable loadings (heat, electric, extreme transitional processes). Such deeds make apparent the hidden element defects, but consequently the latter might become superficial. In this case potentially unreliable elements either fracture or their characteristics unallowably change. Time consumed by such a quality control and potential unreliability diagnostics makes tens and hundreds of hours. Thus, time loss stands in the way of producing the quality and reasonable-cost elements of radio-electronic apparatus. The decrease in the consumed time could be reached at the expense of growing influence-intensity. Hereby the hidden cracks become evident sooner but meanwhile new defects could appear consequently of the extreme intensity. For instance, a thirty-hour test of semiconductor elements at the temperature 150 °C is equivalent to the performance of trials during 168 hours at 125 °C. We should also take into consideration that with increase in temperature, element damages caused by mass uncontrolled inner currents or other effects that are not liable to regulation while element testing under the condition of high temperatures could befall.

The duration of an electronic elements test constitutes a main problem both for manufacturer and consumer. Rather small number of electronic elements refusals could be treated as the issue of insufficient test duration and vice versa their large amount could confirm the excessiveness of a test term. Following from the aforementioned, the highest topicality should be referred to the methods of non-destructive diagnostics of an electronic element inner structure. One of them is the electronic elements diagnostics by dint of elements proper noise that is singled out by rather small time consumption and absence of damage risk for the researched element.

The noise that is unaltered by the operation mode of an element in an electric circuit belongs to the electronic elements proper noise. First of all, it is a thermal noise whose power depends on the temperature of an element. As have been proved, the flicker-noise whose intensity (energetic spectrum $S(f)$) rises with a frequency fall should also be referred to this sort. Flicker-noise is characterized as such that appears in the systems remaining in a thermodynamic disequilibrium. Naturally, we should put some efforts in order to get the system out of the equilibrium, characterized by a certain minimal level of inner energy. The latter is converting into the inner energy additionally accumulated by the system. For instance, in the case of a crystal body, it is the energy of inner strains, dislocations, microcrack surfaces, and etc. Evidently, the flicker-noise is an issue of the presence of extra energy exceeding the minimally possible system energy. Therefore it could be determined through the measured flicker-noise.

The energetic spectrum of the proper noise $S_H(f)$ of the system, remaining in a disequilibrium state and simulating the real electronic elements, is determined: $S_H(f) = \frac{e^{f\tau}}{e^{f\tau}-1} \cdot S_P(f)$, where $S_P(f)$ is the energetic spectrum of system noise in a quasi-equilibrium state; τ is the system relaxation constant; f is the frequency. The dependence of an electronic elements proper noise energetic spectrum on the frequency f is notified in Fig. 6.10.

The hatched area corresponds to the noise energy due to the additionally accumulated inner energy exceeding the energy of the system remaining in a quasi-equilibrium state. This extra energy that could be evoked by the electronic elements structure defects causes the appearance of a disequilibrium system state. The frequencies f_{01}, f_{02}, f_{03} correspond to the values of the frequency f which $S_H(f_{0n}) \approx S_P(f)$.

To wit, f_{0n} is the frequency above which the flicker-component of noise is equal to zero. In the Fig. 6.10 we can observe that f_{0n} depends on the relaxation constant τ: the higher τ, the lower f_{0n}. The value f_{0n} could be determined from the ratio $\frac{e^{f_{0n}\tau n}}{e^{f_{0n}\tau n}-1} \approx 1.001$ (the power of noise at frequency f_{on} is for 0.1 % larger than that at the frequencies $f > f_{on}$), where τ_n is the relaxation constant for the concrete electronic elements type. Research has shown that the relaxation constant of the researched system depends on the characteristics of its inner structure. The value τ could be calculated by means of the experimentally determined frequency characteristics of system proper noise: $\tau = \frac{1}{f_0}$, where f_0 corresponds to the frequency at which: $S_H(f_0) = 1.58 S_P(f)$.

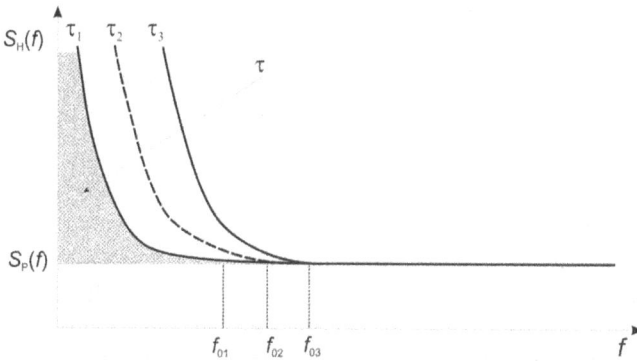

Fig. 6.10. Frequency dependences of an electronic elements proper noise energetic spectrum.

The noise powers are caused by the defectiveness of electronic elements inner structure. In the case of a resistor with the conductive film made from metal-silicon alloy layered on a ceramic pivot with the nominal power of dissipation 0.125 W, the noise power P_{HM} makes $21.9 \cdot 10^{-18}$ J/s within the bandwidth from $f_1 = 10^{-20}$ Hz to $f_2 = 1200$ Hz. For a similar resistor with the film of metal oxide on the surface, power is estimated as $P_{HC} = 39.2 \cdot 10^{-18}$ J/s within the bandwidth from $f_1 = 10^{-20}$ Hz to $f_2 = 2000$ Hz. In the case of high-frequency silicon diode, the similar power is $84.5 \cdot 10^{-18}$ J/s within the bandwidth from $f_1 = 10^{-20}$ Hz to $f_2 = 4500$ Hz (Fig. 6.11). The difference in the values P_H is related to peculiarities of inner structure of the mentioned elements, namely to the number and type of defects. Defects of a resistor structure are evidently caused by the heterogeneity of a conductive film as well as its additional imperfections of the vacancy type. A higher noise level within the range of low frequencies for a high-frequency diode as compared to resistors is explained by the necessity of additional considering the *p-n* junction defects and these of concentrating the less important charge carriers.

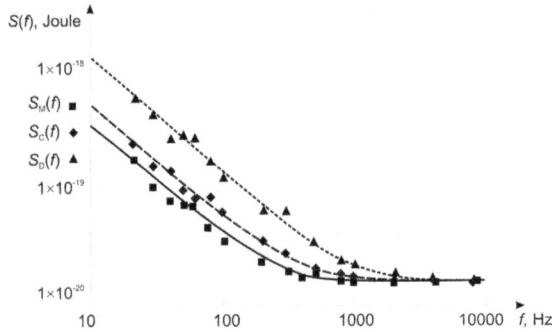

Fig. 6.11. Energetic spectra of resistors proper noise $S_M(f)$, $S_C(f)$ and high-frequency diode $S_D(f)$.

So, in comparison with other non-destructive methods of electronic elements inner structure defectiveness diagnostics, the method of research by dint of electric noise power is multipurpose, and enables detecting the potentially low-reliable elements. With involving the dependence of flicker-noise parameters on the controlled object structure, we could diagnose the electronic elements state and its evolution, especially at the primary stages of defects formation. Owing to the measurement of current fluctuation parameters, method of electronic elements noise spectroscopy could be applied to reliability diagnostics of both analog and digital electronic elements to the same extent. It is based on calculating the frequency above which the flicker-component of noise is equal to zero [68-69]. As follows, passive noise spectroscopy is the non-destructive method of studying the proper electric noise and resultantly of testing the materials and objects of different sizes (bulk, micro- and nano- dimensional) which could be realized the most precisely owing to the usage of NTs as calibrated measuring devices. The highest topicality should be referred to the methods of non-destructive diagnostics of an element inner structure.

Remark. In the case of passive methods, the statistics of studying results could be augmented, e.g. for electric noise. However, for this purpose, we should determine or at least calculate whether the removal of electric power during measuring could not change characteristics, to wit, whether a methodic error-component would not appear, e.g. while measuring the voltage of electric noise of the amplitude 10^{-8} V by a passive method ($P=10^{-16}$ W) any shaking of a pattern that produces in it the energy of 10^{-6} J, would determine the duration of this power removal during measurement as $\sim 10^{10}$ s or ~ 300 years.

6.4. Raman Method in Metrology and Thermometry

Measurement of the solid body surface temperature with taking advantage of a Raman phenomenon is related to one of few methods of primary thermometry being realized with the help of a thermometer whose state equation could be written in an explicit form, avoiding the involvement of unknown constants dependable on temperature. The given method helps to measure the temperature for objects ranged from 100 nm to 100 μm as well as within this from cryogen till mid-high temperatures, which in addition does not demand calibration before measurement, could be distinguished. For instance, temperature measurements are made by exciting the quantum dots with a laser, to obtain their emission spectra.

The similar application of Raman thermometer hits in measuring the temperature and diameter of CNTs. Temperature is controlled externally; very low excitation power on the sample surface is used (<0.5–1 mW). For example, to decrease the different components of the error, the *He-Ne* laser of continuous effect and small power 1…2 mW as well as the *MS3501i* spectrometer are used.

In addition, while studying the thermal conductivity data, Raman spectrometer acts as a thermometer with the resolution ΔT_G, associated with the peculiarities of heat transfer due to changes in phonon scattering [34]. Raman thermometer is a contactless thermometer which due to small diameter of laser beam enables to reduce the size of a thermometring zone to tens μM and lesser (for instance, practicing this thermometer we could study temperature of chip exploitation with a μm-resolution).

Increasing the opportunities of Raman thermometer in terms of researching the different objects (materials) and evolving the method itself requires the elaboration of metrological reliability (error, uncertainty, and etc.) of measurements, their repeatability and reproducibility [66]. We accepted the determination of temperature due to the ratio of intensities of Stokes and anti-Stokes components [67]. As well as the method of thermometry, the given method is based on the usage of the temperature dependence $T = T\left(I_s / I_{as}\right)$:

$$T = \frac{hc\nu_0}{k_B \dfrac{I_s}{I_{as}} - 3k_B \, ln \dfrac{\nu_i - \nu_0}{\nu_i + \nu_0}}, \qquad (6.20)$$

where, the variables are the intensities of Stokes I_s and anti-Stokes I_{as} components of the dispersed radiation; ν_0 and ν_i are the wavenumbers of reflected phonons at the given number of scattered phonons and the used laser bunch accordingly, sm^{-1}, h is the Planck constant, k_B is the Boltzmann constant. As result of performed investigations the CNTs are treated to be standard nanopatterns for testing and calibrating the nanotechnological means. The gained results of experimental research have given possibility of realizing the metrologically correct evaluation of temperature measurement results with considering the peculiarities of both measuring tool and standard nanopattern.

Unfortunately, the structure of spectral lines of CNTs is complex [34], and lines represent a set of mutually superposed sublines. Therefore at processing the experimental results of Raman spectra in order to establish values ν_0; ν_i and beam intensities at these frequencies, arises a significant uncertainty, additional to the known one. It is almost impossible for researchers to evaluate the latter without additional experiments and metrological skill. We have conducted a special study to study this uncertainty, depending on the type of assessment of light re-radiation curves by object. Resultantly, it was developed the software of processing frequency dependencies, which, consequently, enabled the determination of aforementioned uncertainty of temperature independently by experimenter. Here we could take into account the average intensity of the lines or the local maxima/minima of spectral line intensity.

6.4.1. Metrology of Raman Thermometer with Universal Calibration

Nanothermometry is a subfield of thermometry, concerned with the science of measurement at the nanoscale level. Nanometrology has a crucial role in order to produce nanomaterials and devices with a high degree of accuracy and reliability in nanomanufacturing [2]. A challenge in this field including nanothermometry is to develop or create new measurement techniques and standards, to meet the needs of next-generation advanced manufacturing, which will rely on nanometer scale materials and technologies. It is difficult to provide samples on which precision MIs can be calibrated on nanoscale. Calibration standards are important for repeatability to be ensured. It is difficult to select a universal calibration artefact with which we can achieve repeatability on nanoscale. At nanoscale, while calibrating care needs to be taken for the influence of external factors (noise, vibration, etc.) and internal factors such as interaction between the artefact and equipment which can cause significant deviations [68].

Calibration artefacts with full point-by-point uncertainty characterization enable a wide range of calibrated time- and frequency-domain measurements to be performed in industrial laboratories, thus accelerating product development. Nanosized calibration standard is important for gauge repeatability to be ensured. It is difficult to select a universal calibration artefact with which we can achieve repeatability on nanoscale. At nanoscale calibrating care needs to be taken for the influence of external factors (noise, vibration, motion) and internal factors such as interaction between the artefact and equipment which can cause significant deviations. Therefore the producers of nanotechnology means often propose their equipment fitted with nanosized calibration standards crucial to equipment operation.

Those are the common problems of different nanoresearches. However, the most universal Raman method, apt for the direct temperature measurement of micro- and nano-objects could be distinguished [34, 67]. For instance, temperature measurements are made by exciting the quantum dots with a laser, to obtain their emission spectra. Raman shifts of the laser excited nanoparticles based on phonon confinement are used to determine the cell temperature from experimental correlations of particle size with temperature-dependent grain growth. The similar application of Raman thermometer in measuring the temperature and diameter of CNTs is made in [69]. Temperature is controlled externally; very low excitation power on the sample surface is used (<0.5–1 mW). In addition, studying the thermal conductivity data, Raman spectrometer acts as a thermometer with the resolution ΔT_G, associated with the peculiarities of heat transfer due to changes in phonon scattering.

Increasing the opportunities of Raman thermometer in terms of researching the different objects (materials) and evolving the method itself require the elaboration of metrological reliability (error, uncertainty, etc.) of measurements, their repeatability and reproducibility. It is time for realization of the mentioned work, since on the one hand, the unsolved aspects of nanometrology hinder the progress in nanotechnologies, and on the other hand, the nanoobjects, ready to serve as standard patterns, are already designed. The classist of nanotechnologies and meanwhile the researcher of Raman method, prof. K. Dresselhaus deems that the CNTs represent exactly such patterns [34].

The spectrum of CNTs is simple and well reproducible which helps verify the researched patterns, and assures pretty good reproducibility of results in an easy way.

The totality of single-wall CNTs of different sizes and of general mass till 1 g and the square 20 mm² has been chosen for the research. For the trustfulness of the gained results, two independently made measuring sets were designed, and values, averaged due to the period of measurement (24 hours), were represented. The research has been conducted within the range 30...250 °C. The temperature of an object has been gauged additionally by a thermistor with the error not exceeding 0.01 % at sensor linear size till 100 µm. At any temperature, ten spectrum values were gained (for different parts of an object area) so that the averaging of results due to the object area has been performed. Every spectrum is gained by averaging 9 results of measurement at the concrete temperature (averaging due to the temperature). To remove the error of temperature mode drift (the error caused by the additional heating by a laser bunch in the process of measurement), every measurement was conducted with the interval of 2 minutes. This time was sufficient for stabilization in time of the researched object temperature field, caused by the previous heating.

The chain of the averaged temperature values according to the readings of Raman thermometer as well as the control temperature values measured by the specially calibrated thermistor (of 0.1 % precision class) have been fixed, and relative measurement errors calculated. The latter does not exceed 4 % within limits of controlled values of the temperature range 30...250 °C. Considering that the threshold value of an instrumental error of the *MS3501i* spectrophotometer is up to 4 % (passport data for example), the gained results could be treated as satisfactory.

- Determination of temperature due to the ratio Stokes and anti-Stokes components intensities of Raman method.

Temperature gauge due to the ratio of Stokes and anti-Stokes components of scattered light intensities is more prevalent and exact kind of Raman method. The given method of thermometry is based on the usage of characteristics $T = T(I_s; I_{as}; \nu_0)$. Positive feature of the method is the next: a state of the researched surface (micro unevenness, local reflection coefficient, etc.) practically does not affect the results of measurement, since Stocks and anti-Stocks components of scattered light are altering simultaneously.

- Temperature determination due to frequency of object proper radiation shift

Temperature gauge due to radiation shift is simpler and more speed kind of Raman method, and is based on the temperature dependence of the scattered radiation frequency shift [70].

Particularly, this method enables us to perform measurement by using the temperature dependence of shift v_0 of anti-Stokes component of reflected light. With temperature ΔT raising a wavelength of the reflected light (the anti-Stocks component) smoothly reaches a laser wavelength. In the case of Silicium such a shift keeps linearity within the range 300 K...400 K: $v_0[cm^{-1}] = 0.025\Delta T$. Moreover, the error of temperature determination is proportional to that of shifting determination of reflected light frequency. At room temperatures, the suggested method acquires the essential preference with respect to the gauging method by the ratio of intensities, since the intensity of the anti-Stocks line decreases substantially with a fall in temperature from 300...400 K. For investigated CNTs within the range 15...250 °C at the frequencies 1585.6 cm^{-1} ...1576.1 cm^{-1} the given frequency falls down with the temperature (Fig. 6.12) [34, 69].

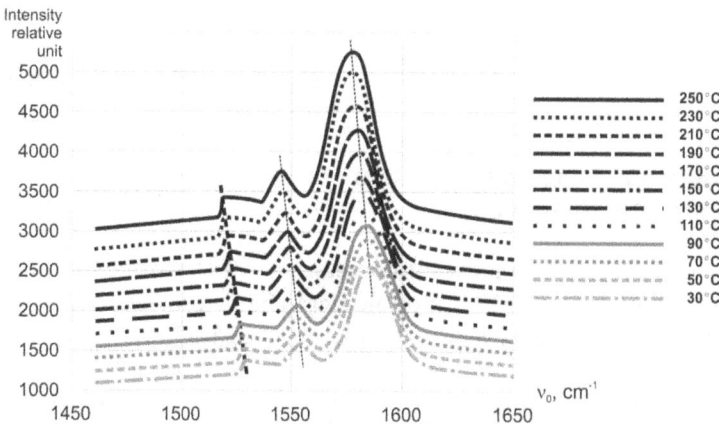

Fig. 6.12. Raman spectrum of CNTs (for He-Ne laser).

Temperature coefficient of frequency change makes –0.041 cm^{-1}/°C. G-peak of intensity for Bi-layer Graphene is fixed at 1580 cm^{-1} ($\lambda_{exc} = 488\ nm$). Frequency downshift (–0.015 cm^{-1}/°C) with T within the range −150...+90 °C is unusual for optical mode since the interatomic distances shorten with T, and lattice contraction leads normally to the upward shift of the frequencies. To compare for graphite at the 4-phonon dispersion, the temperature coefficient of frequency change is up to −0.023 cm^{-1}/°C, and for graphene about −0.035 cm^{-1}/°C.

However, in all cases there exists certain characteristic nonlinearity (Fig. 6.13) at the room temperatures which makes it difficult to use this value for determining the object temperature.

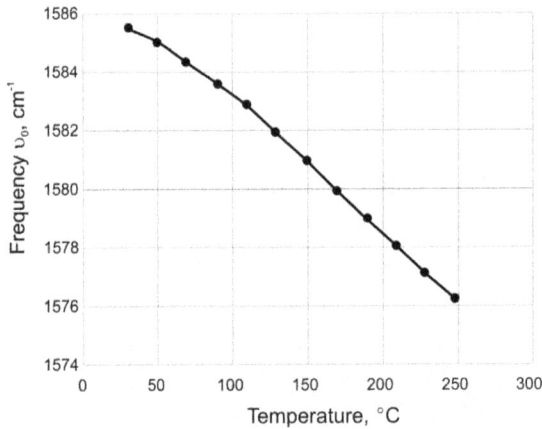

Fig. 6.13. Temperature changes of CNTs frequency shift in Raman method.

Experimental Research of Factors of Influence on the Measurement Results.

Influence factors of Raman thermometer is a variety of factors showing a significant impact on the metrological characteristics of the Raman method and accordingly Raman thermometer [66-67].

- The influence of incident radiation angle. To realize the Raman method in practice, the equipment notified in [34] is employed. The available equipment installed on the optical bench does not imply the performance of measurement at the normal angle of light incidence upon the researched pattern. To determine the optimal angle between the laser beam and the headlet of an optical light-pipe, the series of spectrum measurement are made for three different materials (Artificial diamond, Silicium crystal, CNTs), at temperatures 20 °C, 70 °C, 120 °C. The

274

gained experimental data confirm that optimal angle between a laser beam and a normal to the object should vary within 10...30°, and should not exceed 60° while the error of measurement is considerably rising.

- The influence of durations of exposure and spectrum determination. To determine the efficient time of measurement, the research of different time intervals' influence of object lighting (1.6·0.6·0.1 mm³) by the *He-Ne* laser of continuous effect and with the power of 1 mW at 297 K has been conducted. The velocity of gaining the spectrum and correspondently the velocity of its elaboration depend on the speed of operation of spectrophotometer mechanical components, camera reading velocity and the intensity of gained radiation. For any posture of a spectrophotometer step-engine (during one step of an engine, the 50 nm-wide spectrum band gets into a registering camera), the appropriate time-interval of delay is ascertained in order to determine the optimum duration of the realization of single measurement.

- The general time of measurement. It is restricted by both the total exposure intervals and time needed to change the posture of a diffracting grate (~0.3 s). The need in more time-consuming measurement is explained by the dim intensity of an anti-Stokes component at room temperature. Therefore the time of registration is fixed at the level of 3.0; 3.5; 4.0; 13.0 s. Moreover, the latter is not used furthermore with regard to reaching a photoreceiver saturation limit (within the range $- 250...255$ cm⁻¹). On the other hand, too low level of a useful signal could become the reason for the considerable measurement error. As a result, the optimal measurements are treated to be so with the exposure time 0.01...0.5 s for any step-engine posture.

- The influence of background radiation. The spectra for CNTs have been researched: a) at their complete light screening from any lighting except laser one; b) at the presence of additional lighting of natural light. The conducted investigations have shown that at the increasing of background radiation intensity the capability to identify the peaks of Stokes and anti-Stokes components of a signal is falling (by applying the rapid Furrier transformation in processing results we could gain more precise spectra). The following temperature measurement results are received: namely 29.10 °C – without background lighting; 29.02 °C – with background lighting of 100 lux intensity. This additional error component – 0.08 °C – is practically eliminated at the minimization of the distance between the object and optical headlet.

- The influence of distance between object and photoreceiver. It is most profitable theoretically to place the researched pattern in direct contact with a receiver, but obviously such a design eliminates all the advantages of noncontact measuring methods, that is why the factor of distance "object - photoreceiver" becomes apparent. The error of measurement is estimated (Fig. 6.14), moreover the extreme nature of the dependence is stipulated by the action of two factors – light diffraction and background lighting (100 lux). The latter decrease in distance is impossible without special equipment. However, in the case of passing laser radiation through the glass fiber, the distance could be decreased to the hundredths of a millimeter. Let us notice that the distance "laser - object" practically does not affect the results of measurement which is caused by the parallelism of beams in the bunch.

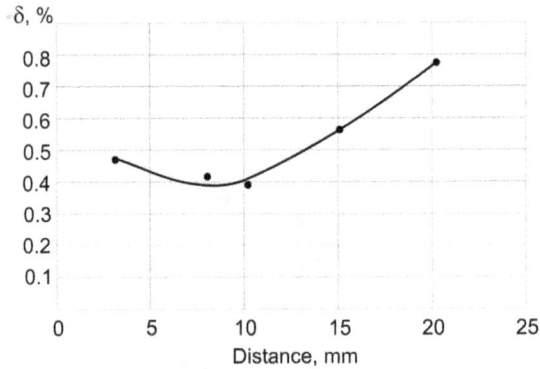

Fig. 6.14. Dependence of a measurement error on the distance between object and entrance into an optical channel of a spectrophotometer (at 30 °C).

- Example of processing the results of experimental investigations. The spectrum is recorded into the memory of computer. Program software PCI-Line has proven insufficient for processing the data. The document of Excel is created for this purpose containing ten spectra for any object. During measurement, the temperature has made 15 °C (288 K). Substituting the correspondent values – the mean of Stokes intensity: $i_s = 77740.538$ of anti-Stokes intensity: $i_{as} = 6953.8785$, laser wavenumber: $v_i = 1579800 \ m^{-1}$, wavenumber of reflected radiation: $v_0 = 52200 \ m^{-1}$, constants of Planck and Boltzmann and the velocity of light – into (6.20), the measuring temperature 287.82 K is gained.

In the case of employing a spectrophotometer with one registration camera and without spatial channels division when $\delta i_s = \delta i_{as} = \delta i_{ChAM}$ the main instrumental error is determined in [67-68] as:

$$\delta T = \frac{1}{Z}\left[2\delta i_{kam} + \frac{6v_0 v_i}{v_i^2 - v_0^2} \cdot \delta v_i + \left(Z + \frac{6v_0 v_i}{v_i^2 - v_0^2}\right) \cdot \delta v_0\right], \qquad (6.21)$$

$$Z = \ln\frac{i_s}{i_{as}} - 3\ln\left(\frac{v_i - v_0}{v_i + v_0}\right) = 2.612402. \qquad (6.22)$$

The error of measuring intensities depends on a camera and is not exceeding $\delta i_{ChAM} = 0.5\,\%$ according to a passport. The value δv_i depends on the precision of laser wavelength determination: $\lambda_L = 62.9974 \pm 0.0003\ nm$; $\delta v_i = \frac{\Delta\lambda_L}{\lambda_L} \cdot 100 = 4.72 \cdot 10^{-5}\,\%$. The error δv_0 depends on the precision of spectrophotometer localization only, and is equal to: $\delta v_0 = \frac{\Delta\lambda_0}{\lambda_0} \cdot 100 = 3 \cdot 10^{-3}\,\%$. As a result, the main component of an instrumental measurement error, calculated due to (6.21), is up to 0.386 %. The additional component of an instrumental measurement error is caused by the instability of feeding voltage (for all devices) and the change in spectrophotometer temperature. Remaining factors stay within a normal value ranges. Let us notice that for the spectrophotometer, the value of an additional CoE does not exceed the value of a main component for any 5 K-change in environmental temperature T_0. As a result, the expression according to which $\delta T_D = 0.195\,\% < 0.2\,\%$ is gained:

$$
\begin{aligned}
&\delta T_D \\
&= \frac{1}{Z}\left[\delta i + \frac{1}{2}\left(Z + \frac{6v_0 v_i}{v_i^2 - v_0^2}\right) \cdot \delta v_0 + \frac{T - 18}{5}\left(z + \frac{6v_0 v_i}{v_i^2 - v_0^2}\right) \cdot \delta v_0\right].
\end{aligned}
\qquad (6.23)
$$

Introducing the correction into the result, i.e. eliminating a systematic CoE, it should be given the duration of one measurement – 8 s. Then the result of measurement is recorded with the predetermined adjustment 0.55 K: $T_{true} = T_{measur} - \Delta_{meth} = 287.82 - 0.55 = 287.27K$. Hereby, the indicated temperature has been measured with a random error not exceeding $\pm 0.6\,\%$ or ± 1.72 K.

- Uncertainty of the gained results. While using Raman thermometer, the temperature is estimated indirectly by means of measuring four values – look equation (6.20): it is determined as a result of an indirect measurement, moreover, by a single observation.

Let us estimate the standard uncertainty of intensity measurement under normal conditions: $u_B(i_{cam}) = \frac{0.5 \cdot 16384}{100 \cdot \sqrt{3}} = 47.3 \, mA$. We should also determine standard uncertainty of measuring the wavenumber for reflected radiation and for laser respectively: $u_B(v_0) = \frac{0.003 \cdot 1136363.63}{100 \cdot \sqrt{3}} = 19.68 \, m^{-1}$; $u_B(v_0) = \frac{4.72 \cdot 10^{-5} \cdot 1579800}{100 \cdot \sqrt{3}} = 0.43 \, m^{-1}$. Relative value of these values makes: $\frac{u_B(v_i)}{v_i} \cdot 100 = \frac{0.43}{1579800} \cdot 100 = 2.72 \cdot 10^{-5} \, \%$ and is treated as negligibly small. For the camera *HS 102H*, the coefficient of influence makes $C_{power} = 0.5$; then the standard uncertainty of a device, caused by the change in the feeding voltage, is determined as $u_B(i_{cam})_{power} = 23.65 \, mA$. Influence of feeding voltage on the wavenumber of reflected radiation for a spectrophotometer which the coefficient of influence makes $C_{power} = 0.5$ is determined as $u_B(v_0)_{power} = 9.84 \, m^{-1}$. It is also expedient to conduct the calculation of the standard spectrophotometer uncertainty related to the change in temperature $u_B(v_0)_T = \frac{3}{5} \cdot 19.68 = 11.808 \, m^{-1}$.

To identify the combined standard uncertainty of the B type results [9], it is necessary to calculate at first the coefficients of results uncertainty influence: $C_{i_S} = -0.00141 \, K/_{mA}$, $C_{i_{AS}} = -0.01584 \, K/_{mA}$, $C_{v_0} = 0.00552 \, m \cdot K$. The facts concerning the correlation of spectrophotometer and detector indices, caused by the change in feeding voltage, are absent. Therefore let us adopt that the appropriate coefficients of influence are characterized by the mean square value $r = 1/_{\sqrt{3}} = 0.577$, and temperature influence takes place on a spectrophotometer only. Then the expression for the combined standard uncertainty of the gained results acquires the following form:

$$
\begin{aligned}
u_{C\,B}(T) \\
= \{ C_{i_S}^2 [u_B^2(i_{kam})_S + u_B^2(i_{kam})_{S.power}] \\
+ C_{i_{AS}}^2 [u_B^2(i_{kam})_{AS} + u_B^2(i_{kam})_{AS.power}] + C_{v_0}^2 \\
\cdot [u_B^2(v_0) + u_B^2(v_0)_{power}] + 2 \cdot |C_{i_S}| \cdot |C_{i_{AS}}| \cdot |C_{v_0}| \quad (6.24) \\
\cdot r \\
\cdot [u_B^2(i_{kam})_{S.power} + u_B^2(i_{kam})_{AS.power} \\
+ u_B^2(v_0)_{power}] \}^{1/2}
\end{aligned}
$$

Substituting the appropriate values, we have gained $u_{CB}(T) = 0.852\,K$. The relative value of the combined standard uncertainty of the temperature measurement results is $\delta_{U_{CB}}(T) = \frac{u_{CB}(T)}{T_{nom}} \cdot 100 = \frac{0.852}{287.27} = 0.2966$. Considering the correlation among the components of feeding voltage influence on indices of detector and spectrophotometer, we could estimate the effective quantity of freedom degrees (at the approximation of this value distribution by the Student distribution with specially selected so-called efficient quantity of freedom degrees) using the equation:

$$
\begin{aligned}
v_{eff} &= u_{CB}^4(T) \\
&\cdot \left\{ \left[(C_{i_{AS}}^4 + C_{i_S}^4) \cdot \left(\frac{u_B^4(i_{kam})}{v_{sp}} + \frac{u_B^4(i_{kam})_{power}}{v_{power}} \right) \right. \right. \\
&\left. + C_{v_0}^4 \left(\frac{u_B^4(v_0)}{v_{sp}} + \frac{u_B^4(v_0)_{power}}{v_{power}} + \frac{u_B^4(v_0)_T}{v_T} \right) \right]^{-1} \\
&\left. + \left[2 \cdot C_{i_{AS}}^2 C_{i_S}^2 C_{v_0}^2 r^2 \left[\frac{u_B^2(i_{kam})_{power} + u_B^4(v_0)_{power}}{v_{power}} + \frac{u_B^4(v_0)_T}{v_T} \right] \right]^{-1} \right\}
\end{aligned}
\tag{6.25}
$$

Considering that the relative uncertainty for spectrophotometer and camera is not prevailing 5 % in normal conditions, and at the change in feeding voltage it is not exceeding 25 %, we could estimate the values of freedom degrees. The temperature influence on a spectrophotometer is determined by the uncertainty 20 %. For normal conditions are: $v_{sp} = 0.5 \cdot \frac{1}{0.05^2} = 200$, for the change in feeding voltage $v_{power} = 0.5 \cdot \frac{1}{0.25^2} = 8$ and for temperature $v_T = 0.5 \cdot \frac{1}{0.2^2} = 12.5$. Substituting the appropriate values into (6.25), we have gained $v_{eff} = 130.6$. For the credence level $P = 0.95$ and the received efficient quantity of freedom degrees $v_{eff} = 130.6$, the value of an expanding coefficient makes $k_{0.95} \approx 1.96$ [10]. Therefore the expanded uncertainty of indirect measurement result makes $U(T) = k_{0.95} \cdot u_{CB}(T) = 1.67\,K$. The relative value of expanded uncertainty of temperature measurement result is $\delta_{U_{CB}}(T) = \frac{U(T)}{T_{tr}} \cdot 100 = \frac{1.67}{287.27} \cdot 100 = 0.58\,\%$.

6.4.2. Elaboration of Raman Method

Countries, that have entered into "nanotechnological era", understand the urgency of advances in optical metrology. It is the exactness and reliability of measurements capable to stimulate the development of relevant industries and to serve as a deterrent. Especially in optical spectroscopy [2] a challenge is to study current measurement techniques and standards, and to meet the needs of next-generation advanced manufacturing, including CPSs. The most universal nanothermometry method, Raman method, apt for the direct temperature measurement of micro- and nano-objects as well as for row of other gauging methods, which in addition does not demand previous calibration.

To increase the opportunities of Raman thermometer in terms of researching the different objects and to evolve the method itself, the elaboration of metrological reliability of measurements, their repeatability and reproducibility is required. Unfortunately the structure of spectrum lines of different substances is complicated [34, 69], and these lines present themselves as totality of mutually imposed sublines. Therefore under processing of experimental results with the aim to find the true values of frequencies as well as beam's intensity of specified wavelengths, the substantial uncertainty appears. Virtually, researchers can not estimate it without conducting extra experiments.

Within Raman method the temperature measuring can be conducted in two ways: by the ratio of intensities of Stokes (I_s) and anti-Stokes (I_{as}) components – see equation (6.20) – or by the frequency shift in spectrum. By the first way there is distinct temperature dependence of ratio of intensities of Stokes (I_s) and anti-Stokes (I_{as}) components of dispersion radiation. Experimental researches had been conducted on calibrated optical bench. Laser *SL03/1* spectrum with wavelength 632 nm is registered by *HS 102H* camera in the regime of vertical binning and is recorded on digital storages with use of PCI-Line program. Obtained results recorded in special file due to internal tools of the PCI-Line program are not enough good to process received information. To simplify the spectra analysis (Fig. 6.16) the laser wavelengths were filtered by notch filter. However the last one was filtered only wavelength 632 nm and lateral harmonics remain in spectrum.

To calculate integral values of the area of Stokes and anti-Stokes intensity in terms of complexity of the waveform, a number of modes of calculating of complex figures integral area were studied with the aim to select the most optimal mode for the studied object.

Therefore computer processing is additionally performed by spectrum's filtration within wavelength of the laser. Obtained result, transformed in frequency area, is shown in the Fig. 6.15. Filtration has been performed by means of ideal band-pass filter with automatically adjustment of bandwidth which is cut off (Fig. 6.16).

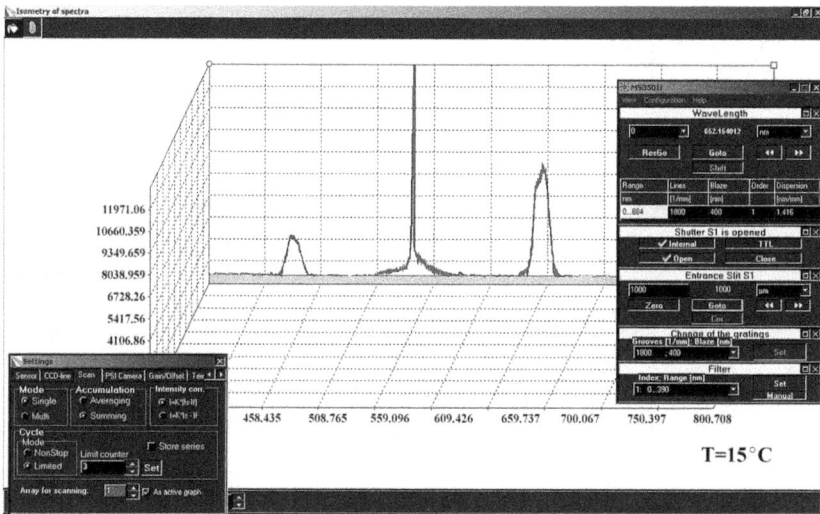

Fig. 6.15. Raman spectrum obtained by water irradiation.

During the mathematical modeling process conducting in MatLab ambience the method of rectangles, method of trapezium, and Monte-Carlo methods have been tested. Every modes uncertainty was calculated for modeled sine function and experimentally measured spectrum. Having conducted analysis of obtained results it was proposed to calculate the integral area of Stokes (I_s) and anti-Stokes (I_{as}) components with help of trapezium method. This method is inherent in minimal uncertainty at high operation speed.

It is necessary to consider separately methodology of determination of Raman shift wavelength. This methodology is considered to be acceptable to determine the main measured parameters in several variants.

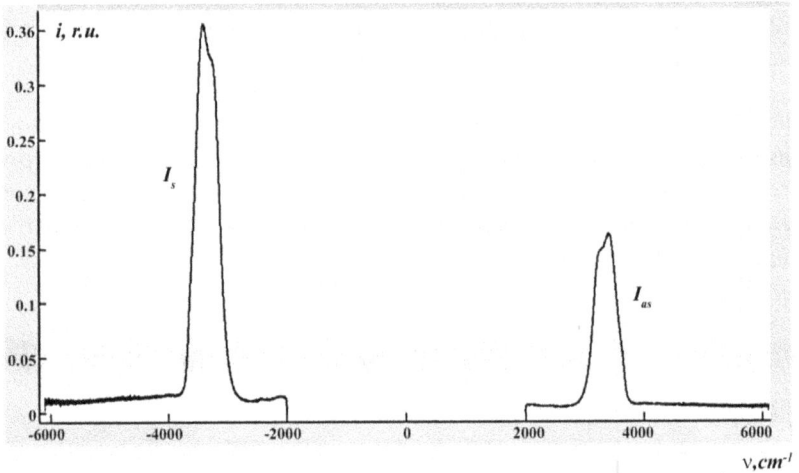

Fig. 6.16. Raman spectrum of water after filtering the major wavelength (632 nm) of laser *SL03/1*.

Firstly let's consider the technique of the determination of Raman frequency shift by maximum value of peak intensity. Further all researches were conducted exceptionally for anti-Stokes component of spectrum as Stokes component is equidistant to major frequency. To check correctness of any methods and modes, the chain of measurements has been performed (100 measurements for concrete temperature point). The maximal value was determined for every measured spectrum.

Ten obtained spectra with corresponding values of maximums (the relative Raman shift frequency is calculated for every one) are shown in the Fig. 6.17. They are received at 18.1 °C.

The uncertainty of temperature measuring through averaging values of Raman wavelength shift of anti-Stokes component, estimated by the maximum of the intensity, is determined in 1.27 %.

Another mode of studying the Raman wavelength shift is the definition of averaged integral value of spectrum's anti-Stokes component. Precondition of other modes' involvement (except determination by maximal value) can be considered possible complication of obtained spectral signal, availability of several maxima and some beforehand unknown widths of particular peaks. Therefore, we have investigated a series of materials with different types of spectra signals.

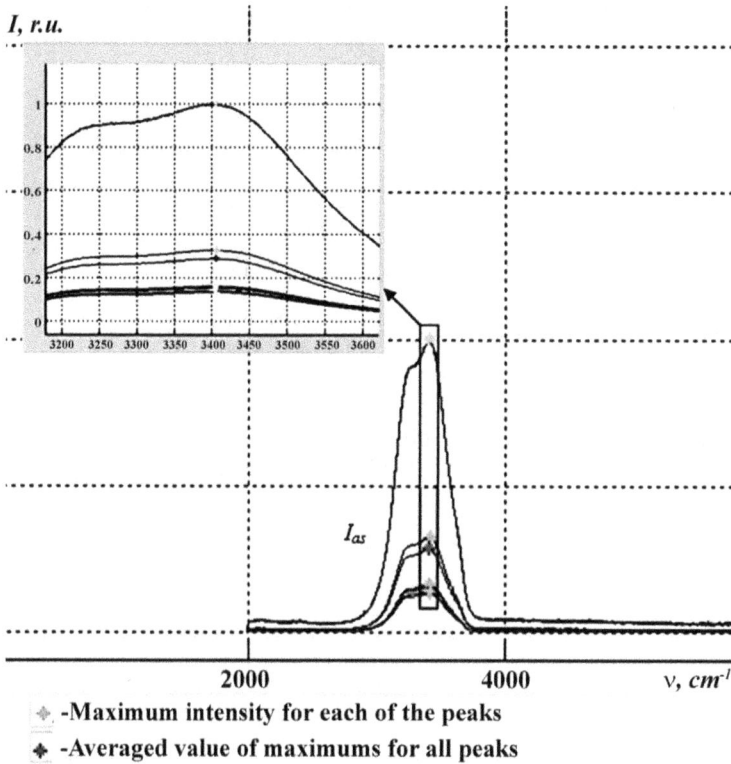

Fig. 6.17. Determination of maxima and their averaged intensity of H_2O anti-Stokes spectra components at 18.1 °C (632 nm).

For instance, in Fig. 6.18 are given 10 anti-Stokes wavelengths of ceramics spectrum (measurement is conducted at 19.68 °C) and corresponding values of their intensities. Evidently, the Raman wavelengths shift determination by the intensity maximum can embody the error linked to the discreteness of Spectrum analyzer of conducted measurements.

At testifying above mentioned peculiarities of spectral studies while analyzing the multimode spectral signal (Fig. 6.19), it was proposed to utilize mode of Raman frequency shift determination by averaged integral value.

The same mode is applied to determine Raman wavelengths shift of water spectra (Fig. 6.20).

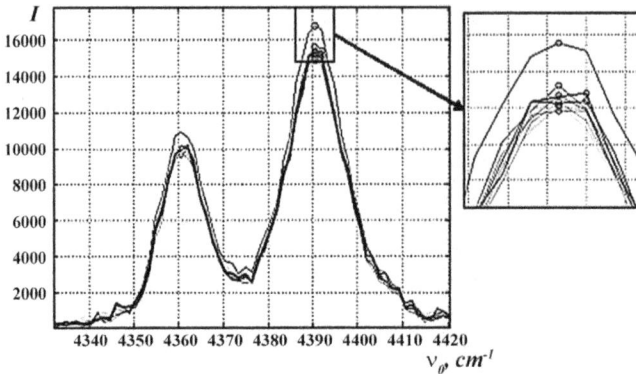

Fig. 6.18. Determination of anti-Stokes intensity maxima of ceramics component at 19.68 °C and λ=632 nm.

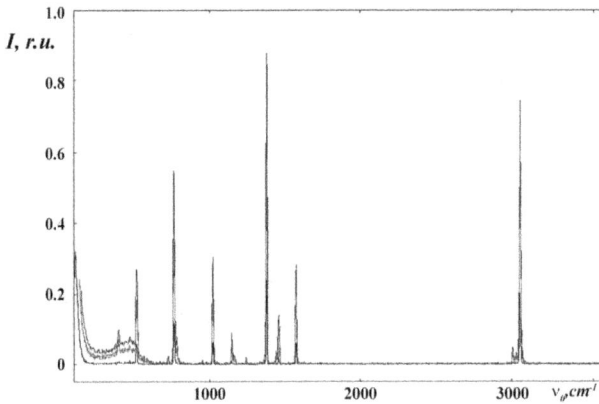

Fig. 6.19. Anti-Stokes spectrum of cyclohexane (sampling of 10 measurements at 20°C and λ= 632 nm).

The uncertainty of temperature measuring through averaged values of anti-Stokes wavelengths shift is determined by averaged integral value and is equal to 1.63 %. Noise reduction in spectrum enables to estimate temperature measuring by Raman method with smaller uncertainty. Share of informative signal seems to be quite significant. To realize this, the spectrum restrictions by the wavelengths and the following selections of anti-Stokes components are introduced. Spectrum limitation is performed by means of MatLab tools. Results of the filtration and determination of averaged value of Raman wavelengths shift are evidenced in Fig. 6.21.

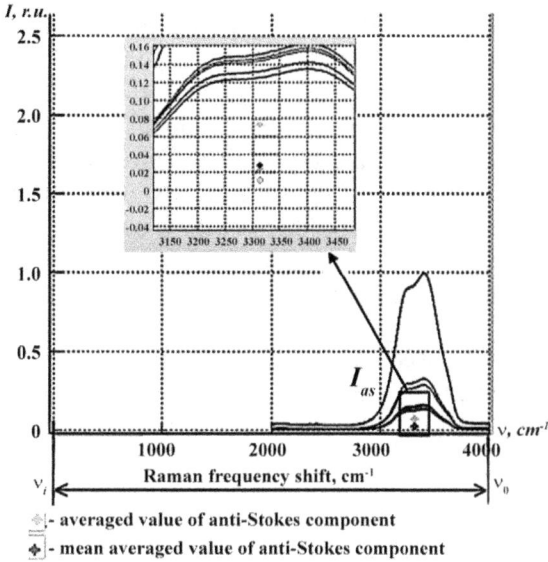

Fig. 6.20. Definition of anti-Stokes frequency shift after averaged integral value of water spectrum at 18.1 °C and λ=632 nm.

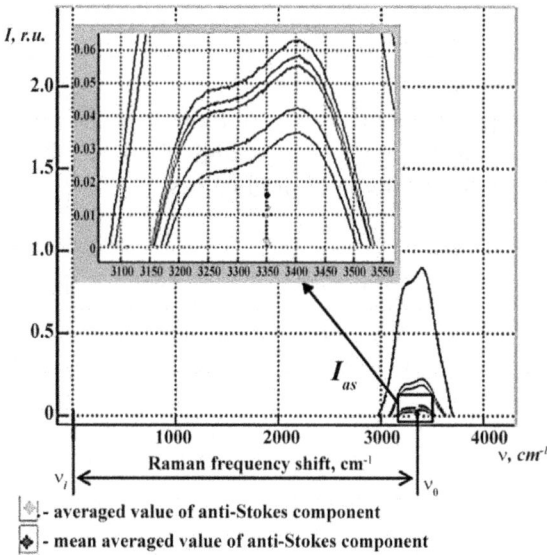

Fig. 6.21. Filtered spectrum of anti-Stokes component of water at 18.1°C and λ=632 nm.

The uncertainty of temperature measuring in the case of restricting Raman wavelengths shift of anti-Stokes component by MatLab tools, in accordance with described way of determination of averaged integral value, is equal to 0.97 %. The current mode of averaging the mean integral values and extracting the desired signal by the frequency has decreased the aforementioned uncertainty versus the mode of determination of Raman wavelengths shift by maximal value of the peak intensity. Therefore it is reasonable further to single out the desired signal from spectrum by filtration.

Receiving the processed results of multiple measurements for every temperature points, we define the dependence of anti-Stokes component averaged values of Raman wavelengths shift for water on temperature as:

$$v_i - v_0 = AT^2 + BT + C, \qquad (6.26)$$

where A= –0.0152 cm⁻¹, B=0.4839 cm⁻¹, C=3309.6 cm⁻¹, Δv is the Raman wavelengths shift, cm⁻¹ ($\Delta v = v_i - v_0$).

From (6.26), the temperature value is obtained:

$$T = \frac{-B \pm \sqrt{B^2 - 4B(A - \Delta v)}}{2C}, \qquad (6.27)$$

The Raman frequency shift curve, or more correctly the anti-Stokes wavelength shift curve (Fig. 6.22), is built for water by applying this equation and experimentally measured values.

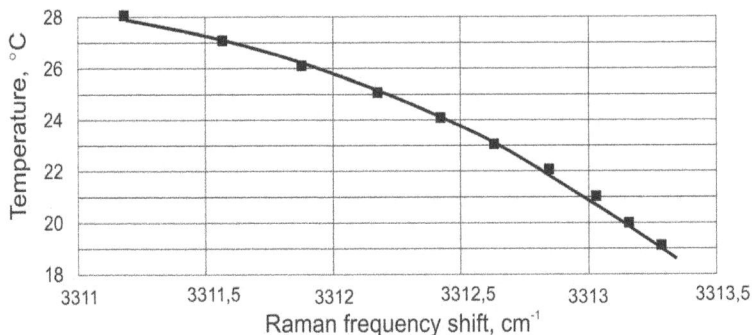

Fig. 6.22. Anti-Stokes frequency shift for water.

6.5. Conclusions

Totality of studied frequency, noise and spectrum methods form the delineated range of works that in one way or another promote the further development of Cyber-Physical Systems.

Method of studying the proper noise, as method of non-destructive testing, lies into the basis of passive noise spectroscopy, and could be precisely realized by application of noise thermometers as calibrated measuring instruments. Diagnostics of CPS elements by the energy spectrum of the stochastic signal, generated by these elements, is inherent in less temporal expenses and absence of damage risk caused with studied affects. This method is universal, since it can be applied to study the seamless and secure systems of different nature. System with small relaxation time becomes less reliable, and vice versa. Minimal value of relaxation time corresponds to the most unreliable system.

Promising method seems to be the Raman spectroscopy. With the help of Raman thermometer, conducted accurate measurement of CNT temperature within the range 30...250 °C have revealed that the tubes may be treated in Thermometry as Standard Nanopatterns or Calibration Artefacts. Proving of this point has been performed, basing on the elaboration of methodology and algorithmic principles of metrological performance improvements. Method of determining the temperature by anti-Stokes frequency shift has been enhanced. A number of modes for improving the accuracy and reducing the measurement uncertainty have included the frequency shift studies by evaluation of: a) maximal value of the peak intensity; b) averaging the mean integral value of anti-Stokes component of spectra on the basis perfecting the hardware (notch-filters) and software for minimizing the noise interference.

References

[1]. Spectrum Sharing Research and Development, *Telecommunications Industry Association (TIA),* Dec. 2013. 11 p., (https://www.tiaonline.org/sites/default/files/pages/SpectrumSharingR&DPaper=10-20-13.pdf).
[2]. K.Gåsvik, Optical Metrology, 3rd Ed., *John Willey & Sons Ltd*, England, 2002.
[3]. Budget Submission to Congress. Advanced Communications Research and Development and Testing, *NIST National Technical Information Service,* FY 2014, 2014.

[4]. S. Yatshyshyn, B. Stadnyk, Ya. Lutsyk, L. Buniak, Handbook of Thermometry and Nanothermometry, *IFSA Publishing*, Spain, 2015.

[5]. H. Seppä, Noise properties in an RF-biased Josephson junction noise thermometer, *J. Appl. Phys.*, Vol. 55, 1984, 1578.

[6]. S. Benz, Boulder Co, Jifeng Qu, H. Rogalia et al, Improvement in the NIST Johnson Noise Thermometry System, *IEEE Transactions on Instrumentation and Measurement*, Vol. 58, Issue 4, 2009, pp. 884-890.

[7]. H. Nyquist, Thermal agitation in conductors, *Physical Review*, Vol. 32, 1927, pp. 110-113.

[8]. F. Gasparyan, Excess Noises in (Bio-) Chemical Nanoscale Sensors, *Sensors & Transducers*, Vol. 122, Issue 11, 2010, pp. 72-84.

[9]. F. N. Hooge, T. G. M. Kleinpennin, L. K. J. Vandamme, Experimental studies of 1/f noise, *Reports on Progress in Physics*, Vol. 44, 1981, pp. 479-532.

[10]. G. R. Mutta et al., Volume charge carrier number fluctuations probed by low frequency noise measurements in InN layers, *Applied Physical Letters*, Vol. 98, 2011, 252104 (3 pages).

[11]. M. P. Anisimov, A. G. Cherevko, Fluctuation phenomena in physical and chemical research, *Nauka*, Novosibirsk, USSR, 1986 (in Russian).

[12]. Yu. Bobalo, Z. Kolodiy, B. Stadnyk, S. Yatsyshyn, Development of Noise Measurements. Part 4. Problems and Methodology, *Sensors & Transducers*, Vol. 153, Issue 6, June 2013, pp. 226-234.

[13]. S. Yatsyshyn, B. Stadnyk, Z. Kolodiy, Research in Nanothermometry, Part 5: Noise thermometry and nature of substance, *Sensors & Transducers*, Vol. 141, Issue 6, June 2012, pp. 8-16.

[14]. H. Hofmann, Advanced nanomaterials, Course support, *Powder Technology Laboratory*, IMX, EPFL, Version 1, September 2009.

[15]. P. Glansdorf, I. Prigogine, Thermodynamic theory of structure, stability and fluctuations, *Wiley*, New York, 1971.

[16]. Ia. Mills et al., The New SI: units and fundamental constants, *Royal Society Discussion Meeting*, 25-28.01.2011, http://www.bipm.org/utils/common/pdf/RoySoc/Ian_Mills.pdf

[17]. G. Bertotti, M. Celasco, F. Fiorello, P. Mazzetti, Study of dislocation dynamics in metals through current noise measurement, *Scripta Metallurgica*, Issue 12, 1978, pp. 943-948.

[18]. B. Stadnyk, Z. Kolodiy, S. Yatsyshyn, Exactness of metallic noise thermometers at low temperatures, *Measuring Equipment and Metrology*, Issue 45, 1989, pp. 8-10 (in Ukrainian).

[19]. Yu. Rumer, Thermodynamics, statistical physics and kinetics, *Nauka*, Moscow, 1977 (in Russian).

[20]. L. M. Franco-Neto, Noise in high electric field transport and low noise field effect of transistor design: the ergodic method, Ph.D Thesis, *Stanford*, 1999, 102 p.

[21]. S. R.de Groot, P. O. Mazur, Non-Equilibrium Thermodynamics, *Dover Publ.*, 1962, 510 p.

[22]. T. Dominiuk, S. Yatsyshyn, About thermodynamic consideration the impact of mechanical stresses and strains on the thermo-emf, *Measuring Equipment and Metrology*, Issue 59, 2002, pp. 66-69 (in Ukrainian).

[23]. Ya. Lutsyk, B. Stadnyk, S. Yatsyshyn, Drift of thermoelectric properties and electromechanochemical noise in thermometric converters, *Journal of Thermoelectricity*, 2, 2004, pp.75-81.

[24]. Michael Berger, Selective gas sensing with pristine graphene, *Nanowerk Nanotechnology Spotlight,* April 26, 2012, http://www.nanowerk.com/spotlight/spotid=25043.php

[25]. H. Rao, G. Bosman, Simultaneous low-frequency noise characterization of gate and drain currents in AlGaN/GaN high electron mobility transistors, *Journal of Applied Physics,* Vol. 106, 2009, pp. 103712- 103715.

[26]. S. Jacyszyn, B.Stadnyk, J.Lucyk, Efekty szumowe w termometrii, Pomiary, automatyka, kontrola, Issue 7/8, 2003. pp. 15-17 (in Polish).

[27]. S. Yatsyshyn, B. Stadnyk, Ya. Lutsyk et al, Research in Nanothermometry. Part 8: Summary, *Sensors & Transducers,* Vol. 144, Issue 9, September 2012, pp. 1-15.

[28]. I. Novikov, Defects of crystalline structure, *Metallurgy*, Moscow, 1975. 208 p. (in Russian).

[29]. I. Novikov, Thermodynamic aspects of the deforming and destroying, in Physical and mechanical properties of metals, *Nauka,* Moscow, 1976, pp. 170-176 (in Russian).

[30]. S. Kutateladze, V. Borishanski, Handbook of heat transfer, *Gostehizdat,* Leningrad, 1959, 348 p.(in Russian).

[31]. Ultrasound, Red. of I .Golamin, *Metallurgy,* Moscow, 1979, 332 p. (in Russian).

[32]. Ch. Kittel, Introduction to Solid State Physics, 8th Ed., *John Wiley & Sons*, Nov. 2004, 704 p.

[33]. H. Postma, T. Teepen, Zhen Yao, C. Dekker, 1/F noise in carbon nanotubes, in *Proceedings of the 36th Rencontres de Moriond,* France, 2001, 4 p.

[34]. M. Dresselhaus, G. Dresselhaus, R. Saitoc, A. Joriod, Raman spectroscopy of carbon nanotubes, *Physics Reports,* Vol. 409, Issue 2, March 2005, pp. 47–99.

[35]. B. Levin, Theoretical basis of statistical radio engineering, *Sov. Radio,* Vol. 2, Moscow, 1975 (in Russian).

[36]. A. van der Ziel, Noise in Solid State Devices and Circuits, *Wiley*, 1986.

[37]. M. Buckingham, Noise in electronic devices and systems, *Wiley-Halsted,* 1983.

[38]. A. V. Savateev, Noise thermometry, *Energy,* Leningrad, 1987 (in Russian).

[39]. I. Mykytyn. Noise filtering with the use of fast Fourier transformation, *Bulletin of Lviv Polytechnic. Automation, Measurement and Control,* 551, 2006, pp. 37-42 (in Ukrainian).

[40]. G. Zhigalski, Nonequilibrium $1/f^\gamma$ -noise in conducting films and contacts, Successes of physical sciences, Vol. 173, Issue 5, 2003, pp. 465-490 (in Russian).

[41]. L. Larikov, Healing of defects in metals, *Naukova Dumka,* Kyiv, 1980 (in Russian).

[42]. B. I. Stadnyk, S. P. Yatsyshyn, Electromechanochemical Noise in Temperature Converters, *Journal of Thermoelectricity*, 1, 2003, pp.56-64.

[43]. S. Yatsyshyn, B. Stadnyk, Accuracy and metrological reliability enhancing of thermoelectric transducers, *Sensors & Transducers,* Vol. 123, Issue 12, December 2010, pp. 69-75.

[44]. Z. Kolodiy, B. Stadnyk, S. Yatsyshyn, Development of Noise Measurements. Part 3. Passive Method of Electronic Elements Quality Characterization, *Sensors & Transducers,* Vol. 152, Issue 5, May 2013, pp. 164-168.

[45]. T. J. Kobayashi, Bayesian Information Processing in Stochastic Biological Systems, in *Proceedings of the 22nd International Conference on Noise Fluctuations,* June 24-28, 2013, Corum de Montpellier, France, p. 137.

[46]. B. Guillet, S. Wu, L. Mechin, C. Adamo, D. G. Schlom, J. M. Routour, Correlation of 1/f noise with DC electrical properties in $La_{0.7}Sr_{0.3}MnO_3$ thin films, in *Proceedings of the 22nd International Conference on Noise Fluctuations,* June 24-28, 2013, Corum de Montpellier, France, p. 43.

[47]. A. Meddah, T. Kadri, Stochastic Behavior of a Bridge Under Moving Vehicles with Irregular Deck, in *Proceedings of the 22nd International Conference on Noise Fluctuations,* June 24-28, 2013, Corum de Montpellier, France, p. 13.

[48]. C. G. Theodorou, N. Fasarakis, T. Hoffman, T. Chiarella, G. Ghibaudo, C. A. Dimitriadis, Origin of the low-frequency noise in-channel Fin FETs, *Solid-State Electronics,* Vol. 82, Apr. 2013, pp. 21-24.

[49]. E. Sergeev, A. Knapek, L. Grmela, J. Šikula, Noise Diagnostic Method of Experimental Cold Field-Emission Cathodes, in *Proceedings of the 22nd International Conference on Noise Fluctuations,* June 24-28, 2013, Corum de Montpellier, France, p. 38.

[50]. F. A. Levinzon, K. J. Vandamme, Comparison of 1/f noise in jfets and mosfets with several figures of merit, *Fluctuation and Noise Letters,* Vol. 10, Issue 4, 2011, pp. 447–465.

[51]. C. Varotsos, I. Melnikova, M. Efstathiou, C. Tzanis, 1/f noise in the UV solar spectral irradiance, *Theoretical & Applied Climatology,* Vol. 111, Issue 3/4, Feb. 2013, pp. 641-648.

[52]. A. V. Desherevsky, A. A. Lukk, A. Y. Sidorin, G. V. Vstovsky, S. F. Timashev, Flicker-noise spectroscopy in earthquake prediction research, *Natural Hazards and Earth System Sciences,* Vol. 3, 2003, pp. 159-164.

[53]. M. Zahid Hossain, S. Rumyantsev, M. Shur, A. Balandin, Reduction of 1/f noise in graphene after electron-beam irradiation, *Applied Physics Letters,* Vol. 102, Issue 15, 2013, pp.153512-153512-5.

[54]. E. Shmelev et al., Complexes of spatially multistable defects as the source of 1/f noise in devices, *Fluctuation & Noise Letters*, Vol. 12, Issue 1, 2013, pp. 1-13.

[55]. G. Leroy, L. Yang, J. Gest, L. K. J. Vandamme, Research on the properties of ZnO films by 1/f noise measurement, in *Proceedings of the 22nd International Conference on Noise Fluctuations*, June 24-28, 2013, Corum der Montpellier, France, p. 48.

[56]. F. G. Aliev, J. P. Cascales, F. Bonell, S. Andrieu, Band edge noise spectroscopy, in *Proceedings of the 22nd International Conference on Noise Fluctuations*, June 24-28, 2013, Corum de Montpellier, France, p. 66.

[57]. T. Fink, H. Bluhm, Noise Spectroscopy Using Correlations of Single-Shot Qubit Readout, *Physical Review Letters*, Vol. 110, Issue 1, 2013, pp. 010403-1-010403-5.

[58]. L. K. J. Vandamme, How useful is Hooge's empirical relation, in *Proceedings of the 22nd International Conference on Noise Fluctuations*, June 24-28, 2013, Corum de Montpellier, France, p. 17.

[59]. G. Litak, Yu. Polyakov, S. Timashev, R. Rusinek, Dynamics of stainless steel turning: Analysis by flicker-noise spectroscopy, *Physica A: Statistical Mechanics and its Applications*, Vol. 392, Issue 23, 2013, pp. 6052-6063.

[60]. S. F. Timashev, Flicker-Noise spectroscopy in analysis of chaotic fluxes in distributed dynamical dissipative systems, *Russian Journal of Physical Chemistry*, Vol. 75, Issue 10, 2001, pp. 1742-1749 (in Russian).

[61]. P. Wagner, T. Aichinger, T. Grasser, M. Nelhiebel, Possible Correlation between Flicker Noise and Bias Temperature Stress, in *Proceedings of the 20th International Conference on Noise Fluctuations*, 2009, pp. 621–624.

[62]. Sh. Qinghui, L. Guanxiong, D. Teweldebrhan, A. Balandin, Noise in Bilayer Graphene Transistors, *Electron Device Letters*, Vol. 30, Issue 3, 2009, pp. 288–290.

[63]. Z. Kolodiy, A. Kolodiy, Informational Noise's Entropy, *IEEE Journal of Electrical Engineering*, Vol. 2, Issue 2, 2014, pp. 92-95.

[64]. Z. A. Kolodiy, Flicker-Noise of Electronic Equipment: Sources, Ways of Reduction and Application, *Radioelectron. Commun. Syst.*, Vol. 53, Issue 8, 2010, pp. 412–417.

[65]. A. Diniz et al., Contemporary theories of 1/f noise in motor control, *Human Movement Science*, Vol. 30, Issue 5, October 2011, pp. 889–905.

[66]. B. Stadnyk, S. Yatsyshyn, O. Seheda, Yu. Kryvenchuk, Metrology of Cyber-Physical Systems. Part 8. Elaboration of Raman Method, *Sensors & Transducers*, Vol. 189, Issue 6, 2015, pp. 116-120.

[67]. B. Stadnyk, S. Yatsyshyn, O. Seheda, Metrology of Temperature Transducer based on Raman Effect, *Sensors & Transducers*, Vol. 117, Issue 6, June 2010, pp. 78-84.

[68]. B. Stadnyk, S. Yatsyshyn, O. Seheda, Research in Nanothermometry, Part 6: Metrology of Raman thermometer with universal calibration artefacts, *Sensors & Transducers*, Vol. 142, Issue 7, July 2012, pp. 1- 9.

[69]. G. Gouadec, Ph. Colomban, Raman Spectroscopy of Nanomaterials: How Spectra Relate to Disorder, Particle Size and Mechanical Properties, *Progress in Crystal Growth and Characterization of Materials,* Vol. 53, Issue 1, 2007, pp. 1-56.

[70]. A. N. Magunov, Laser Thermometry of Solids, *Fizmatlit*, Moscow, 2001 (in Russian).

7.

Qualimetric Estimation of CPSs and Their Products

V. Motalo, B. Stadnyk and S. Yatsyshyn

Qualimetric measurements are the indirect measurement of product quality level, the value of which is found by processing the results of its properties measurements according to multidimensional scaling methodology [1]. The level of product quality is relative characteristics of products quality, based on a comparison of estimated values of products quality with the basic values of the relevant parameters.

So, qualimetric measurement methodology allows determining the level of NG quality, which is a comprehensive assessment of its quality, taking into account all the factors that affect it.

Hydrocarbon Gases are the natural mixture of aliphatic hydrocarbons of different structure, located in the pores and cavities of rocks, dissolved in the oil and stratified waters, or dispersed in the soil. Because of their energy and chemical valuables Gases are related to strategic feedstocks, which led to their wide spreading in both housing and municipal sector, in many industries, and attracted a particular attention to Gas as an object research. Importance of gas analysis today is emphasized so that such issues as the supplying volume and price, possibility of gas mining from sources to consumers have become the elements of foreign policy. Key and urgent issues in this analysis are the gas quality requirements and establishing the gas prices, depending on energy value, which in turn requires advanced and reliable methods of determining the quality. Despite the certain attention to the gas quality assessment, a number of challenges of metrological and technical supports, regulatory evaluation of the gas quality are still emerging.

In most European countries the main qualitative characteristic of NG is the Wobbe Index, which is a function of the calorific value and relative density of gas. According to the current standard ISO 13686:1998 [2]

NGs are divided by Wobbe Index into two varieties: H-gas and L-gas. For the 1st one Wobbe Index constitutes 48.36...57.87 MJ/m³, and for L-gas – 41.28...47.38 MJ/m³. In return, these groups are divided into subgroups, depending on which is established the price of gas for residential consumers. For example, in Germany the NG of group H is divided into subgroups E and ES, and the NG of group L into subgroup on EI and LL; in France there are eight similar subgroups E1, ..., E4 and L1, ..., L4.

LNG is a natural hydrocarbon gas that at normal pressure 0.101325 MPa and temperature of environment 20 °C remains in a gaseous state, but drop in temperature to −160 °C and the pressure increase turns into reduced in 600 times by volume liquid, which facilitates its storage and transport. Large enterprises are increasingly used method of storing liquefied petroleum gas at atmospheric pressure and low temperature. Application of this method is accomplished by refrigeration, which reduces the vapor pressure of LNG. Liquefied propane gas may be kept at −42 °C and atmospheric pressure, resulting in reduced calculation pressure when defining the wall thickness of tanks. It is sufficient that the wall withstood only the hydrostatic pressure of the stored product.

LPG is the propane-butane gas mixture, which by the normal range of pressure and ambient temperature (0.101325 MPa; 20 °C) is in a gaseous state, but at the certain changes in conditions: slight increase in pressure and a sustainable temperature, or a slight decreasing temperature and constant pressure – passes into the liquid state.

The raw materials for the manufacturing the LPG are associated with the petroleum gases and gases of oil and condensate stabilization process as well as a small extent of NG. LPG major components are propane and butane. LPG has a broad field of usage: in municipal consumption, as a fuel for internal combustion engines, as a chemical raw materials and so on.

In spite of certain differences between different types of gases, evaluating their quality expedient has to be based on a generalized methodology. Methods of determining the gas quality should be comprehensive and calculate the value of its most important characteristics, such as caloricity, humidity, density, composition, availability and quantity of non-flammable and harmful components and so on. Also, appropriate to carry out the sorting of gas by its quality level and to establish appropriate prices depending on its energy value.

Basing on the analysis of existing standards and normative documents, which are governed the requirements for NG physical and chemical properties and for procedures of their determination, it has been established below the structure problems of principles, regulations, and technical support in determining the NG quality as well as methodology to solve them. The basic task of modern qualimetry of hydrocarbon gases, effective solution which is appropriate to carry out on gas at theoretical, practical and legislative levels, is formulated further. Consider the following: development of underpinning pillars of the integrated NG quality evaluation and development of methods for determining the quality of studied NG; research of evaluation methods accuracy of gas quality in order to ensure the unity of qualimetric measurements; development of methods and software for automating measurements of NG quality indicators in accordance with required accuracy and selection of MIs necessary for their implementation.

7.1. Qualimetry of Natural Gas as Energy Source

NG, which is highly energy intensive and chemically valuable raw material, undoubtedly belongs to the strategic products. This led to its widespread use in daily life, as well as in many industries. Energetic and chemical values of NG create the considerable attention of multidisciplinary specialists concerning gas as an object of study. However, the question of determining the gas quality practically is not investigated, although it is very important. Currently, the largest area of NG use is in industrial and communal sector, namely the use of gas as an energy source. Therefore, during the continuous growth of world prices for natural gas, demands for gas quality as an energy carrier are becoming increasingly urgent as well as the need to take into account the energy value of gas while setting a price for it.

Thus, there are a number of problems in the sphere of Qualimetry of NG, solutions of which are very important and resulted in theme below.

- Analysis of Current State of Natural Gas Quality Assessment as Energy Source. In Ukraine practically the only criterion of gas quality today is its moisture, which is reflected in the current documents as dew point temperature of the gas on moisture (or "dew point") Θ_p °C and characterizes transport meeting specifications of gas and its ability to ensure trouble-free operation of gas supply systems.

As it can be seen from a brief analysis of the current state of NG quality assessment, currently there is no generally accepted methodology for determining the quality of NG as the energy source, taking into account all the factors that affect its energy value. Main objectives of the research are:

• Selecting the quality evaluation concept of natural gas as the energy source;

• Analysis of physical-chemical properties impact of natural gas on its energy value and choice of indexes range for determining the gas quality as energy source;

• Conducting of experimental studies of natural gas samples according to developed evaluation methodology of its quality as an energy source.

The concept choice of methodology creation of natural gas quality evaluation as an energy source is conditioned by the following considerations.

- Methodology of Natural Gas Quality Evaluation as Energy Source. For a comprehensive evaluation of the natural gas quality, namely gas determination respectively to methodology of qualimetric measurements, there is needed to follow the next steps:

1) Analysis, systematization and choice of natural gas quality parameters to determine its quality level;

2) Setting of a base sample of natural gas as physical standard in determining its quality;

3) Formation of the basic profile of natural gas quality $\Pi_{P,b}$ based on single absolute basic indexes of its quality $P_{b,i}$;

4) Formation of the estimated profile of natural gas quality $\Pi_{P,e}$, or based on absolute individual evaluated quality indexes $P_{e,i}$;

5) Determination of the single-relative basic $K_{b,i}$ and evaluated $K_{e,i}$ natural gas quality parameters;

6) Definition of the validity factors m_i of individual indexes of natural gas quality;

7) Formation of the basic profile of natural gas quality $\Pi_{K,b}$, namely virtual extent of natural gas, and estimated profile of natural gas quality $\Pi_{K,e}$;

8) Determination of the correlation coefficients $r_{K_{e,i}K_{e,j}}$ between the single-estimated relative indexes of natural gas quality $K_{e,i}$ and $K_{e,j}$ $(i \neq j)$;

9) Selection of multidimensional scaling model aiming to determine the level of natural gas Q;

10) Definition of the level of natural gas quality Q according to methodology of multidimensional scaling.

- Analysis and Choice of Natural Gas Quality Parameters. It is proclaimed in [3] that products quality is the degree to which the set of own characteristics of product meets the requirements. Natural gas, as well as any other product, has certain properties. Products attribute their objective features that can be manifested during its creation, production, operation or consumption. For practical evaluation of the products, the quality indexes are used. Products quality index is the quantitative description of one or more properties of products which characterizes its quality that is examined due to certain conditions of its creation, operation or use.

Let us consider the natural gas properties in relation to their effect on its quality as an energy source. The primary numerical assessment of product quality is the single quality product index, which characterizes one of its properties. According to the shape of the product properties reflection, the quality indexes are divided into absolute P_i and relative K_i, and on index function at determining the level of product quality – at estimation, respectively, $P_{e,i}$ and $K_{e,i}$ and basic $P_{b,i}$ and $K_{b,i}$. Absolute index of product quality $P_i, i = 1,2, ..., n$, where n is the number of properties) describes the individual properties of products and numerically is equal to the ith products property p_i and is expressed in its units. In the process of evaluating the quality of products it is directly measurable value. The relative index of product quality $K_i, i = 1,2, ..., n$, where n is the number of properties) describes certain properties of the product in the form of a ratio between the values of certain property, displayed its absolute indexes, and its dimensionless quantity.

Concerning the effect on the value of production quality Q, quality indexes are divided into two groups. In the first group of quality indexes increase of the value of absolute single quality index $P_{e,i}$ causes the improvement of the investigated product Q quality and, therefore, increase of the unit relative quality index $K_{e,i}$, and in the second group reducing the value of the unit absolute quality index $P_{e,i}$ causes the improvement of the investigated product Q quality and, respectively, reducing unit relative quality index $K_{e,i}$. When setting the range of product quality indexes to determine the level of its quality, analysis of current NG is taken into account, and there are regulated requirements for the properties of products, reference literature and experimental results of specific products. Based on the authors' analysis of the references [4-7] that characterize its quality as energy source, the basic physical and chemical properties of NG are defined, previously divided into two groups.

The first group includes indexes of gas, increasing values of which leads to the increase in the quality of gas as an energy source Q: low heating value H_H, MJ/m^3; Wobbe B number, MJ/ m^3 and gas density ρ, kg/m^3. The second group includes indexes of gas, increasing values of which leads to a decrease in quality of gas as an energy source Q: gas moisture W, g/m^3; content of non-combustible mixture in gas (C_{N_2}, % nitrogen and carbon dioxide C_{CO_2}, %) and harmful components (hydrogen sulphide C_{N_2S}, g/m^3 and mercaptan sulfur C_{CH_4S}, g/m^3). Hydrogen sulfide although it is combustible, but toxic and corrosive component, assigned to the second group of gas quality indexes. Therefore, when analyzing the natural gas quality as an energy source, we consider eight of its properties, namely eight quality indexes $P_i, i = 1,2,3, ..., n ; n = 8$.

- Setting the Base Sample of Natural Gas as Physical Standard. Essence of any measurement is the comparison of the measured value with the measure, which reproduces and/or maintains a physical quantity of given size. In the Qualimetry such means of comparison, that standard of the quality of products may be a base (reference, standard) sample of the product. In accordance to definition, a standard sample is the standard as a substance or material with the established, as a result metrological certification, values of one or more quantities characterizing the properties or composition of this substance or material [8-9]. The base is called the sample of products that meets advanced scientific and technical achievements in prescribed period.

Regarding qualimetric measurements, the measured value of which is the level of quality Q of the researched products, to reflect product quality profile P is used, which is a single set of quantitative indexes of production quality [1, 10].

Thus, the base sample of product in numerical terms, may be described as a set of numerical values of unit basic absolute indexes of products quality $P_i, i = 1,2,...,n$, where n is the number of individual indexes which is equal to the number of products properties p_i), which form the basic quality profile $\Pi_{P,b} = \{P_{b,1}; P_{b,2}; 3; ...; P_{b,n};\}$. Also taking into account the separation of products properties concerning their impact on the quality level of Q into two groups, the basic profile $\Pi_{P,b}$ of product quality is formed of two parts:

$$\Pi_{P,b} = \{P_{b,1}; P_{b,2}; P_{b,3}; ...; P_{b,n}\}$$
$$= \{P_{b,1}; P_{b,2}; ...; P_{b,l}\} + \{P_{b,l+1}; P_{b,l+2}; ...; P_{b,n}\}, \tag{7.1}$$

where l is the number of individual quality indexes of the first group of natural gas, namely indexes, which increasing leads to improvement of its quality Q; $n-l$ is the number of individual quality parameters of the second group of natural gas, namely indexes, which increasing leads to reducing of its quality Q.

Numerical value of each element $P_{b,i}$ of basic quality profile $\Pi_{P,b}$, set, according to [10], as accepted reference value, or value that is used as an agreed standard for comparison and which is defined as:

a) Theoretical or established value, based on the scientific principles;

b) Attributed or certified value, based on experimental data of some national or international organizations;

c) Agreed or certified value, based on joint experimental work conducted by scientific or engineering staff;

d) Mathematical expectation of measured value, namely the average value of measurement results - just in case when a), b) and c) are not available.

Thus, the basic quality profile $\Pi_{P,b} = \{P_{b,1}; P_{b,2}; P_{b,3}; ...; P_{b,n};\}$, the numerical values of each of the elements $P_{P,b} = 1,2,3,...,n$ are defined according to the above mentioned rules and are established as reference, is a numerical characteristic of physical standard of products quality, namely the base sample.

In particular, it should be noted that for the procedures implementation of qualimetric measurement it is not necessarily to produce physical sample of studied product. It is enough to form a base quality profile $\Pi_{P,b}$, on the basis of which the basic quality profile $\Pi_{K,bw}$ is formed, from the appropriate relative weighted basic indexes of quality $K_{bw,i}$, namely virtual standard of quality of products, which is used in process of determining the level of studied product quality.

Naturally, basing on the above mentioned analysis we form the basic quality profile $\Pi_{P,b}$, for determining the quality of NG as an energy source.

- Formation of Base Profile of Natural Gas Quality. Base profile of natural gas quality as an energy source $\Pi_{P,b}$, which is a numerical characteristic of physical standard of gas quality, namely the base sample of investigated gas, we form from the unit basic absolute indexes of its quality $P_{P,b} = 1,2,3,\dots,n; n = 8$, corresponding to 8 major, above defined gas properties. Also, for implementation of the methodology for determining the level of NG quality taking into account its properties division according to their impact on level of gas quality into 2 groups, the basic profile of gas quality $\Pi_{P,b}$ according to (7.1) is formed of 2 parts:

$$
\begin{aligned}
\Pi_{P,b} &= \{P_{b,1}; P_{b,2}; P_{b,3}\} + \{P_{b,4}; P_{b,5}; P_{b,6}; P_{b,7}; P_{b,8}\} \\
&= \{P_{b,H_H}; P_{b,B}; P_{b,\rho}\} \\
&\quad + \{P_{b,W}; P_{b,C_{N_2}}; P_{b,CO_2}; P_{b,CH_2S}; P_{b,CH_4S}\},
\end{aligned}
\tag{7.2}
$$

where $P_{b,1} = P_{b,H_H}$ is the absolute basic single-quality index of NG, which is equal to the base value of the specific volume of heat of gas combustion (gas calorific value) $P_{H,b}$, MJ/m^3; $P_{b,2} = P_{b,B}$ is the absolute basic single-quality index of NG, which is equal to the base value of the Wobbe number B_b, MJ/m^3; $P_{b,3} = P_{b,\rho}$ is the absolute basic single-quality index of NG which is equal to the base value of the gas density ρ_b, kg/m^3; $P_{b,4} = P_{b,W}$ is the absolute basic single-quality index of NG which is equal to the base value of gas humidity W_b, g/m^3; $P_{b,5} = P_{b,C_{N_2}}$ is the absolute basic single-quality index of NG equal to the base value of the concentration of nitrogen in the gas mixture $C_{N_2,b}$ %; $P_{b,6} = P_{b,CO_2}$ is the absolute basic single-quality index of NG which is equal to

300

the base value of concentration in the gas mixture of carbon dioxide $C_{CO_2,b}$ %; $P_{b,7} = P_{b,C_{H_2S}}$ is the absolute basic single-quality index of NG which is equal to the base value of content of hydrogen sulphide in the gas mixture $C_{H_2S,b}$ g/m^3 ; $P_{b,8} = P_{b,CH_4S}$ is the absolute basic single-quality index of NG which is equal to the base value of content in the gas mixture of mercaptan sulfur $C_{CH_4S,b}$ g/m^3.

Basic indexes of product quality $P_{b,i}$: these are indexes of products quality, values of which are taken as the basis for comparative assessment of its quality. Overall, the base value of quality products index $P_{b,i}$, may be one of the normalized values - regulated, nominal or boundary which is set depending on the index group. In the first group of gas quality indexes, when higher value of index leads to the raising of its quality, we accept that $P_{b,i} = P_{i,max}$, where $P_{i,max}$ is the maximum possible value of the ith index of gas quality that meets the highest value meaning of its quality Q. In the second group of gas quality indexes when lower value of index leads to the raising of its quality, we accept that $P_{b,i} = P_{i,min}$, where $P_{i,min}$ is the minimum possible value of i-th index of gas quality that meets the highest value to its quality Q (it is desirable that $P_{i,min} \rightarrow 0$). Also in both groups of the gas quality indexes, maximum allowable values of the $P_{i,allow}$, are set, but its content is different. In the first group natural gas does not meet the established requirements in case $P_{b,i} \leq P_{i,allow}$, and in the second group $P_{b,i} \geq P_{i,allow}$.

The numerical values of base quality indexes of NG $P_{b,i}$, are set based on the analysis of current ND; there are regulated requirements for gas property values and the methods of measurement, based on the analysis the relevant references in accordance with [9]. Also, when setting numerical values of the basic gas quality indexes we are considered the specific requirements of each particular analysis of its quality.

Estimated quality profile $\Pi_{P,e}$ of studied NG is formed from unitary estimated absolute indexes of its quality $P_{e,i} = 1,2,3, \dots, n; n = 8$ corresponding to 8 of its most important properties p_i, set above. Taking into consideration the gas separation properties on their impact on the level of quality into two groups, estimated quality profile of NG $\Pi_{P,e}$ as a source of energy as a basic quality profile as $\Pi_{P,b}$, are formed of two parts:

$$\Pi_{P,e}$$
$$= \left\{P_{e,1}; \ P_{e,2}; P_{e,3}\right\} + \left\{P_{e,4}; \ P_{e,5}; \ P_{e,6}; P_{e,7}; \ P_{e,8}\right\}$$
$$= \left\{P_{e,H_H}; \ P_{e,B}; P_{e,\rho} \right\} \tag{7.3}$$
$$+ \left\{P_{e,W}; \ P_{e,C_{N_2}}; \ P_{e,CO_2}; \ P_{e,C_{H_2S}}; P_{e,CH_4S}\right\},$$

where $P_{e,1} = P_{e,H_H}$ is the estimated overall single-quality NG index, which is equal to the base value of the specific volume heat of gas combustion of lower (gas calorific value) $H_{H,e}$, MJ/m^3; $P_{e,2} = P_{e,B}$ is the absolute estimated single-NG quality index which is equal to estimated value of the Wobbe number B_e, MJ /м^3; $P_{e,3} = P_{e,\rho}$ is the absolute estimated single-NG quality index which is equal to the estimated value of gas density ρ_e, kg/m^3; $P_{e,4} = P_{e,W}$ is the absolute estimated single-NG quality index which is equal to estimated value of gas humidity W_e, g/m^3; $P_{e,5} = P_{e,C_{N_2}}$ is the absolute estimated single-NG quality index which is equal to the estimated value of the concentration of nitrogen in gas mixture $C_{N_2 e}$ %; $P_{e,6} = P_{e,CO_2}$ is the absolute estimated single-NG quality index which is equal to estimated value of concentration in the gas mixture of carbon dioxide $C_{CO_2,e}$ %; $P_{e,7} = P_{e,C_{H_2S}}$ is the absolute estimated single-NG quality index which is equal to estimated value content of hydrogen sulphide in the gas mixture $C_{H_2S,e}$ g/m^3; $P_{e,8} = P_{e,CH_4S}$ is the absolute estimated single-NG quality index which is equal to estimated value of content in the gas of mixture mercaptan sulfur $C_{CH_4S,e}$ g/m^3.

Numerical values of estimated NG quality indexes $P_{e,i} = 1,2,3,\dots,n; n = 8$ are determined experimentally by measuring the properties of gas p_i according to current normative documents that include regulated requirements for gas property values and methods of their measurement.

- Defining Single Relative Basic $K_{b,i}$ and Estimated the $K_{e,i}$ of Natural Gas Quality Indexes. Value of single-evaluated relative quality indexes $K_{e,i}$ is always located within $0 \leq K_{e,i} \leq 1$, but depends on group of indexes concerning the impact on their gas quality; it is calculated by different methods.

In the first group of quality indexes the improving the gas quality level Q is caused by increasing of single estimated absolute quality indexes $P_{e,i}$, and, consequently, increasing of the single estimated relative quality indexes $K_{e,i}$. Thus, in the first group of quality indexes - basic values of gas relative quality $K_{b,i} \to 1$, theoretically, $K_{b,i} = 1$, and the numerical values of the estimated relative quality indexes $K_{e,i}$ are calculated by:

$$K_{e,i} = \frac{P_{e,i}}{P_{i,max}} = \frac{P_{e,i}}{P_{b,i}}, \ P_{e,i} \le P_{b,i} \tag{7.4}$$

In the second group of quality indexes the improving the gas quality level Q is caused by decreasing of the single estimated absolute quality indexes $P_{e,i}$, and, consequently, decreasing of the single estimated relative quality indexes $K_{e,i}$. Thus, in the second group of quality indexes - basic values of gas relative quality $K_{b,i} \to 0$, theoretically $K_{b,i} = 0$, and the numerical values of estimated relative quality indexes $K_{e,i}$ are calculated:

$$K_{e,i} = \frac{P_{e,i} - P_{i,min}}{P_{i,allow} - P_{i,min}} = \frac{P_{e,i} - P_{b,i}}{P_{i,allow} - P_{b,i}}, \ P_{e,i} \le P_{b,i} \tag{7.5}$$

where $P_{i,allow}$ is the maximum allowable index value, above which $P_{e,i} > P_{i,allow}$ provides that gas is defective.

- Formation of Basic Quality Profile and Estimated Quality Profile of Natural Gas. Under the basic $\Pi_{P,b}$ and estimated $\Pi_{P,e}$ product quality profiles, generated in accordance with (7.2)-(7.3), we form the basic $\Pi_{K,b}$ and estimated $\Pi_{K,e}$ profiles of products quality with appropriate relative basic $K_{b,i,1}$ and $K_{b,i,2}$ and evaluated quality indexes $K_{e,i,1}$; $K_{e,i,2}$ of the 1st and 2nd groups, calculated by (7.4) - (7.5):

$$
\begin{aligned}
\Pi_{K,b} \\
= \left\{ K_{b,1,1}; \ K_{b,2,1}; \ ...; \ K_{b,l,1} \right\} \\
+ \left\{ K_{b,l+1,2}; \ K_{b,l+2,2}; \ ...; \ K_{b,n,2} \right\} \\
= \left\{ K_{b,H_H}; \ K_{b,B}; \ P_{b,\rho} \right\} \\
+ \left\{ K_{b,W}; \ K_{b,C_{N_2}}; \ K_{b,CO_2}; \ K_{b,CH_2S}; \ K_{b,CH_4S} \right\}
\end{aligned}
\tag{7.6}
$$

$$
\begin{aligned}
\Pi_{K,e} \\
= \left\{ K_{e,1,1}; \ K_{e,2,1}; \ ...; \ K_{e,l,1} \right\} \\
+ \left\{ K_{e,l+1,2}; \ K_{e,l+2,2}; \ ...; \ K_{e,n,2} \right\} \\
= \left\{ K_{e,H_H}; \ K_{e,B}; \ K_{e,\rho} \right\} \\
+ \left\{ K_{e,W}; \ K_{e,C_{N_2}}; \ K_{e,CO_2}; \ K_{e,CH_2S}; \ K_{e,CH_4S} \right\}
\end{aligned}
\tag{7.7}
$$

Numerical values of relative base quality indexes in the first group of indexes $K_{b,i} \to 1, i = 1,2, \dots, l$, in general $K_{b,i} \leq 1$ and in extreme $K_{b,i} = 1$, in the second group - $K_{b,i} \to 0, i = l+1, l+2, \dots, n$, in general $K_{b,i} \geq 0$, and in the extreme variant $K_{b,i} = 0$.

- Formation of virtual quality standard of natural gas and weighted estimated gas quality profile. By regarding the basic $\Pi_{K,b}$ and estimated $\Pi_{K,e}$ product quality profiles, formed in accordance with (7.6) and (7.7), we assess a weighted basic $\Pi_{K,bw}$ and an estimated $\Pi_{K,ew}$ product quality profiles of the respective weighted relative basic $K_{bw,i,1}$ and $K_{bw,i,2}$ and estimated $K_{ew,i,1}$ and $K_{ew,i,2}$ quality indexes of the first and second groups, which are key elements in the procedure of qualimetric measurements:

$$
\begin{aligned}
\Pi_{K,bw} &= \left\{ K_{bw,1,1}; \ K_{bw,2,1}; \dots; \ K_{bw,l,1} \right\} \\
&+ \left\{ K_{bw,l+1,2}; \ K_{bw,l+2,2}; \dots; \ K_{bw,n,2} \right\} \\
&= \left\{ K_{bw,H_H}; \ K_{bw,B}; P_{bw,\rho} \right\} \\
&+ \left\{ K_{bw,W}; \ K_{bw,C_{N_2}}; \ K_{bw,CO_2}; \ K_{bw,C_{H_2}S}; K_{bw,CH_4S} \right\};
\end{aligned}
\tag{7.8}
$$

$$
K_{bw,i,1} = K_{b,i,1} \cdot m_{i,1}; \quad K_{bw,i,2} = K_{b,i,2} \cdot m_{i,2};
\tag{7.9}
$$

$$
\begin{aligned}
\Pi_{K,ew} &= \left\{ K_{ew,1,1}; \ K_{ew,2,1}; \dots; \ K_{ew,l,1} \right\} \\
&+ \left\{ K_{ew,l+1,2}; \ K_{ew,l+2,2}; \dots; \ K_{ew,n,2} \right\} \\
&= \left\{ K_{ew,H_H}; \ K_{ew,B}; K_{ew,\rho} \right\} \\
&+ \left\{ K_{ew,W}; \ K_{ew,C_{N_2}}; \ K_{ew,CO_2}; \ K_{ew,C_{H_2}S}; K_{ew,CH_4S} \right\}
\end{aligned}
\tag{7.10}
$$

$$
K_{ew,i,1} = K_{e,i,1} \cdot m_{i,1}; \quad K_{ew,i,2} = K_{e,i,2} \cdot m_{i,2};
\tag{7.11}
$$

Weighted basic profile of product quality $\Pi_{K,bw}$, formed from single relative weighted basic indexes of quality $K_{bw,i}, i = 1,2,3, \dots, n$ is a virtual product quality standard (in our case a virtual gas quality standard) used in procedure of qualimetric measurements for determining the quality level Q of studied product.

Normalized weight coefficients $m_{i,1}$ and $m_{i,2}$ of single-quality indexes $K_{e,1}$ of the first $K_{e,i,1}$ and second $K_{e,i,2}$ groups of the studied product must satisfy the condition of their normalization:

$$\sum_{i=1}^{l} m_{i,1} + \sum_{i=l+1}^{n} m_{i,2} = 1 \qquad (7.12)$$

- Determination of Weight Coefficients m_i of Single Indexes of Natural Gas Quality. Choice of method for determining weight coefficients m_i of single-quality indexes $K_{e,i}$ of the studied product depends on the type of product and available data on properties. To implement procedures of qualimetric measurements one of the known quality control methods can be used [11]: expensive method, method of regressive dependencies, equivalent ratio method, expert method, as well as mixed or combined methods.

Validity coefficients m_i important of unitary quality parameters of NG, are advisable to determine by the method of boundary and nominal values because due to current normative documents, which regulate requirements for gas properties and methods of measurement, we may set individual values of single absolute quality parameters of gas, in particular, nominal $P_{i,nom}$ and allowed $P_{i,allow}$ values. The values of coefficients determining of the validity m_i of individual quality indexes of the investigated natural gas are provided below.

- Multidimensional Scaling Model in Measuring Gas Quality. To standard the level quality of investigated natural gas Q, namely for full implementation of qualimetric measurements procedures, it is necessary to provide comparing the estimated weighted profile of gas quality $\Pi_{K,ew}$ with weighted basic profile $\Pi_{K,bw}$, namely as with virtual quality standard. To compare profiles of quality $\Pi_{K,ew}$ and $\Pi_{K,bw}$, we apply methodology of multidimensional scaling - one of the sections of mathematical statistics, the subject of research of which is the data processing on pairwise similarities, connections and relationship between objects that are analyzed in order to present these objects in the form of points of some multidimensional space [10]. This technique allows the comparison of individual estimated weighted relative and basic quality indexes $K_{ew,i}$ and $K_{bw,i}$, followed by the construction of scales, which determined the value of these indexes, to one-dimensional scale to define the level of quality Q of the studied products.

Choice of implementation model of multidimensional scaling we perform based on the analysis of presence/absence of statistical correlation between the estimated absolute single-quality indexes of studied gas $P_{e,i}$ and therefore between the estimated relative gas quality indexes $K_{e,i}$.

In case of statistically noncorelated single-quality products indexes to compare weighted profile of investigated products $\Pi_{K,ew}$ with reasonable quality basic profile $\Pi_{K,bw}$, namely with standard of virtual studied products quality, we use weighted Euclidean model of individual differences on which function of differences $\Delta\Pi$ between profiles $\Pi_{K,ew}$ and $\Pi_{K,bw}$ are defined according to:

$$
\Delta\Pi = \left\{ \sum_{i=1}^{l} \left(K_{ew,i,1} - K_{bw,i,1}\right)^2 + \sum_{i=l+1}^{n} \left(K_{ew,i,2} - K_{bw,i,2}\right)^2 \right\}^{1/2}
$$
$$
= \left\{ \sum_{i=1}^{l} m_{i,1}^2 \cdot \left(K_{e,i,1} - K_{b,i,1}\right)^2 + \sum_{i=l+1}^{n} m_{i,2}^2 \cdot \left(K_{ew,i,2} - K_{bw,i,2}\right)^2 \right\}^{1/2} \qquad (7.13)
$$

In the case of correlation between individual quality indexes of the studied gas, to determine its quality level we use three-modal model of multidimensional scaling, which allows to consider correlation between the estimated single-gas quality indexes $K_{O,i}$ and $K_{O,j}$, $i \neq j$, $i = 1,2,3, \dots, n$ and the defined function of differences $\Delta\Pi$ between profiles $\Pi_{K,ew}$ and $\Pi_{K,bw}$:

$$
\Delta\Pi
$$
$$
= \left\{ \sum_{i=1}^{l} \left(K_{e,i,1} - K_{b,i,1}\right)^2 + \sum_{i=l+1}^{n} \left(K_{ew,i,2} - K_{bw,i,2}\right)^2 \right.
$$
$$
\left. + 2 \sum_{i=1}^{n-1} \sum_{j=i+1}^{n} m_i m_j \left(K_{e,i} - K_{b,i}\right) \cdot \left(K_{e,j} - K_{b,j}\right) \cdot r_{K_{e,i}} r_{K_{e,j}} \right\}^{1/2}
$$
$$
= \left\{ \sum_{i=1}^{l} m_{i,1}^2 \cdot \left(K_{e,i,1} - K_{b,i,1}\right)^2 + \sum_{i=l+1}^{n} m_{i,2}^2 \cdot \left(K_{ew,i,2} - K_{bw,i,2}\right)^2 \right. \qquad (7.14)
$$
$$
\left. + 2 \sum_{i=1}^{n-1} \sum_{j=i+1}^{n} m_i m_j \left(K_{e,i} - K_{b,i}\right) \cdot \left(K_{e,j} - K_{b,j}\right) \cdot r_{K_{e,i}} r_{K_{e,j}} \right\}^{1/2}
$$

Here m_i and m_j are the normalized weight coefficients of single-quality indexes $K_{e,i}$ and $K_{e,j}$; $i \neq j, i = 1,2,3, \dots, n$; $r_{K_{e,i},K_{e,j}}$ are correlation coefficients between the estimated single-relative indexes of product quality $K_{e,i}$ and $K_{e,j}$, $i \neq j$.

- Determination of Correlation Coefficients Between Single Estimated Relative Indexes of Natural Gas Quality. We define correlation coefficients $r_{K_{e,i},K_{e,j}}$ between estimated single-relative product quality indexes $K_{e,i}$ and $K_{e,j}$, $i \neq j$ as the ratio of covariance $R_{K_{e,i},K_{e,j}}$ of indexes $K_{e,i}$ and $K_{e,j}$, $i \neq j$ for product evaluations of their standard deviations $s_{K_{e,i}}$ and $s_{K_{e,j}}$:

$$
r_{K_{e,i},K_{e,j}} = \frac{R_{K_{e,i},K_{e,j}}}{S_{K_{e,i}} \cdot S_{K_{e,j}}}
$$

$$
= \frac{\sum_{\chi=1}^{\eta} \left(K_{e,i_\chi} - \overline{K}_{e,i} \right) \cdot \left(K_{e,i_\chi} - \overline{K}_{e,i} \right)}{\sqrt{\sum_{\eta=1}^{\eta} \left(K_{e,i_\chi} - \overline{K}_{e,i} \right)^2 \cdot \sum_{\chi=1}^{\eta} \left(K_{e,i_\chi} - \overline{K}_{e,j} \right)^2}}, \qquad (7.15)
$$

where $\overline{K}_{e,i}$ and $\overline{K}_{e,j}$ are the results of measurement indexes $K_{e,i}$ and $K_{e,j}$, $i \neq j$, which are subject to the normal distribution of measuring experimental results K_{e,i_χ} and K_{e,j_χ}, $\chi = 1,2,3, \dots, n$, we calculate as the mean values of the samples:

$$
\overline{K}_{e,i} = \frac{1}{\eta}\sum_{\chi=1}^{\eta} K_{e,i_\chi} ; \ \overline{K}_{e,j} = \frac{1}{\eta}\sum_{\chi=1}^{\eta} K_{e,j_\chi} \qquad (7.16)
$$

- Synthesis of Measurement Scale of Natural Gas Quality. To determine the quality level of the investigated gas pursuant to the qualimetric measurement methodology, it is necessary to build one-dimensional scale of gas quality level. In accordance with generally accepted international definition the measurements scale is an ordered set of values of certain kind used for ranking by the size of magnitude of this kind.

Basing on (7.13) or (7.14) the function $\Delta\Pi$ of deviations meaning between profiles of quality $\Pi_{K,ew}$ and $\Pi_{K,bw}$ we build scale of determining the level of NG quality Q:

$$
Q = 1 - \Delta\Pi , or \ Q = (1 - \Delta\Pi) \cdot 100\% \qquad (7.17)
$$

So, importance of natural gas quality level is determined by developed technique varies from 0 to 1, which is convenient and methodologically correct for practice of evaluating the gas quality, and is consistent with above mentioned definition of measuring scale. To assess the quality of natural gas is proposed the method based on deviation function $\Delta\Pi$ and gas quality Q that allows sorting of gas in terms of its quality and, respectively, setting the different prices for various gas parties.

- Methods of Natural Gas Quality Measuring with Multiple Observations. Result of measurement of NG quality Q in the case of measurement with multiple observations provided at their results normal distribution we define the mean value $\bar{Q}$ of the sample from obtained values of individual measurements of quality $Q_\chi, \chi = 1,2,3, \dots, \eta$, where η is the number of individual measurements, calculated by $Q_\chi = 1 - \Delta\Pi_\chi$:

$$Q = \bar{Q} = \frac{1}{\eta} \sum_{\chi=1}^{\eta} Q_\chi = \frac{1}{\eta} \sum_{\chi=1}^{\eta} (1 - \Delta\Pi_\chi), \qquad (7.18)$$

where $\Delta\Pi_\chi$ is the value of differences function $\Delta\Pi$ for single observation. If using mean-weighted Euclidean model of individual differences in defining the natural gas level of quality Q, the value of differences function $\Delta\Pi_\chi$ is calculated with (7.13) at $\chi = 1,2,3, \dots, \eta$ by:

$$\Delta\Pi_\chi$$
$$= \sqrt{\sum_{i=1}^{l} m_{i,1}^2 \cdot \left(K_{e,i,1_\chi} - K_{b,i,1}\right)^2 + \sum_{i=l+1}^{n} m_{i,2}^2 \cdot \left(K_{e,i,\ \chi} - K_{b,i,2}\right)^2}, \chi \qquad (7.19)$$
$$= 1,2,3, \dots, \eta$$

and in the case of three-modal model usage of multidimensional scaling to find the natural gas quality level Q, the function of differences $\Delta\Pi_\chi$ is determined with (7.14) at $\chi = 1,2,3, \dots, \eta$ by:

$$\Delta\Pi_\chi$$
$$= \left\{ \sum_{i=1}^{l} m_{i,1}^2 \cdot \left(K_{e,i,1_\chi} - K_{b,i,1}\right)^2 + \sum_{i=l+1}^{n} m_{i,2}^2 \cdot \left(K_{e,i,2_\chi} - K_{b,i,2}\right)^2 \right.$$
$$\left. + 2 \sum_{i=1}^{n-1} \sum_{j=i+1}^{n} m_i m_j \left(K_{e,i_\chi} - K_{b,i}\right) \cdot \left(K_{e,j_\chi} - K_{b,j}\right) \cdot r_{K_{e,i}} r_{K_{e,j}} \right\}^{1/2}, \chi \qquad (7.20)$$
$$= 1,2,3, \dots, \eta,$$

where $K_{e,i,1_\chi}$ is the χ-meaning ($\chi = 1,2,3, \dots, \eta$) i^{th} single estimated relative gas quality index of the first group $K_{e,i,1}, i = 1,2, \dots, l$, where l is number of indexes of the first group); $K_{e,i,2_\chi}$ is the χth meaning ($\chi = 1,2,3, \dots, \eta$) i^{th} single estimated relative gas quality index of the second group $K_{e,i,2}, i = l + 1, l + 2, \dots, n$, where $n - l$ is the number of parameters of the second group; n is the number of all investigated parameters of natural gas); here K_{e,i_χ} is the χth meaning ($\chi = 1,2,3,\dots,\eta$) of the first single estimated relative index of natural $K_{e,i}, i = 1,2,3,\dots,n$ of the entire set of indexes; K_{e,j_χ} is the χth meaning of ($\chi = 1,2,3,\dots,\eta$) j^{th} estimated single-relative quality index of NG $K_{e,j}, i \neq j$, $i = 1,2,3,\dots,n$ from entire set of indexes.

Number of single indexes of natural gas quality n, taken to evaluate its quality, depends on the type of the investigated gas. In general, the developed technique of determining the level of NG quality respectively to qualimetric measurement methodology can be applied at ≥ 2, but we must mention that at $n = 2$ this technique can be used, on the occasion of the products proportion of quality indexes differences:

$$\left(K_{e,i,1} - K_{b,i,1}\right) \cdot \left(K_{e,j,2} - K_{b,i,2}\right) \qquad (7.21)$$

and their validity ratios $m_{i,1} \approx m_{i,2}$, that is on condition:

$$m_{i,1} \cdot \left(K_{e,i,1} - K_{b,i,1}\right) \approx m_{i,2} \cdot \left(K_{e,i,2} - K_{b,i,2}\right). \qquad (7.22)$$

If specified condition is not fulfilled, for example if:

$$m_{i,1} \cdot \left(K_{e,i,1} - K_{b,i,1}\right) < \frac{1}{3}m_{i,2} \cdot \left(K_{e,i,2} - K_{b,i,2}\right), \qquad (7.23)$$

then the first term in (7.13) for function of differences $\Delta\Pi$ is negligibly small compared to the 2nd member and it is advisable to standard-gas quality determining by Quality Score as $K_{e,i,2}$ [11].

Providing the necessary metrological reliability of measurement results of individual quality indexes of studied NG at small amounts of sample values for these parameters in each case we need to achieve by choosing this method of measurement and MIs that would provide the appropriate exactness of the obtained measurement result, in particular, its uncertainty, which in turn would have ensured the reliability of measurement results even at the disposable measurements.

- Study of Natural Gas Quality. Input information for defining NG quality as energy source by methodology of qualimetric measurements were the results of experimental studies of natural gas samples from deposits of Lviv region, conducted in laboratory of management "Lvivgasmining" of "Naftogas of Ukraine". Range of indexes for determining the NG quality as energy source was set by aforementioned analysis. Definition of gas components was performed by chromatograph. Measuring and processing of data were carried out by method with requirements regulated by [7, 12]. Dew point temperature at humidity Θ_P, °C was measured by hygrometer, and absolute gas humidity W, g/m^3 depending on the measured dew point of gas at humidity Θ_p, °C was defined. Calculation of specific volume heat combustion H_H, MJ/m^3; density ρ, kg/m^3 and Wobbe number, MJ/m^3 were conducted respectively [7] basing on gas composition and mass concentrations of hydrogen sulfide C_{H_2S} g/m^3, mercaptan sulfur C_{CH_2S} g/m^3 in the gas mixture. The results of the conducted analysis of gas samples are given in Table 7.1.

Table 7.1. Results of the experimental defining of natural gas quality as an energy source

No	Title of the index	Absolute indexes				Relative indexes		Weight coefficients	Relative weighted indexes	
		$P_{max,i};$ $P_{allow,i}$	$P_{mix,i}$	$P_{b,i}$	$P_{e,i}$	$K_{b,i}$	$K_{b,i}$	m_i	$K_{bW,i}$	$K_{ew,i}$
First group of indexes										
1	Heat of combustion < H_H, MJ / m^3	45.00	-	45.00	39.65	1	0.881	0.235	0.235	0.207
2	Wobbe number B, MJ/m^3	54.50	-	54.50	48.35	1	0.877	0.224	0.224	0.191
3	Density ρ, kg/m^3	0.960	-	0.960	0.820	1	0.854	0.122	0.122	0.104
Second group of indexes										
4	Absolute humidity W, g/m^3	0.110	0	0	0.085	0	0.773	0.218	0	0.169
5	Conc-on of nitrogen C_N , %	1.400	0	0	1.134	0	0.810	0.071	0	0.058
6	Conc-on of carbon dioxide C_{CO_2},%	0.450	0	0	0.272	0	0.604	0.051	0	0.031
7	Mass conc-on of hydrogen sulphide C_{H_2S}, g/m^3	0.020	0	0	0.002	0	0.100	0.039	0	0.004
8	Mass concentration of mercaptan sulfur C_{CH_2S}, g/m^3	0.036	0	0	0.005	0	0.139	0.040	0	0.006

Remark. Dimensions of values, provided in table, are taken for standard conditions, namely to pc=0.101325 MPa; T_c = 293.15

NG quality level as energy source Q is defined in two stages. First, in accordance to (7.12) the function deviations envisaged that ΔP is equal to $\Delta P = 0.187$, and further due to (7.16), NG quality level Q is defined as $Q=0.813$ or 81.3 %. For this Q the level of NG quality is estimated quite high. This is due to the proximity of absolute values of single-quality evaluated gas indexes $P_{e,i}$, and estimated relative individual quality gas indexes $K_{e,i}$, to the basic values of these indexes - $P_{b,i}$ and $K_{b,i}$ respectively.

It becomes possible to improve the gas quality by applying such technological operations as gas drying and gas mixture cleaning from noncombustible and harmful components. Then theoretical value of 5 single-absolute last evaluated parameters of gas quality $P_{e,i}$, namely the second group indexes, and respectively 5 last single-quality evaluated gas indexes $K_{e,i}$, may be reduced to zero. Dimensions of deviations function $\Delta P = 0.05$, and the quality of gas $Q = 0.95$ or 95 % are close to the base value. But this augmentation of gas quality could lead to significant increase in gas prices.

7.2. Foundations of Objective Qualimetry

CPS development programs have been suggested by scientists in automatics, software, networks. However, their effective development is impossible without taking into account the metrological aspects of CPS designing, constructing and operating. Therefore current NIST program [13] focuses on involving metrological science to resolve some CPS-problems.

For instance, the program [14] focuses on four areas which are closely interrelated: material characterization, real-time process control, process and product qualification, and systems integration. In the process and product qualification area, the program establishes foundations for equivalence-based qualification of materials, processes, and parts used in AM by developing novel test methods and protocols for round robin testing, as well as generating trusted data for sharing among the AM stakeholders.

However, the continuous development of CPSs, designed schematically at least, by above-mentioned documents, comes up against a number of difficulties due to the following circumstances. Firstly, assign a priori that CPS creating [15] on the basis of industrial Internet opportunities envisages the free conjunction and matching of wares from different countries of different not fully estimated quality. As the current set of standards in different countries is insufficient to describe successfully the means and tools of modern scientific technology, it is constantly evolving and improving. Secondly, laboratories and leading research centers based on modern productive machinery are not able to be tested by virtue of their own complexity and problems of delivery to certified laboratories. Thirdly, unique and newly created machinery often requires self-verification and standardization of metrology facilities to ensure the quality work.

- Development of Qualimetry in Additive Manufacturing and Related Fields. The basis of any measurement is comparison of the measured value with a reference or standard that retains and reproduces a physical quantity of the certain dimension. Specificity of qualimetric measurement is the absence of specific physical measures of the quality of particular products. Available basic product samples not always correspond to the metrological requirements that are applied to measures. Therefore not always is possible methodologically to compare the studied products with the basic sample that, in fact, constitutes the main problem of these measurements implementation.

- Conceptual Principles. To ensure methodological implementation of qualimetric measurement procedures is suggested virtual measure of product quality, which is a theoretical analogue of corresponding physical measure, i.e. the reference sample of the studied product [3]. To construct the product quality virtual measure [16] is applied the basic provisions of virtual instrumentation technology as one of modern information technologies. Its essence lies in a computer program that simulates real physical instrumentation, measuring systems and control, as also the set theory of mathematics. Thus, virtual measure of product quality corresponds to mathematical expressed and software backed real physical measure.

On the other hand, the product quality is defined by a set of the different origin's properties. So, virtual measure of product quality is a plurality of some arbitrary objects, united by certain properties common for these units. Such objects in Qualimetry are individual absolute P_i and relative K_i indicators of products quality $P_i = 1,2,3, ... , n$, where n is the number of individual indicators, which equals to number of properties p_i). Proceeding from this analysis, synthesis of virtual measure of product quality is carried out on dot sets theory basis in n-dimensional Euclidean space, in which are made the product quality assessing. That is due to different physical nature and various dimensions of individual absolute indicators of product quality P_i. The latter are the points on corresponding coordinate axes of multidimensional space. Hereby, scales on certain i-coordinate axes are different and determined by weight coefficients m_i of appropriate individual quality indicators P_i.

- Methodology of qualimetric measurements realization in mechanical manufacture. At resolving theoretical and practical problems of qualimetric measurement in quality evaluation of construction materials we have developed a comprehensive methodology for determining quality by considering properties and internal structure impact on decisive indicator of the quality, namely by using the measurement uncertainties of gained results.

For this purpose primarily there were analyzed characteristics of construction materials quality and was chosen quality evaluation method; then was established a decisive indicator of quality for their strength evaluation considering the relationship between this indicator and structure-sensitive properties; ultimately was studied this indicator measurement method taking into account the substance inner structure as well as and the method of assessment of aforementioned indicator measurement accuracy.

In examination were also taken into account that most important characteristics of constructional materials include elastic properties in particular modulus of elasticity. That is due to the next. Modulus is related with energy of crystal lattice and is a measure of atomic bonds so can be applied in materials research particular in study of substance structure, phase transformations and more. In addition, the elastic modulus is used in mechanics of solids at designing the details and units of various machines as makes it possible to predict the material response on applied load.

In materials science an ultrasonic pulse method for measuring mechanical properties is widespread; it provides the highest exactness. The method is based on the dependence of longitudinal elastic ultrasonic waves speed v in tested material on its modulus of elasticity and density:

$$E = rv^2, Pa, \qquad (7.24)$$

where ρ is the material density, kg/m³; v is the measured waves velocity in a sample, m/s. Authors of the work take into account v only the instrumental error, they adopt that $\delta_E = \delta_{E,instr}$, which in accordance with (7.24) equals to:

$$d_{E,instr} = d_r + 2d_u, \%. \qquad (7.25)$$

The measurement error consists of two constituents: instrumental one $\delta_{E,instr}$ and another methodical one $\delta_{E,imet}$, or:

$$\delta_E = \delta_{E,instr} + \delta_{E,met} \qquad (7.26)$$

and $\delta_{E,met}$ can be equal to few per cent ($\gg \delta_{E,instr}$), making the main contribution to the value of total error δ_E of the mechanical properties measurement.

Thus, we have developed a technique of correcting until excluding the methodical CoE while ultrasound method application. Behind it, simultaneously to measuring the velocity of UWs it should be measured one of following structure-sensitive characteristics of sample. For instance, these can be considered specific conductivity γ for conductive materials, relative dielectric permeability ε for dielectrics or relative magnetic permeability μ for ferromagnets. Resulting corrected value of elasticity modulus of studied pattern is determined taking into account these additionally measured parameters. The appropriate equations for calculating the adjusted modulus values E_M of different construction materials as a primary indicator of quality for the developed technique are presented in Table 7.2.

Evaluation for measurement accuracy of corrected values of construction materials elasticity modulus was carried out by means of assessing the measurements uncertainty, i.e. values of modulus E_M, of studied samples. For this purpose technique of defining uncertainty $u(E_M)$ of elastic modulus E_M received value was developed. Its peculiarities consist of the following. First, modulus measurement result E_M in the case of measurements with multiple observations on condition the normal distribution of these observations $E_{M,i}$ are calculated as mean sampling value $\bar{E}_M$, obtained from individual modulus measurements $E_{M,i}, i = 1,2,3, \dots, \eta$, where η is the number of individual measurements:

$$E_M = \bar{E}_M = \frac{1}{\eta}\sum_{i=1}^{\eta} E_{M,i} \qquad (7.27)$$

Table 7.2. Corrected values of elastic modulus of materials.

No	Kind of sample substance	Equation of corrected values of elastic modulus E_M, Pa
1.	Conductive materials	$E_M = \rho_0 v^2 \dfrac{\gamma_0}{\gamma}$
2.	Dielectrics	$E_M = \rho_0 v^2 \dfrac{\varepsilon_0}{\varepsilon}$
3.	Ferromagnets	$E_M = \rho_0 v^2 \dfrac{\mu_0}{\mu}$

Remark 1. Here ρ_0 is the theoretical density of defect-free material, kg/m^3; γ is the real electrical conductivity of the sample substance with structure defectiveness, 1/(Ω·m); γ_0 is the theoretical electric conductivity of defect-free material, 1/(Ω·m); ε is the real relative dielectric permeability of sample substance with structure defectiveness; ε_0 is the theoretical relative dielectric permeability of defect-free material, μ is the real relative permeability of the sample with defectiveness; μ_0 is the theoretical relative permeability of defect-free material.

The latter are determined by the formulas given in Table 7.2 dependently on the kind of studied sample substance. Second definition of elastic modulus values $E_{M,i}$, of each of n measurement results, as can be seen from the formulas (Table 7.1), depending on the type of studied samples, are indirect measurements. Third, as for sustainable measuring conditions of experiment obtained results of individual measurements of modulus E_M, are equally accurate, then the standard uncertainty of type B of elasticity modulus E_M measurement result is found on the basis of uncertainty analysis of only one measurement result $E_{M,i}$ from obtained η results.

Thus, the combined standard uncertainty $u(E_M)$ of elastic modulus measurement result E_M of construction materials by ultrasound pulsed method consists of standard uncertainty of type A $u_A(E_M)$ that is determined statistically by processing a sequence of n observations $E_{M,i}$ of studied material:

$$u_A(E_M) = \sqrt{\frac{1}{n(n-1)} \sum_{i=1}^{n} (E_{M,i} - \bar{E}_M)^2} \qquad (7.28)$$

and combined standard uncertainty of type B $u_{cB}(E_M)$, Pa, which is defined by processing the results of indirect measurements.

Determination of combined standard uncertainty of type B $u_{cB}(E_{M,i})$, Pa, of elastic modulus measurement result $E_{M,i}$, Pa of ith observation is carried out by analyzing the general relation:

$$E_{M,i} = \rho_0 v_i^2 \frac{x_0}{x_i}, \qquad (7.29)$$

where x_0 and x_i are the generalized designations of studied material's structurally-sensitive properties. They are applied to correct methodical error of modulus measurement. Similarly are used specific conductivity values γ_0 and γ_i while studying the electrically conductive materials, and relative dielectric permeability ε_0 and ε_i in the case of dielectric materials research. The same concerns relative magnetic permeability μ_0 and μ_i at ferromagnetics research.

Combined standard uncertainty $u_c(E_M)$ of investigated material modulus E_M measurement result is defined as the square root of sum of the squares of predetermined standard uncertainty of type A $u_A(E_M)$ and the combined standard uncertainty of type B $u_{cB}(E_{M,i})$:

$$u_c(E_M) = \sqrt{u_A^2(E_M) + u_{cB}^2(E_{M,i})}, Pa \qquad (7.30)$$

The uncertainties' budget of studied material modulus is given in Table 7.3.

7.3. Objective Qualimetry on the Basis of Thermodynamics

Assessment of products quality is a complex multifactorial problem, within which is difficult to evaluate the role and relative weight of each factor as well as to expressed in physical units the objective numerical value of it or to validate this factor, and finally to determine certain characteristics. At best, the result is expressed as the number reasonably combining all these influence factors. That is a Subjective Qualimetry. It is exploited while comparing the similar products of the same destination from different manufacturers. However, no one can prove conclusively the correctness of the choice of those or other factors that affect the assessment.

Table 7.3. Uncertainties' budget of studied material elastic modulus E_M.

Input value x_j	Standard uncertainty $u(x_j)$	Evaluation type, distribution law	Coefficient of sensitivity $C_j = \partial f / \partial x_j$	Product of $|C_j| \cdot u(x_i)$
ρ_0, kg/m^3	$u_B(\rho_0) = \frac{q_{quan,\rho_0}}{2\sqrt{3}}$, kg/m^3	Type B, uniform	$C_{\rho_0} = v_i^2 \frac{x_0}{x_i}$, m^2/s^2	$C_{\rho_0} \cdot u_{cB}(\rho_0)$, Pa
v_i, m/s	$u_{cB}(v_i)$, m/s	Combined, normal	$C_{v_i} = 2v_i^2 \rho_0 \frac{x_0}{x_i}$, kg/(m^2·s)	$C_{v_i} \cdot u_{cB}(v_i)$, Pa
$x_0, 1_x$	$u_B(\rho_0) = \frac{q_{quan,x_0}}{2\sqrt{3}}, 1_x$	Type B, uniform	$C_{x_0} = \frac{\rho_0 v_i^2}{x_i}$, kg/$(m \cdot s^2 \cdot 1_x)$	$C_{x_0} \cdot u_B(x_0)$, Pa
$x_i, 1_x$	$u_{cB}(x_i), 1_x$	Combined, normal	$C_{x_i} = \frac{\rho_0 v_i^2 x_0}{x_0^2}$, kg/$(m \cdot s^2 \cdot 1_x)$	$C_{x_i} \cdot u_{cB}(x_i)$, Pa
Output value	Evaluation type, distribution law	Combined standard uncertainty		
$E_{M,i}$, Pa	Combined, normal law	$u_{cB}(E_{M,i}) = \{C_{\rho_0}^2 u_B^2(\rho_0) + C_{v_i}^2 u_{cB}^2(v_i) + C_{x_0}^2 u_B^2(x_0) + C_{x_i}^2 u_{cB}^2(x_i)\}^{1/2}$, Pa		

From the metrological point of view, no one can guarantee absence of correlation for separate factors among themselves that negative influence on the obtained results. As this comparison becomes in a certain degree subjective, it would be performed only for products of very similar appointment, e.g. vehicles of the same sector, power, etc. In terms of metrological-qualimetric approach this is the result of only researcher's subjective approach concerning correct choice of coordinates (major characteristics, by which the studied object is estimated). Indeed, when selected coordinates are somehow associated (in metrological sense correlated) and when correlativity is not considered in the obtained results at least by correlation coefficients, the final results become unreliable. Mutual correlation of certain parameters is often explicit, often evident, but hardly argued. For instance, returning to cars' estimation, we can note that quality of outer covering that forms such qualitative parameter as the appearance car, is however involved in fuel economy. So, the mentioned parameters could consider to be significantly correlated with the high correlation coefficient.

Previously on the basis of studying factors influencing the performance of thermoelectric sensors, to evaluate their metrological quality we suggested [17] to carry out methodologically correct selection of uncorrelated factors. The basis for this implementation was taken similar to adopted in thermodynamics [18]. The latter ensures correctness of choice the determining factors – characteristics – of being evaluated product or process as a set of unrelated variables (measurands in metrology) denominated as thermodynamic forces and flows. This thermodynamics approach is able to provide an Objective Qualimetry to create multidimensional space with guaranteed independent and uncorrelated coordinates. Currently in phenomenological thermodynamics are known [18] 6 independent degrees of freedom. They cover almost all physical phenomena eligible to describe the characteristics of arbitrary objects, conjugated by certain properties for estimating goals. For instance, in [19] we disclosed the path how to tie aforementioned Young's modulus through superficial freedom degree of the basic equation of thermodynamics with thermo-EMF drift of thermoelectric thermometer manufactured from cermet materials. Obtained results fully coincide with given above in Tables 7.1 and 7.2 expressions for the modulus of elasticity.

Should be noted that on the basis of empirical experience Qualimetry works with linear algebraic expression of type:

$$U = aX + bY + \cdots + kZ, \qquad (7.31)$$

where a, b, c are the coefficients; X, Y, Z are the studied properties, or coordinated which number is defined by researcher. Since operating in n-dimensional Euclidean space, the proper establishing the size of vector, which module P is determined from the following expression:

$$P^2 = aX^2 + bY^2 + \cdots + kZ^2, \qquad (7.32)$$

can serve as integer giving the assessment, especially when coordinates are specified independent by involving thermodynamic approach. In thermodynamics expression of type aX^2, where X is thermodynamic force concerning the certain freedom degree, corresponds to the dimension of energy. Therefore at correct choice of studied properties in (7.32) the received result is equivalent to P^2 under that dimension.

Thus, proposed approach with proper choice of coordinates describing the relevant properties of studied object leads to absolutely correct its qualimetric assessment, which coincides with the thermodynamic evaluation on the energetic basis.

318

Remark. In thermodynamics due to equations of state it appears the possibility to choose any parameters of the system as independent variables. At a certain choice of independent variables there are always functions of the system that are successful for studying various processes. These thermodynamic functions are named as thermodynamic potentials or characteristic functions if they satisfy the following requirements: a) They have to be additive and unique functions of the system state; b) For a particular set of physical variables their derivatives are inherent in simple and clear physical meaning; c) Under certain conditions the thermodynamic function in equilibrium is characterized by extremum.

7.4. Conclusions

For comprehensive quality assessment of complex products produced by Cyber-Physical Systems and comparing between similar products of the same destination from different manufacturers it could be offered a newly established field of metrology that is specified as Objective Qualimetry. In contrast to traditional Subjective Qualimetry the latter is able to realize the selection of metrologically uncorrelated row of determining characteristics of these products and, thus, to identify products of higher quality.

Determination of critical characteristics, especially those that do not correlate with a number of other characteristics, is crucial for paying the maximum attention to metrological assurance as well as to validation of these particular characteristics for Cyber-Physical System specific type or for its final product, and can be achieved only on the basis of thermodynamically justified involvement. As result, the reliability and perfection of smart and flexible operation of mentioned systems could be permanently improved.

References

[1]. V. P. Motalo, Problems of metrological assurance of qualimetrical measurements, *Measuring Equipment and Metrology,* No 68, 2008, pp. 190 -195 (in Ukrainian)
[2]. International standard Natural gas – Quality designation: ISO 13686:1998. Implemeted 01.01.83, *International Organization for Standardization,* Geneva, 1983, 49 p.

[3]. B. Stadnyk, V. Motalo, Evaluation of Qualimetrical Measurements Quality Based on the Uncertainty Concept, *Pomiary. Automatyka. Kontrola,* Warszawa, Vol. 59, Issue 9, 2013, S. 950-953.

[4]. D. Katz, R. Cornell, R. Kobayashi, Handbook Natural Gas Engineering, *McGraw Hill,* New York, 1959.

[5]. The Properties of Gases and Liquids, By B. E. Poling, J. M. Prausnitz, J. P. O'Connell, *Engineering and Transportation,* 5th Edition, 2000.

[6]. ISO 13686:2013, Natural gas – Quality designation.

[7]. ISO 13686:2013, Natural gas – Calculation of calorific values, density, relative density and Wobbe index from composition.

[8]. JCGM 200:2012 (E/F), International vocabulary of metrology - Basic and general concepts and associated terms.

[9]. ISO 5725-1-2003, Accuracy of measurement methods and results. Part 1. General principles and definitions.

[10]. M. L. Davison, Multidimensional scaling, Multidimensional Scaling, *Wiley,* New York, 1983.

[11]. P. E. Green, J. Frank, Jr. Carmone, Sc. M. Smith, Multidimensional Scaling: Concepts and Applications, *Allyn & Bacon,* Boston, 1989.

[12]. ISO 6974-2:2001, Natural gas. Determination of composition with defined uncertainty by gas chromatography. Part 2. Measuring system characteristics and statistics for processing of data.

[13]. NIST National Technical Information Service, FY 2014, Budget Submission to Congress, 2014.

[14]. ASTM International, Technical Committee F42, Additive Manufacturing Technologies, http://www.astm.org/COMMITTEE/F42.htm

[15]. Future Project Industry 4.0, *VDE,* online, https://www.vde.com/en/dke/std/Pages/industry40.aspx

[16]. P. E. Green, J. Frank, Jr. Carmone, Sc. M. Smith, Multidimensional Scaling: Concepts and Applications, *Allyn & Bacon,* Boston, 1989.

[17]. B. Stadnyk, S. Yatsyshyn, Accuracy and metrological reliability enhancing of thermoelectric transducers, *Sensors & Transducers,* Vol. 123, Issue 12, December 2010, pp. 69-75.

[18]. P. Glansdorf, I. Prigogine, Thermodynamic theory of structure, stability and fluctuations, *Wiley,* New York, 1971.

[19]. S. Yatsyshyn, B. Stadnyk, Ya. Lutsyk, Impact of porosity on the changes of thermo-EMF of thermometric materials, *Bulletin of National University Lviv Polytechnic, Automation, Measurement and Control,* Issue 530, 2005, pp. 22-28 (in Ukrainian).

Index

A

accuracy threshold, 201
acoustic
    detectors, 26
    resonance thermometers, 235
actuators precision, 108
ad hoc networks, 46
adaptability, 36
ADC, 34, 44, 49, 147, 158, 161
additive
    error correction, 147
    manufacturing, 60
ageing problems, 36
air traffic control, 37
alarm systems, 23
amendments, 211
application servers, 57
atomic standards, 118
autonomous smart sensors, 44
Avogadro number, 193

B

ballistic galvanometer, 202
beam splitter, 64
bi-layer Graphene, 283
Boltzmann constant, 111, 119,
    120, 122, 127, 242, 279
built-in automatic calibration, 130
built-in self-calibration, 130
bus interface, 44

C

calibration, 30
    artefacts, 30
    period, 108

procedures, 109
surface, 66
calibrators, 30
carbon nanotubes, 30
cold-junction temperature, 44
computational elements, 19
computer tomography, 61
conductance quantum, 112
conductive film, 276
conjugated material, 61
content management systems, 57
Coriolis
    effect, 221
    mass flowmeter, 221
correlation amplifier, 260
coulomb blockade thermometer,
    122
cyber-physical systems, 19, 24

D

data acquisition, 25
    system, 34, 45. 160
data processing algorithms, 55
data rate, 46
Debye model, 195
digital intelligent filters, 253
direct
    manufacturing, 24
    measurement, 109
disequilibrated noise, 263
distributed
    control, 47
    robot garden, 37
    sensor network, 46
    smart sensors, 45
distribution law, 142
dynamic error, 203, 224

E

Einstein model, 195
electric current measurement, 144
electrical
    conductivity, 30, 110
    noise, 261
    resistance, 110
electronic commutators, 153
embedded
    microprocessors, 23
    software, 31
    systems, 19, 35
emergency detectors, 26
energy spectrum, 269
entropy, 121
envelope function, 63
equilibrated noise, 263
error
    correction, 50
    budget, 149
Ethernet, 45
explosion detectors, 26
Extensible Markup Language
    (XML), 58
external calibration, 130
extraordinary calibration, 130

F

Faraday constant, 111
fire
    alarm sensors, 45
    detectors, 26
    sensors, 23
firmware, 114
Flicker noise, 239, 257
Flicker noise spectroscopy, 270
flicker-component fluctuations,
    273
flicker-noise parameters, 29
flow velocity, 221
Fluctuation-dissipation theorem,
    186

Fourier transformation, 29, 253
fractional processes, 58
fractional-order signal processing,
    58
free electron gas primary
    thermometer, 122
frequency gauges, 234
frequency-domain measurements,
    30
fuzzy logics, 26

G

gas
    pipelines, 37
    sensor, 193
    thermosensitive substance, 193
gateway, 47
Gaussian spectrum, 63
glass flask, 124
global
    information network, 36
    network systems, 26
graphene, 112
    nanopatterns, 110

H

Hamon network, 31, 116
harmonic tones amplitude, 264
heat
    energy balance, 95
    flux transducers, 237
heterogeneous networks, 25
hostile environment affect, 74
hydrocarbon gas, 303, 304

I

Industrial Revolution 4.0, 21
Industry 4.0, 24
inexactness, 219

infrared radiometers, 237
input signal inverting method, 158
instrumental error, 51, 198, 204, 224
intelligent fire detectors, 23
intelligent
    instrumentation, 25
    temperature sensor, 44
intercalibration interval, 108
interferogram
    envelope, 64
    segment, 65
interferometer, 64
interferometric pattern, 63, 65
invariance provision, 156
irreversible thermodynamics, 186
isotopic composition, 124

J

Johnson noise thermometer, 119
Johnson-Nyquist equation, 120
Johnston junctions array, 126
Josephson
    constant, 111
    effect, 127, 237
    element, 123, 237
    junction noise thermometer, 123, 237
    junctions, 125

K

Klitzing constant, 125

L

leakage current, 114, 161
Lightweight Directory Access
    Protocol (LDAP), 58
low-frequency interference, 158

M

Magnetic
    method of thermometry, 121
    resonance, 111
    susceptibility, 121
manufacturing intelligence, 24
Markov approach, 43
mat matrix, 72
measurement subsystems, 42
measuring
    instruments, 49, 108
    subsystems, 34
mechanical strain, 76
methodical error, 189, 210
method of
    errors automatic correction, 147
    thermometry, 120
metrological
    calibration, 109
    failures, 108
    reliability, 33, 108
    software, 52, 57, 108
    subsystems, 31, 32
    verifications, 138
middleware, 48, 57
mobile devices, 30
monochromatic interferometry, 63
Monte Carlo
    simulation, 43
    methods, 291
multi-level network, 130
multiplicative error, 184
multipoint temperature
    monitoring, 27
multi-sensing, 44

N

nanomanufacturing, 188
nanomaterials, 280
nanopatterns, 30
nanothermodynamics, 241
nanothermometry, 120, 123, 279

method, 28
natural gas quality, 307, 319
navigation, 37
nested-chopper amplifiers, 158
network topology, 46
net-zero building, 27
neutron sensors, 26
noise
    immunity, 260
    interference, 260
    measurement, 194, 251
    method, 29
    method of thermometry, 120
    metrology, 238
    thermometers, 29, 32
non-equilibrium thermodynamics,
    185
non-invasive diagnostics, 267
nonlinear errors, 184
nuclear quadrupole resonance
    thermometer, 234
Nyquist
    equation, 249
    formula, 205

O

optical
    metrology, 28, 233
    networks, 25
    sensors, 26
    thermometry, 120

P

Peltier effect, 110
periodic calibration, 108, 129
photocurrent measurement, 199
physical quantity unit, 110
piezo crystal plate, 235
Planck constant, 111, 123, 127
242, 279
plasticity criterion, 250

p-n junction, 257
polarity switch, 159
precision
    measurements, 109
    threshold, 201
pressure
    calibration service, 30
    sensor, 36
primary
    thermometry, 118, 120
probability distribution, 145
product quality, 315, 323
programmable logic arrays, 52
propane-butane gas mixture, 304
Pt-Rh thermocouples, 225
pyrometer, 97, 237
    radiation, 98

Q

QoP, 108
QoS, 46, 108
qualimetry, 34
quantization error, 110
quantum
    dots, 36
    Hall Effect resistance
    standards, 110
quartz
    crystal vibrator, 235
    thermometer, 234

R

radiometric thermometer, 235
Raman
    method, 28, 278, 289
    method of thermometry, 121
    spectrometer, 278
    thermometer, 278, 284, 290
    thermometry, 198
    error, 192

reference software, 55
relaxation time, 271
remote
    calibration, 130, 133
    error correction, 162
resistance
    sensors, 225
    thermotransducers, 72
root-mean-square approximation, 60

S

Schottky equation, 256
security, 53
segmental approximation, 63
    method, 65
selective laser sintering, 61
self-adaptation, 132
self-adjustment, 42, 176
self-calibration, 42, 109, 118
self-correction, 132
self-diffusion coefficient, 250
self-regulatory network, 36
self-validation, 42, 118
self-verification, 22, 42, 118
sensitive balance, 201
sensor
    sensors array, 150
    networks, 45
    node monitoring, 37
    node, 47, 49
shunt resistance, 144
signal
    conditioning unit, 34
    processing, 44
Simple Object Access Protocol, 58
smart
    actuators, 20
    buildings, 24, 27, 81
    energy-effective houses, 37
    energy-efficient house, 85
    fire detector, 23
    grid, 26, 23
    grid sensor, 47

smart
    health-care system, 27
    house, 82
    machines, 24
    manufacturing, 20, 24
    measuring instruments, 42, 49
    meters, 48
    networks, 24
    sensors, 25, 42, 44, 50
    sensors grids, 27
    temperature sensor, 27. 44
    transducers, 42
    transport, 24
smartphones, 37
software metric, 51
solid body surface temperature, 121
spatial problems, 36
spectrophotometer, 281
spectrum
    access system, 28
    metrology, 27, 233
state standards of electrical resistance, 112
statistical thermodynamics, 186
Stefan-Boltzmann law, 97, 98
stochastic
    signals, 268
    systems, 189
stress gradient, 76
S-type
    diode, 235
    thermocouple, 227
surface
    profile measurement, 63
    reconstruction, 63
    reconstruction method, 68
    topology measurement, 63
symmetrical observations, 217
systematic error, 109, 210

T

temperature
    coefficient, 283

compensation, 49
correcting, 51
measurement, 193
measuring standard, 118
scale, 119
sensors, 32, 95
test signals method, 154
thermal
equilibrium, 123
noise, 239, 255, 263
response time, 226
shock, 264
thermocouples, 44, 50
thermodynamic
disequilibrium, 32
equilibrium, 124, 248
temperature, 119, 120, 123
thermoelectric
sensors, 34
thermotransducer, 207
thermometric material, 225
thermosensitive
materials, 122
transducer, 44
thermotransducers, 71, 187, 203
tolerance, 188
tomography systems, 51
transient
noise process, 32
resistance, 31
transposition method, 156
tunnel junction arrays, 122

U

ultrasonic pulse method, 324
uncertainty
analysis, 114

of measurement result, 184

V

validation, 21
verification, 21
methods, 141
verified measuring instrument,
109

W

Watt balance, 111
web servers, 57
Wheatstone bridge, 30, 112
white-light interferogram, 64
wireless
networking, 37
sensor networks, 48
sensors, 47

X

X-rays crystal density, 111

Y

Young's modulus, 329

Z

zero error, 210